THE CHEMISTRY AND PHYSICS OF ENGINEERING MATERIALS

Volume 1: Modern Analytical Methodologies

THE CHEMISTRY AND PHYSICS OF ENGINEERING MATERIALS

MATERIALS

Volume 1: Modern Analytical Methodologies

Edited by
Alexandr A. Berlin
Roman Joswik
Nikolai I. Vatin

Reviewers and Advisory Members
A. K. Haghi
Gennady E. Zaikov

APPLE
ACADEMIC
PRESS

Apple Academic Press Inc.
3333 Mistwell Crescent
Oakville, ON L6L 0A2 Canada

Apple Academic Press Inc.
9 Spinnaker Way
Waretown, NJ 08758 USA

First issued in paperback 2021

Exclusive worldwide distribution by CRC Press, a member of Taylor & Francis Group
No claim to original U.S. Government works

ISBN 13: 978-1-77463-128-7 (pbk)
ISBN 13: 978-1-77188-741-0 (hbk)

The Chemistry and Physics of Engineering Materials (2-volume set)
ISBN 13: 978-1-77188-742-7 (hbk)

Library and Archives Canada Cataloguing in Publication

The chemistry and physics of engineering materials / edited by Alexandr A. Berlin, Roman Joswik, Nikolai I. Vatin ; reviewers and advisory members, A.K. Haghi, Gennady E. Zaikov.

Reissued with one fewer chapter from each volume with revised indexes. Includes biblio-graphical references and indexes. Contents: Volume 1. Modern analytical methodologies. ISBN 978-1-77188-741-0 (v. 1 : hardcover)--ISBN 978-0-429-45360-1 (PDF)

1. Materials--Mechanical properties. 2. Strength of materials. 3. Metallurgical analysis. 4. Polymers--Structure. 5. Chemistry, Technical. I. Berlin, Al. Al., 1940-, editor II. Joswik, Roman, editor III. Vatin, Nikolai I. (Nikolai Ivanovich), editor

| TA405.C54 2018 | 620.1'1292 | C2018-901927-1 |

CIP data on file with US Library of Congress

CONTENTS

LIST OF CONTRIBUTORS

M. I. Abdullin
Bashkir State University, Ufa, 450077, Russia, E-mail: ProfAMI@yandex.ru

O. M. Alekseeva
Emanuel Institute of Biochemical Physics, Russian Academy of Sciences, ul. Kosygina, 4, Moscow, 119334; Fax: (499) 137-41-01

Sogrina Darya Alexandrovn
Moscow State University of Food Production, 125080, Moscow, Volokolamskoye shosse, 11. E-mail: kaf.vms@rambler.ru

L. A. Badykova
Institute of Organic Chemistry, Ufa Scientific Center, Russian Academy of Sciences, Ufa, Bashkortostan, Russia; Bashkir State Medical University, Ufa, Bashkortostan, Russia; E-mail: badykova@mail.ru

A. A. Basyrov
Bashkir State University, Ufa, 450077, Russia

Alexey Benda
Moscow State University of Printing Arts, 127550 Moscow, Pryanishnikova Street, 2a, Russian Federation

V. I. Berendyaev
L.Ya.Karpov Institute of Physical Chemistry, Moscow, Russia

S. B. Bibikov
Russian Academy of Sciences, Moscow, Russia

S. A. Bogdanova
Department of Plastics Technology, Kazan State Technological University, Kazan, Russia

A. A. Bogomazova
Sterlitamak branch of Bashkir State University, 49 Lenina avenue, 453103 Sterlitamak, Russia

S. N. Bondarenko
Volzhsky Polytechnical Institute, branch of Volgograd State Technical University, Volzhsky, Volgograd Region, Russia; E-mail: mystery_21_12@mail.ru; www.volpi.ru

A. V. Bychkova
Russian Academy of Sciences, Moscow, Russia

Ana M. T. D. P. V. Cabral
Faculty of Pharmacy, University of Coimbra, 3000-295 Coimbra, Portugal, E-mail: acabral@ff.uc.pt

Guipeng Cai
Department of Chemistry and Fire Retardant Research Facility, Marquette University PO Box 1881, Milwaukee, WI 53201, USA

V. V. Chernova
Bashkir State University, Russia, Republic of Bashkortostan, Ufa, 450074, St. Zaki Validi, 32

Margarita E. Dzodzikova
North Ossetian State Nature Reserve, D.1, Chabahan Basiev St., Alagir, North Ossetia-Alania, 363000, Russia; Tel.: +7-918-822-42-69; E-mail: dzodzikova_m@mail.ru

B. Dzyadevych
National Forestry University of Ukraine, General Chuprynky St. 103, 79057 Lviv, Tel. (380–322) 237-80-94

V. N. Erokhin
Emanuel Institute of Biochemical Physics, Russian Academy of Sciences, ul. Kosygina, 4, Moscow, 119334; Fax: (499) 137-41-01

D. V. Feoktistov
Institute of Organic Chemistry, Ufa Scientific Center, Russian Academy of Sciences, Ufa, Bashkortostan, Russia; Bashkir State Medical University, Ufa, Bashkortostan, Russia

Ryszard Fiedorow
Adam Mickiewicz University, Faculty of Chemistry, Grunwaldzka 6, Poznań, Poland

A. B. Gilman
N.S. Enikolopov Institute of Synthetic Polymeric Materials of RAS

A. B. Glazyrin
Bashkir State University, Ufa, 450077, Russia

M. D. Goldfein
Saratov State University named after, N.G. Chernyshevskiy, Russia, E-mail: goldfeinmd@mail.ru

N. G. Grigor'eva
Institute of Petrochemistry and Catalysis of RAS, 141 pr. Oktyabria, 450075, Ufa, Russia

N. Guarrotxena
Instituto de Ciencia y Tecnología de Polímeros (ICTP), Consejo Superior de Investigaciones Científicas (CSIC). C/Juan de la Cierva 3, 28006n Madrid, Spain, E-mail: nekane@ictp.csic.es

Y. Hnatyshyn
National Forestry University of Ukraine, General Chuprynky St. 103, 79057 Lviv, Tel. (380–322) 237-80-94, E-mail: hnat_ya@inbox.ru

Daniel Horák
Institute of Macromolecular Chemistry, Academy of Sciences of the Czech Republic, Heyrovskeho Sq. 2, 162 06 Prague 6, Czech Republic

Peter Jurkovič
VIPO, Partizánske, Slovakia; E-mail: upolnovi@savba.sk

V. F. Kablov
Volzhsky Polytechnical Institute, branch of Volgograd State Technical University, Volzhsky, Volgograd Region, Russia; E-mail: mystery_21_12@mail.ru; www.volpi.ru

K. A. Kalinova
Volzhsky Polytechnical Institute (branch) Volgograd State Technical University, 42a Engelsa Street, Volzhsky, 404121, Russia

A. N. Kazakova
Ufa State Petroleum Technological University, 1 Kosmonavtov Str., 450062 Ufa, Russia; Tel: (347) 2420854, e-mail: nocturne@mail.ru

N. A. Keibal
Volzhsky Polytechnical Institute, branch of Volgograd State Technical University, Volzhsky, Volgograd Region, Russia; E-mail: mystery_21_12@mail.ru; www.volpi.ru

T. Sh. Khakamov
Institute of Organic Chemistry, Ufa Scientific Center, Russian Academy of Sciences, Ufa, Bashkortostan, Russia; Bashkir State Medical University, Ufa, Bashkortostan, Russia

V. G. Kochetkov
Volzhsky Polytechnical Institute (branch) Volgograd State Technical University, 42a Engelsa Street, Volzhsky, 404121, Russia

V. Kochubey
National University "Lviv Polytechnic," Lviv

Alexander Kondratov
Moscow State University of Printing Arts, 127550 Moscow, Pryanishnikova Street, 2a, Russian Federation, Tel: +7 963 605 70 27; E-mail: apk@newmail.ru

N. V. Kostenko
Volzhsky Polytechnical Institute (branch) Volgograd State Technical University, 42a Engelsa Street, Volzhsky, 404121, Russia

A. L. Kovarsky
Russian Academy of Sciences, Moscow, Russia

N. I. Kozhevnikova
Saratov State University, 410012 Saratov, Russian Federation

N. V. Kozhevnikov
Saratov State University, 410012 Saratov, Russian Federation, E-mail: KozevnikovNV@info.sgu.ru

A. V. Krementsova
Emanuel Institute of Biochemical Physics, Russian Academy of Sciences, ul. Kosygina, 4, Moscow, 119334; Fax: (499) 137-41-01; E-mail: akrementsova@mail.ru

O. S. Kukovinets
Bashkir State University, Ufa, 450077, Russia

E. I. Kulish
Bashkir State University, Russia, Republic of Bashkortostan, Ufa, 450074, St. Zaki Validi, 32

B. I. Kutepov
Institute of Petrochemistry and Catalysis of RAS, 141 pr. Oktyabria, 450075, Ufa, Russia

V. G. Leontiev
Russian Academy of Sciences, Moscow, Russia

Victor M. M. Lobo
Department of Chemistry, University of Coimbra, 3004–535 Coimbra, Portugal

Hieronim Maciejewski
Poznań Science and Technology Park, A. Mickiewicz University Foundation, Rubież 46, Poznań, Poland

Milan Marônek
Slovak Academy of Sciences, Polymer Institute of the Slovak Academy of Sciences, 845 41 Bratislava, Slovakia

Ján Matyašovský
VIPO, Partizánske, Slovakia; E-mail: upolnovi@savba.sk

Ivan Michalec
Slovak Academy of Sciences, Polymer Institute of the Slovak Academy of Sciences, 845 41 Bratislava, Slovakia

N. N. Mikhailova
Ufa State Petroleum Technological University, 1 Kosmonavtov Str., 450062 Ufa, Russia; Tel: (347) 2420854, e-mail: nocturne@mail.ru

V. V. Molokin
Department of Plastics Technology, Kazan State Technological University, Kazan, Russia

R. Kh. Mudarisova
Institute of Organic Chemistry, Ufa Scientific Center, Russian Academy of Sciences, Ufa, Bashkortostan, Russia; Bashkir State Medical University, Ufa, Bashkortostan, Russia

L. N. Nikitin
Russian Academy of Sciences, Moscow, Russia

F. F. Niyazi
Doctor of Chemical Sciences, Professor, Head of the Department "Fundamental chemistry and chemical technology", South-West State University, 305040, Kursk, street October 50, 94, E-mail: farukhniyazi@yandex.com

Igor Novák
Department of Welding and Foundry, Faculty of Materials Science and Technology in Trnava, 917 24 Trnava, Slovakia

O. M. Novopoltseva
Volzhsky Polytechnical Institute (branch) Volgograd State Technical University, 42a Engelsa Street, Volzhsky, 404121, Russia, www.volpi.ru; e-mail: nov@volpi.ru

A. N. Ozerin
N.S. Enikolopov Institute of Synthetic Polymeric Materials of RAS

Vitalii Patsula
Institute of Macromolecular Chemistry, Academy of Sciences of the Czech Republic, Heyrovskeho Sq. 2, 162 06 Prague 6, Czech Republic

D. A. Provotorova
Volzhsky Polytechnical Institute, branch of Volgograd State Technical University, Volzhsky, Volgograd Region, Russia; E-mail: mystery_21_12@mail.ru; www.volpi.ru

G. Z. Raskildina
Ufa State Petroleum Technological University, 1 Kosmonavtov Str., 450062 Ufa, Russia; Phone (347) 2420854, E-mail: nocturne@mail.ru

Ana C. F. Ribeiro
Department of Chemistry, University of Coimbra, 3004–535 Coimbra, Portugal, Tel: +351-239-854460; Fax: +351-239-827703; E-mail (corresponding author): anacfrib@ci.uc.pt; luisve@gmail.com; avalente@ci.uc.pt; vlobo@ci.uc.pt

E. G. Rozantsev
Saratov State University named after, N.G. Chernyshevsky

B. M. Rumyanyzev
Russian Academy of Sciences, Moscow, Russia

A. V. Sazonova
Candidate of Chemical Sciences, Lecturer of the Department "Fundamental chemistry and chemical technology", South-West State University, 305040, Kursk, street October 50, 94, E-mail: ginger313@mail.ru

V. A. Semenov
Emanuel Institute of Biochemical Physics, Russian Academy of Sciences, ul. Kosygina, 4, Moscow, 119334; Fax: (499) 137-41-01

Ladislav Šoltés
Institute of Experimental Pharmacology of the Slovak Academy of Sciences, 845 41 Bratislava, Slovakia

O. N. Sorokina
Russian Academy of Sciences, Moscow, Russia

Rostyslav Stoika
Institute of Cell Biology, National Academy of Science of Ukraine, Drahomanov St. 14/16, 79005 Lviv, Ukraine

I. F. Tuktarova
Bashkir State University, Russia, Republic of Bashkortostan, Ufa, 450074, St. Zaki Validi, 32, E-mail: tuktarova_irina@rambler.ru

R. R. Usmanova
Ufa State Technical University of Aviation; 12 Karl Marks str., Ufa 450000, Bashkortostan, Russia; E-mail: Usmanovarr@mail.ru

Artur J. M. Valente
Department of Chemistry, University of Coimbra, 3004–535 Coimbra, Portugal

Marian Valentin
Department of Welding and Foundry, Faculty of Materials Science and Technology in Trnava, 917 24 Trnava, Slovakia

Luís M. P. Veríssimo
Department of Chemistry, University of Coimbra, 3004–535 Coimbra, Portugal

Ananiev Vladimir Vladimirovich
Moscow State University of Food Production, 125080, Moscow, Volokolamskoye shosse, 11. kaf. vms@rambler.ru

Agata Wawrzyńczak
Adam Mickiewicz University, Faculty of Chemistry, Grunwaldzka 6, Poznań, Poland

Charles A. Wilkie
Department of Chemistry and Fire Retardant Research Facility, Marquette University PO Box 1881, Milwaukee, WI 53201, USA, E-mail: charles.wilkie@marquette.edu

M. Yu. Yablokov
N.S. Enikolopov Institute of Synthetic Polymeric Materials of RAS

V. Yalechko
National Forestry University of Ukraine, General Chuprynky St. 103, 79057 Lviv, Tel. (380–322) 237-80-94, E-mail: V0l0dyMyR@ukr.net

A. E. Zaikin
Department of Plastics Technology, Kazan State Technological University, Kazan, Russia

G. E. Zaikov
N.M. Emanuel Institute of Biochemical Physics, Russian Academy of Sciences, 4 Kosygin str., Moscow 119334, Russia; Bashkir State University, Ufa, 450077, Russia; E-mail: chembio@sky.chph.ras.ru

Beata A. Zasonska
Institute of Macromolecular Chemistry, Academy of Sciences of the Czech Republic, Heyrovskeho Sq. 2, 162 06 Prague 6, Czech Republic

S. S. Zlotsky
Ufa State Petroleum Technological University, 1 Kosmonavtov Str., 450062 Ufa, Russia; Phone (347) 2420854, E-mail: nocturne@mail.ru

LIST OF ABBREVIATIONS

AAS	atomic absorption spectroscopy
AFM	atomic force microscope
AMPA	2,2'-azobis(2-methylpropionamidine) dihydrochloride
AMS	antibiotic amikacin sulfate
APP	ammonium polyphosphate
APTES	(3-aminopropyl)triethoxysilane
ATH	alumina trihydrate
ATR FTIR	attenuated total reflectance Fourier transform infrared spectroscopy
BPEI	branched polyethylenimine
CA	carbonization agent
CB	carbon black
CCPGQY	charge carrier photogeneration quantum yield
CFT	cefotaxime
CHT	chitosan
CNR	chlorinated natural rubber
CNT	carbon nanotubes
CP	citrus pectin
CTC	charge-transfer complex
DLS	dynamic light scattering
DMEM	Dulbecco's Modified Eagle's Medium
DSC	differential scanning calorimetry
DTG	differential thermogravimetry
DVM	digital voltmeter
EDA	electron donor-acceptor
EMR	electron magnetic resonance
EPDM	ethylene-propylene-diene terpolymer
EVA	ethylene-vinyl-acetate copolymer
FR	fire retardant
GLC	gas-liquid chromatography
HDPE	high density polyethylene
HFI	hyperfine interaction
HIPS	high impact polystyrene

HRC	heat release capacity
IFR	intumescent fire retardant
LBL	layer by layer technique
LDH	layered double hydroxide
LDPE	low density polyethylene
LOI	limiting oxygen index
MCC	microscale combustion calorimeter
MCPBA	meta-chloroperbenzoic
MH	magnesium hydroxide
MMT	montmorillonite
MNP	magnetic nanoparticles
MR	magnetic resonance
MRI	magnetic resonance imaging
PA6	polyamide 6
PAEI	polyalkanetherimide
PDID	perylenediimide derivative
PDMAAm	poly(N, N)-dimethylacrylamide
PE	polyethylenes
PEG	poly(ethylene glycol)
PER	pentaerythritol
PES	photoelectric sensitivity
PHRR	peak heat release rate
PMMA	poly(methyl methacrylate)
PP	polypropylene
PV	photovoltaic
PVA	polyvinyl alcohol
SAN	poly(acrylonitrile-costyrene)
SAXS	small-angle X-ray scattering
SEM	scanning electron microscope
TEM	transmission electron microscopy
TEOS	tetraethylorthosilicate
TG	thermogravimetry
THR	total heat release
TMOS	tetramethylorthosilicate
TTI	time to ignition
XRD	x-ray diffraction
ZP	zeta potential
ZrP	α-zirconium phosphate

PREFACE

The collection of topics in the two-volume publication is to reflect the diversity of recent advances in this field with a broad perspective which may be useful for scientists as well as for graduate students and engineers. This new book presents leading-edge research from around the world in this dynamic field.

Diverse topics published in this book are the original works of some of the brightest and most well-known international scientists in two separate volumes.

In the first volume, modern analytical methodologies are presented here.

The first volume offers scope for academics, researchers, and engineering professionals to present their research and development works that have potential for applications in several disciplines of engineering and science. Contributions range from new methods to novel applications of existing methods to provide an understanding of the material and/or structural behavior of new and advanced systems.

In the second volume, limitations, properties and models are presented. These two volumes:

- are collections of articles that highlight some important areas of current interest in recent advances in chemistry and physics of engineering materials
- give an up-to-date and thorough exposition of the present state-of-the-art of chemical physics
- describe the types of techniques now available to the chemist and technician, and discuss their capabilities, limitations and applications.
- provide a balance between chemical and material engineering, basic and applied research.

We would like to express our deep appreciation to all the authors for their outstanding contributions to this book and to express our sincere gratitude for their generosity. All the authors eagerly shared their experiences and expertise in this new book. Special thanks go to the referees for their valuable work.

ABOUT THE EDITORS

Alexandr A. Berlin, DSc
Director, Institute of Chemical Physics, Russian Academy of Sciences, Moscow, Russia
Professor Alexandr A. Berlin, DSc, is the Director of the N. N. Semenov Institute of Chemical Physics at the Russian Academy of Sciences, Moscow, Russia. He is a member of the Russian Academy of Sciences and many national and international associations. Dr. Berlin is world-renowned scientist in the field of chemical kinetics (combustion and flame), chemical physics (thermodynamics), and chemistry and physics of oligomers, polymers, and composites and nanocomposites. He is a contributor to 100 books and volumes and has authored over 1000 original papers and reviews.

Roman Joswik, PhD
Director, Military Institute of Chemistry and Radiometry, Warsaw, Poland
Roman Joswik, PhD, is the Director of the Military Institute of Chemistry and Radiometry in Warsaw, Poland. He is a specialist in the field of physical chemistry, chemical physics, radiochemistry, organic chemistry, and applied chemistry. He has published several hundred original scientific papers as well as reviews in the field of radiochemistry and applied chemistry.

Nikolai I. Vatin, DSc
Director of Civil Engineering Institute, Saint-Petersburg State Polytechnical University, Chief of Construction of Unique Buildings and Structures Department
Nikolai I. Vatin, DSc, is the Chief Scientific Editor of the *Magazine of Civil Engineering*, and Editor of *Construction of Unique Buildings and Structures*. He is specialist in the field of chemistry and chemical technology. He has published several hundred scientific papers (original and review) and several volumes and books.

A. K. Haghi, PhD

Member of the Canadian Research and Development Center of Sciences and Cultures (CRDCSC), Montreal, Quebec, Canada; Editor-in-Chief, International Journal of Chemoinformatics and Chemical Engineering; Editor-In-Chief, Polymers Research Journal

A. K. Haghi, PhD, holds a BSc in urban and environmental engineering from the University of North Carolina (USA); a MSc in mechanical engineering from North Carolina A&T State University (USA); a DEA in applied mechanics, acoustics and materials from Université de Technologie de Compiègne (France); and a PhD in engineering sciences from Université de Franche-Comté (France). He is the author and editor of 165 books as well as 1000 published papers in various journals and conference proceedings. Dr. Haghi has received several grants, consulted for a number of major corporations, and is a frequent speaker to national and international audiences. Since 1983, he served as a professor at several universities. He is currently Editor-in-Chief of the "*International Journal of Chemoinformatics and Chemical Engineering* and *Polymers Research Journal*" and on the editorial boards of many international journals. He is a member of the Canadian Research and Development Center of Sciences and Cultures (CRDCSC), Montreal, Quebec, Canada.

Gennady E. Zaikov, DSc

Head of the Polymer Division, N. M. Emanuel Institute of Biochemical Physics, Russian Academy of Sciences, Moscow; Professor, Moscow State Academy of Fine Chemical Technology and Kazan National Research Technological University, Russia

Gennady E. Zaikov, DSc, is the Head of the Polymer Division at the N. M. Emanuel Institute of Biochemical Physics, Russian Academy of Sciences, Moscow, Russia, and Professor at Moscow State Academy of Fine Chemical Technology, Russia, as well as Professor at Kazan National Research Technological University, Kazan, Russia. He is also a prolific author, researcher, and lecturer. He has received several awards for his work, including the Russian Federation Scholarship for Outstanding Scientists. He has been a member of many professional organizations and on the editorial boards of many international science journals.

CHAPTER 1

CHEMICAL MODIFICATION OF SYNDIOTACTIC 1,2-POLYBUTADIENE

M. I. ABDULLIN, A. B. GLAZYRIN, O. S. KUKOVINETS, A. A. BASYROV, and G. E. ZAIKOV

Bashkir State University, Ufa, 450077, Russia,
E-mail: ProfAMI@yandex.ru

CONTENTS

ABSTRACT

Results of chemical modification of syndiotactic 1,2-polybutadiene under various chemical reagents are presented. Influence of the reagent nature both on the reactivity of $>C=C<$ double bonds in syndiotactic 1,2-polybutadiene macromolecules at its chemical modification and on the composition of the modified polymer products is considered basing on the analysis of literature and the authors' own researches.

1.1   INTRODUCTION

Obtaining polymer materials of novel or improved properties is considered a major direction in modern macromolecular chemistry [1]. Researches aimed at obtaining polymers via chemical modification methods have been quite prominent along with traditional synthesis methods of new polymer products by polymerization or polycondensation of monomers.

Polymers with unsaturated macrochains hold much promise for modification. The activity of carbon-carbon double bonds as related to many reagents allows introducing substituents of different chemical nature in the polymer chain (heteroatoms including). Such modification helps to vary within wide limits physical and chemical properties of the polymer and render it new and useful features.

Syndiotactic 1,2-polybutadiene obtained by stereospecific butadiene polymerization in complex catalyst solutions [2–7] provides much interest for chemical modification.

In contrast to 1,4-polybutadiens and 1,2-polybutadiens of the atactic structure, the syndiotactic 1,2-polybutadiene exhibit thermoplastic properties combining elasticity of vulcanized rubber and ability to move to the viscous state at high temperatures and be processed like thermoplastic polymers [8–11].

The presence of unsaturated $>C=C<$ bonds in the syndiotactic 1,2-PB macromolecules creates prerequisites for including this polymer into various chemical reactions resulting in new polymer products. Unlike 1,4-polybutadiens, the chemical modification of syndiotactic 1,2-PB is insufficiently studied, though there are some data available [12–15].

A peculiarity of the syndiotactic 1,2-PB produced nowadays is the presence of statistically distributed *cis*- and *trans*-units of 1,4-diene polymerization [16, 17] in macromolecules along with the order of 1,2-units at polymerization of butadiene-1,3. Their content amounts to 10–16%. Thus, by its chemical structure, syndiotactic 1,2-polybutadiene can be considered as a copolymer product containing an orderly arrangement of 1,2-units and statistically distributed 1,4 polymerization units of butadiene-1,3:

1

Taking into account syndiotactic 1,2-PB microstructures and the presence of $>C=C<$ various bonds in the polydiene macromolecules, the influence of some factors on the polymer chemical transformations has been of interest. The factors in question are determined both by the double bond nature in the polymer and the nature of the substituent in the macromolecules.

In the paper the interaction of the syndiotactic 1,2-PB and the reagents of different chemical nature as ozone, peroxy compounds, halogens, carbenes, aromatic amines and maleic anhydride are considered.

A syndiotactic 1,2-PB with the molecular weight $M_n = (53–72) \times 10^3$; $M_w/M_n = 1,8–2,2$; 84–86% of 1,2 butadiene units (the rest being 1,4-polimerization units); syndiotacticity degree 53–86% and crystallinity degree 14–22% was used for modification.

1.2   OZONATION

At interaction of syndiotactic 1,2-PB and ozone the influence of the inductive effect of the alkyl substituents at the carbon-carbon double bond on the reactivity of double bonds in 1,2 and 1,4-units of butadiene polymerization is vividly displayed. Ozone first attacks the most electron-saturated inner double bond of the polymer chain. The process is accompanied by the break in the $>C=C<$ bonds of the main chain of macromolecules and a noticeable decrease in the intrinsic viscosity and molecular weight of the polymer (Fig. 1.1) at the initial stage of the reaction (functionalization degree $\alpha<10\%$) [17–20]. Due to the ozone high reactivity, partial splitting of the vinyl groups is accompanied by spending double bonds in the main polydiene chain.

[η] , dl/g
$M_n \cdot 10^{-3}$
2,0
1,5
1,0
0,5
0
80
70
60
50
40
30
20
10
0
0 2 4 6 8 10 12 14 16
α, %
1
2

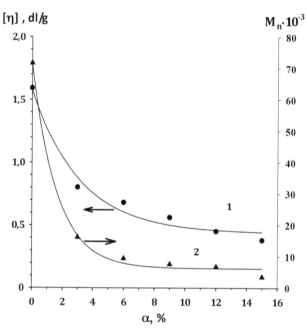

FIGURE 1.1 Dependence of the intrinsic viscosity [η] (1) and the average molecular weight M_n (2) of the formyl derivative of the syndiotactic 1,2-PB from the functionalization degree of the α polymer (with chloroform as a solvent, 25°C).

However, it does not affect the average molecular weight of the polymer up to the functionalization degree of 15% (Fig. 1.1). Depending on the chemical nature of the reagent used for the decomposition of the syndiotactic 1,2-PB ozonolysis products (dimethyl sulfide or lithium aluminum hydride) [17–20], the polymer products containing aldehyde or hydroxyl groups are obtained (Scheme 1).

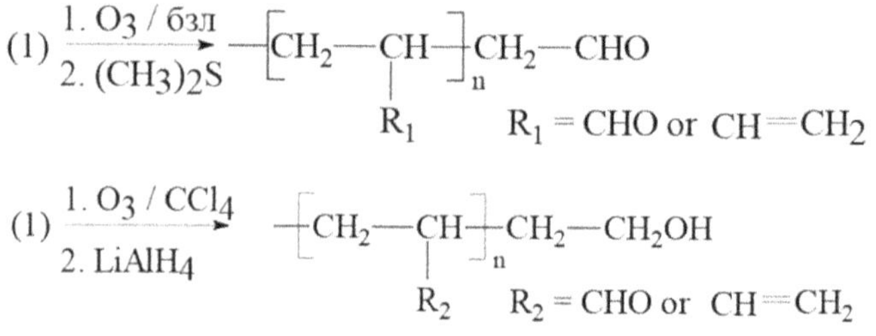

(1)
1. O_3 / бзл
2. $(CH_3)_2S$
$-[CH_2-CH]_n-CH_2-CHO$
R_1
R_1 = CHO or CH=CH_2
(1)
1. O_3 / CCl_4
2. $LiAlH_4$
$-[CH_2-CH]_n-CH_2-CH_2OH$
R_2
R_2 = CHO or CH=CH_2

SCHEME 1

The structure of the modified polymers is set using IR and NMR spectroscopy [17]. The presence of C-atom characteristic signals connected with aldehyde (201.0–201.5 ppm) or hydroxyl (56.0–65.5 ppm) groups in ^{13}C NMR spectra allows identifying the reaction products.

Thus, the syndiotactic 1,2-PB derivatives with oxygen-contained groupings in the macromolecules may be obtained via ozonation. Modified 1,2-PB with different molecular weight and functionalization degrees containing hydroxi- or carbonyl groups are possible to obtain by regulating the ozonation degree and varying the reagent nature used for decomposing the products of the polymer ozonation.

1.3 EPOXIDATION

Influence of the double bond polymer nature in the reaction direction and the polymer modification degree is vividly revealed in the epoxidation reaction of the syndiotactic 1,2-PB, which is carried out under peracids (performic, peracetic, meta-chloroperbenzoic, trifluoroperacetic ones) [21–27], tert-butyl hydroperoxide [21–23] and other reagents [28–30] (Table 1.1).

TABLE 1.1   Influence of the epoxidizing agent on the functionalization degree α of syndiotactic 1,2-PB and the composition of the modified polymer*

Epoxidizing agent	α, mol. %	Content in the modified polymer, mol.%			
		Epoxy groups		$>C=C<$bonds	
		1,2-units	1,4-units	1,2-units	1,4-units
**R^1COOH / H_2O_2	11.0–16.0	-	11.0–16.0	84.0	0–5.0
**R^2COOOH	32.1	16.1	16.0	67.9	-
**R^3COOOH	34.6	18.6	16.0	65.4	-
Na_2WO_4 / H_2O_2	31.0	15.0	16.0	69.0	-
Na_2MoO_4 / H_2O_2	23.7	7.7	16.0	76.3	-
$Mo(CO)_6 / t$-BuOOH	18.0	18.0	-	66.0	16.0
NaClO	16.0	-	16.0	84.0	-
$NaHCO_3 / H_2O_2$	16.0	-	16.0	84.0	-

* the polymer with the content of 1,2- and 1,4-units with 84 and 16%, respectively;

** where R^1 – H–, Me–, Et–, $CH_3CH(OH)$–; R^2 – м–ClC_6H_4; R^3 – CF_3–.

Depending on the nature of the epoxidated agent and conditions of the reaction [21–27], polymer products of different composition and functionalization degree can be obtained (Table 1.1).

As established earlier [21–24, 26], at interaction of syndiotactic 1,2-PB and aliphatic peracids obtained in situ under hydroxyperoxide on the corresponding acid, the $>C=C<$ double bonds in 1,4-units of macromolecules are mainly subjected to epoxidation (Scheme 2). Higher activity of the double bonds of 1,4-polymerization units in the epoxidation reaction is revealed at interacting syndiotactic 1,2-PB and sodium hypochlorite as well as percarbonic acid salts obtained in situ through the appropriate carbonate and hydroperoxide [31–35].

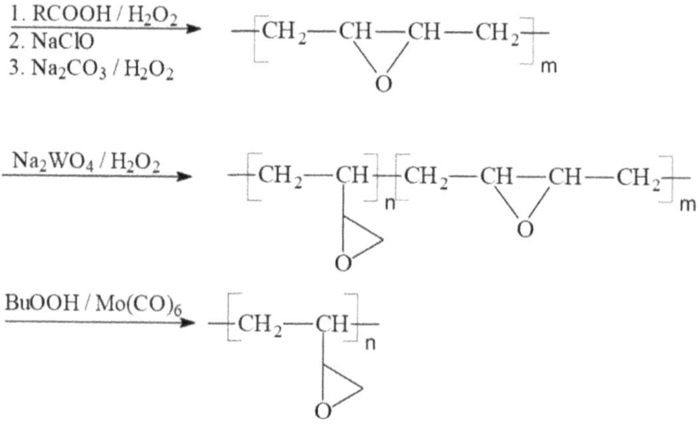

SCHEME 2

It should be noted that the syndiotactic 1,2-PB epoxidation by the sodium hypochlorite and percarbonic acid salts carried out in the alkaline enables to prevent the disclosure reactions of the epoxy groups and a gelation process of the reaction mass observed at polydiene epoxidation by aliphatic peracids [31–32]. The functionalization degree of 1,2-PB in the reactions with the stated epoxidizing agents ($\alpha \leq 16\%$, Table 1.1) is determined by the content of inner double bonds in the polymer.

To obtain the syndiotactic 1,2-PB modifiers of a higher degree of functionalization (up to 35%) containing oxirane groups both in the main and side chain of macromolecules (Scheme 2) it is necessary to use only active epoxidizing agents meta-chloroperbenzoic (MCPBA), trifluoroperacetic acids (TFPA) [23], and metal complexes of molybdenum and tungsten, obtained by reacting the corresponding salts with hydroperoxide) [25] (Table 1.1). From the epoxidizing agents given a trifluoroperacetic acid is most active (Fig. 1.2) [23, 26].

However, at the syndiotactic 1,2-PB epoxidation by the trifluoroperacetic acid, a number of special conditions are required to prevent the gelation of the reaction mass, namely usage of the base (Na_2HPO_4, Na_2CO_3, et al.) and low temperature (less than 5°C) [23].

At reacting the syndiotactic 1,2-PB and the catalyst complex [t-BuOOH – Mo(CO)$_6$] a steric control at approaching the reagents to the double polymer bond is carried out [21–24]. This results in participation of less active but more available vinyl groups of macromolecules in the reaction (Table 1.1, Scheme 2).

Thus, modified polymer products with different functionalization degrees (up to 35%) may be obtained on the syndiotactic 1,2-PB basis according to the epoxidizing agent nature. The products in question contain oxirane groups in the main chain (with aliphatic peracids, percar-

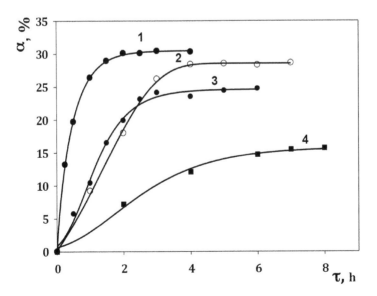

FIGURE 1.2 Influence of the peracid nature on the kinetics of oxirane groups accumulation at the syndiotactic 1,2-PB epoxidation: 1 – TFPA ([Na_2HPO_4] / [TFPA] = 2; 0°C); 2 – MCPBA (50°C); 3 – Na_2MoO_4/H_2O_2 (55°C); 4 – HCOOH / H_2O_2 (50°C).

bonic acid salts, and NaClO as epoxidizing agents), in the side units of macromolecules [t-BuOOH – Mo(CO)$_6$] or in 1,2- and 1,4-units (TFPA, MCPBA, metal complexes of molybdenum and tungsten).

1.4   HYDROCHLORINATION

In adding hydrogen halides and halogens to the $>C=C<$ double bond of 1,2-PB, the functionalization degree of the polymer is mostly determined by the reactivity of the electrophilic agent. Relatively low degree of polydiene hydrochlorination (10–15%) at interaction of HCl and syndiotactic 1,2-PB [16, 36, 37] is caused by insufficient reactivity of hydrogen chloride in the electrophilic addition reaction by the double bond (Table 1.2). Due to this, more electron-saturated $>C=C<$ bonds in 1,4-units of butadiene polymerization are subjected to modification.

The process is intensified at polydiene hydrochlorination under AlCl$_3$ due to a harder electrophile H$^+$[AlCl$_4$]$^-$ formation at its interaction with HCl [37]. In this case double bonds of both 1,2- and 1,4-polidiene units take part in the reaction (Scheme 3):

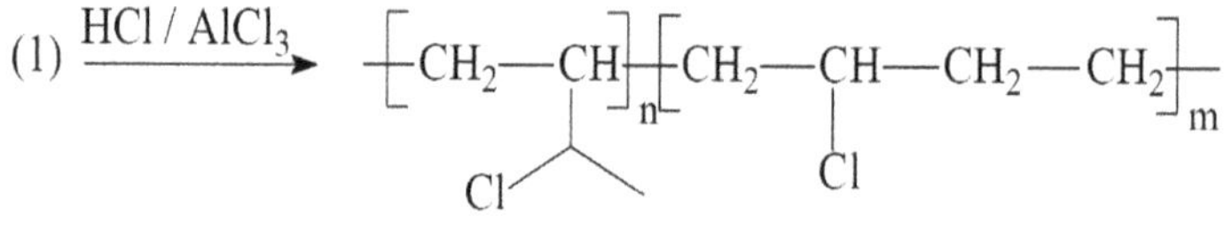

$$(1)\ \xrightarrow{\text{HCl / AlCl}_3}\ \left[\text{CH}_2\text{—CH}\right]_n\left[\text{CH}_2\text{—CH—CH}_2\text{—CH}_2\right]_m$$
$$\underset{\text{Cl}}{\big|} \qquad\qquad \underset{\text{Cl}}{\big|}$$

SCHEME 3

Usage of the catalyst AlCl$_3$ and a polar solvent medium (dichloroethane) allows speeding up the hydrochlorination process (Table 1.2) and obtaining polymer products with chlorine contained up to 28 mass.% and the functionalization degree α up to 71% [37].

By the ^{13}C NMR spectroscopy method [37] it is established that the $>C=C<$ double bond in the main chain of macromolecules is more active at 1,2 PB catalytic hydrochlorination. Its interaction with HCl results in the formation of the structure (a) (Scheme 4). At hydrochlorination of double bonds in the side chain the chlorine atom addition is controlled by formation of the most stable carbocation at the intermediate stage. This results in the structure (b) with the chlorine atom at carbon β-atom of the vinyl group (Scheme 4):

TABLE 1.2 Influence of the Syndiotactic 1,2-PB Hydrochlorination Conditions on the Functionalization Degree α and Chlorine Content in the Modified Polymer

Solvent	Reaction time, h	Chlorine content in the polymer, mass. %	α, %	Polymer out-put, %
chloroform*	24	4.1	10.5	92.0
dichloroethane*	24	5.9	15.1	90.3
chloroform dichloro-ethane	24	12.6	32.2	91.1
	14	18.8	47.9	92.9
dichloroethane	18	25.9	66.3	95.1
dichloroethane	24	27.9	71.2	96.7

*w/o catalyst.

(20–25°C; HCl consumption – 0.2 mol/ h per mol 1,2-PB; [AlCl$_3$]=5 mass. %).

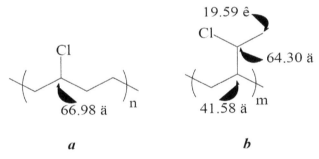

SCHEME 4

1.5 HALOGENATION

Effective electrophilic agents like chlorine and bromine easily join double carbon-carbon bonds [16, 38–40] both in the main chain of syndiotactic 1,2-PB and in the side chains of macromolecules (Scheme 5):

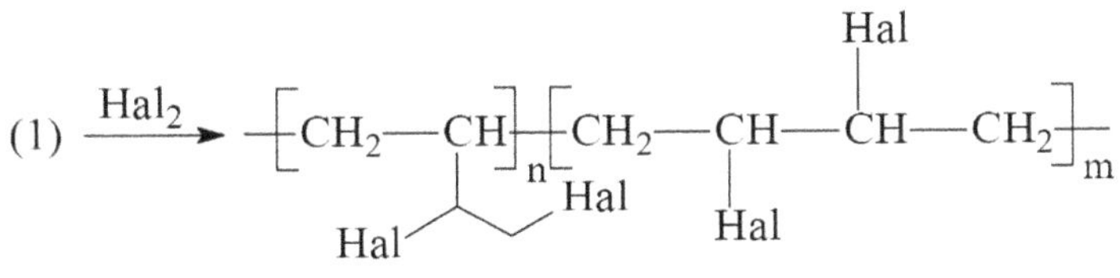

Hal = Cl, Br

SCHEME 5

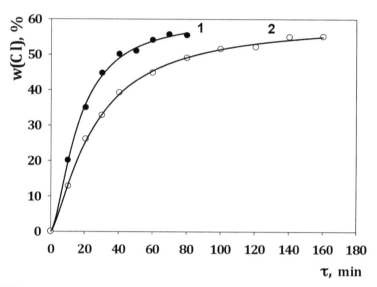

FIGURE 1.3   Kinetics of syndiotactic 1,2-PB chlorination. Chlorine consumption: *1*–1 mol/h per mol of syndiotactic 1,2-PB; *2*–2 mol/h per mol of syndiotactic 1,2-PB; 20°C, with CHCl$_3$ as a solvent.

The reaction proceeds quantitatively: syndiotactic 1,2-PB chlorine derivatives with chlorine w(Cl) up to 56 mass.% ($\alpha \sim$ 98%) (Fig. 1.3) and syndiotactic 1,2-PB bromo- derivatives with bromine up to 70 mass. % ($\alpha \sim$ 94%) are obtained.

According to the ^{13}C NMR spectroscopy, polymer molecules with dihalogen structural units (Scheme 6) and their statistic distribution in the macro chain serve as the main products of syndiotactic 1,2-PB halogenation [16, 40].

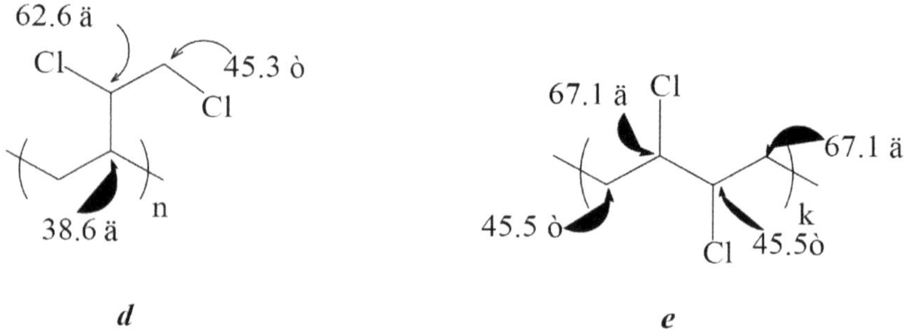

SCHEME 6

1.6 DICHLOROCYCLOPROPANATION

There is an alternative method for introducing chlorine atoms into the 1,2-PB macromolecule structure, namely a dichlorocyclopropanation reaction. It is based on generating an active electrophile agent in the reaction mass at dichlorocarbene modification, which is able to interact with double carbon-carbon polymer bonds [41–43]. The syndiotactic 1,2-PB dichlorocyclopropanation is quite effective at dichlorocarbene generating by Macoshi by the chloroform reacting with an aqueous solution of an alkali metal hydroxide. The reaction is carried out at the presence of a phase transfer catalyst and dichlorocarbene addition in situ to the double polydiene links [44–46] according to Scheme 7:

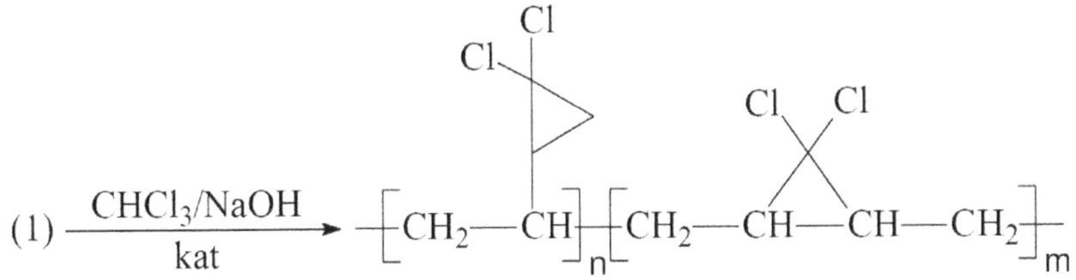

SCHEME 7

The ^{13}C NMR spectroscopy results testify the double bonds dichlorocyclopropanation both in the main chain and side chains of polydiene macromolecules (Scheme 8).

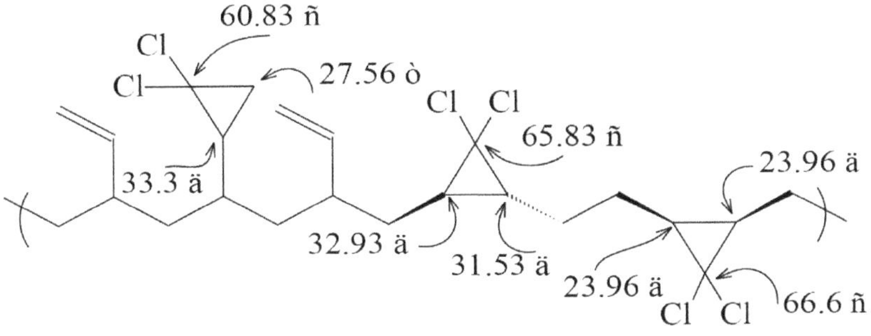

SCHEME 8

Cis- and *trans-*double bonds in the 1,4-addition units [45] are more active in the dichlorocarbene reaction.

The polymer products obtained contain chlorine up to 50 mass.% which corresponds to the syndiotactic 1,2-PB functionalization degree ~97%, that is, in the reaction by the Makoshi method full dichlorocarbenation of unsaturated >C=C< polydiene bonds is achieved [23, 45].

1.7 INTERACTION WITH METHYLDIAZOACETATE

Modified polymers with methoxycarbonyl substituted cyclopropane groups [47–51] are obtained by interaction of syndiotactic 1,2-PB and carb methoxycarbonyl generated at a catalytic methyldiazoacetate decomposition in the organic solvent medium (Scheme 9):

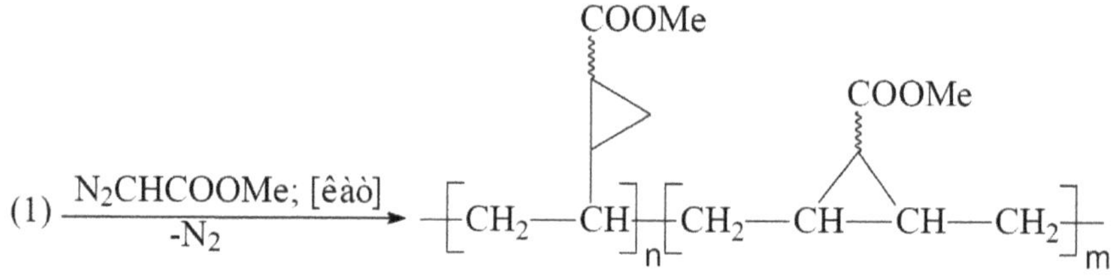

SCHEME 9

The catalytic decomposition of alkyldiazoacetates comprises the formation of the intermediate complex of alkyldiazoacetate and the catalyst [51, 52]. The generated alcoxicarbonylcarbene at further nitrogen splitting is stabilized by the catalyst with the carbine complex formation [51, 52], interaction of which with the alkene results in the cyclopropanation products (Scheme 10):

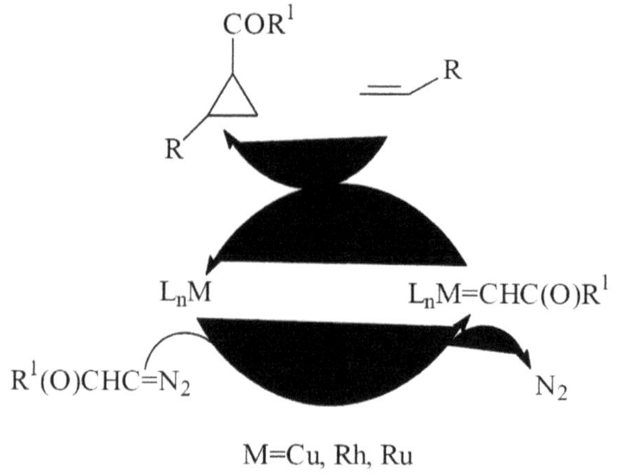

M=Cu, Rh, Ru

SCHEME 10

The output of the cyclopropanation products is determined by the reactivity of the $>C=C<$ double bond in the alkene as well as the stability and reactivity of the carbine complex $L_nM=CH(O)R^1$ which fully depend on the catalyst used [52].

The catalysts applied in the syndiotactic 1,2-PB cyclopropanation range as follows: $Rh_2(OAc)_4$ (α=38%) > [Cu OTf]·0,5 C_6H_6 (α=28%) > $Cu(OTf)_2$ (α=22%) [51].

By the ^{13}C NMR spectroscopy methods it is established that in the presence of copper (I), (II) compounds, double bonds both of the main chain and in the side units of syndiotactic 1,2-PB macromolecules are subjected to cyclopropanation whereas at rhodium acetate $Rh_2(OAc)_4$ mostly the $>C=C<$ bonds in the 1,4- addition units undergo it [51].

Thus, catalytic cyclopropanation of syndiotactic 1,2-PB under methyldiazoacetate allows obtaining polymer products with the functionalization degree up to 38% and their macromolecules containing cyclopropane groups with an ester substituent. The determining factor influencing the cyclopropanation direction and the syndiotactic 1,2-PB functionalization degree is the catalyst nature. By using catalysts of different chemical nature it is possible to purposefully obtain the syndiotactic 1,2-PB derivatives containing cyclopropane groups in the main chain (with rhodium acetate as a catalyst) or in 1,2- and 1,4-polydiene units (copper compounds), respectively.

Along with the electronic factors determined by different electron saturation of the $>C=C<$ bonds in 1,2- and 1,4-units of polydiene addition and the catalyst nature used in modification, the steric factors may also influence the reaction and the syndiotactic 1,2-PB functionalization degree. The examples of the steric control may serve the polydiene reactions with aromatic amines and maleic anhydride apart from the above considered epoxidation reactions of syndiotactic 1,2-PB by *tret*-butyl hydroxyperoxide.

1.8   INTERACTION WITH AROMATIC AMINES

Steric difficulties prevent the interaction of double bonds of the main chain of syndiotactic 1,2-PB macromolecules and aromatic amines (aniline, N, N-dimethylaniline and acetanilide). In the reaction with amines

[17, 20, 23] catalyzed by Na[AlCl$_4$] the vinyl groups of the polymer enter the reaction and form the corresponding syndiotactic 1,2-PB arylamino derivatives (Scheme 11):

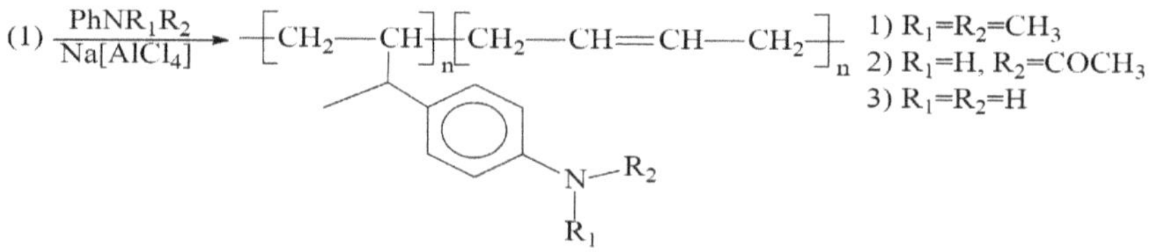

SCHEME 11

From the NMR spectra analysis it is seen that the polymer functionalization is held through the β-atom of carbon vinyl groups [17].

Introduction of arylamino groups in the syndiotactic 1,2-PB macromolecules leads to increasing the molecular weight M_w (Fig. 1.4) and the size of macromolecular coils characterized by the mean-square radius of gyration $(R^2)^{1/2}$ [17, 20].

The results obtained indicate to the intramolecular interaction of monomer units modified by aromatic amines with vinyl groups of polydiene macromolecules at syndiotactic 1,2-PB modification. This leads to the formation of macromolecules of the branched and linear structure (Scheme 12):

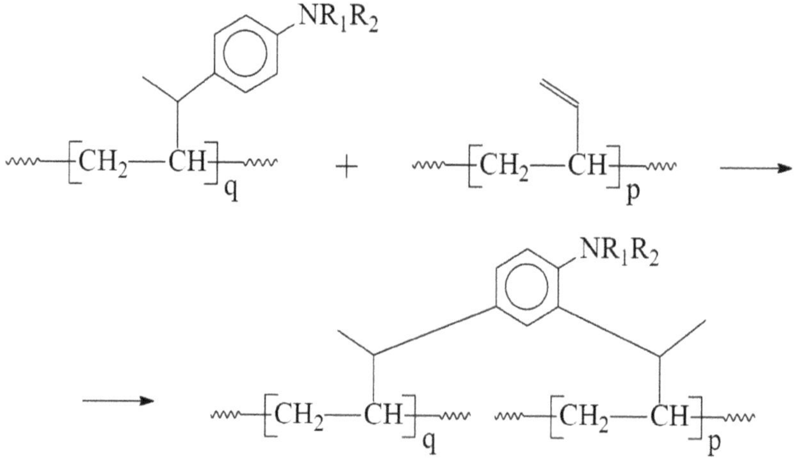

SCHEME 12

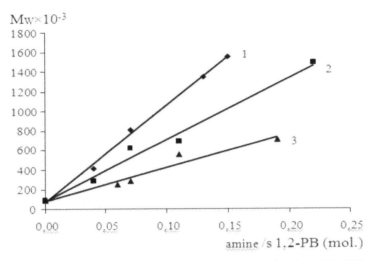

FIGURE 1.4 Influence of the aromatic amine nature on the molecular weight (M_w) of the polymer modified by: 1 – acetanilide; 2 – N, N-dimethylaniline; 3 – aniline.

Steric difficulties determined by the introduction of bulky substituents in the polydiene units ("a neighbor effect" [53, 54]) does not allow to obtain polymer products with high functionalization degree as the arylamino groups in the modified polymer does not exceed 8 mol.%. At the same time secondary intermolecular reactions are induced in the synthesis process involving arylamino groups of the modified macromolecules and result in the formation of linear or branched polymer products with high molecular weight.

1.9   INTERACTION WITH MALEIC ANHYDRIDE

The polymer products with anhydride groups are synthesized by thermal adding (190°C) of the maleic anhydride to the syndiotactic 1,2-PB [23, 24] (Scheme 13):

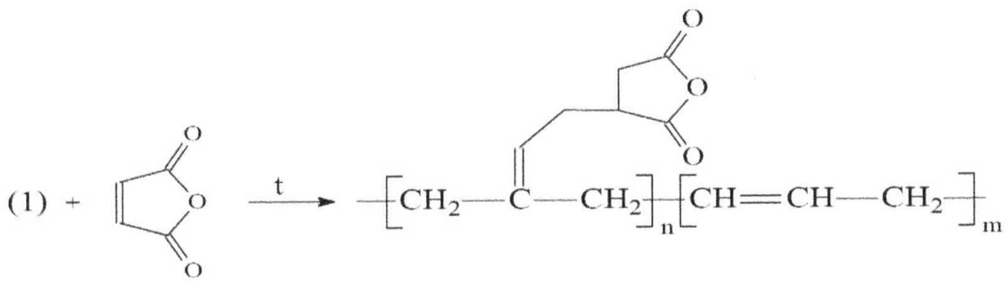

SCHEME 13

The ^{13}C NMR spectroscopy results show that the maleic anhydride addition is carried out as an end-reaction [55] by the vinyl bonds of the polymer without the cycle disclosure and the double bond is moved to the β-carbon atom of the vinyl bond [23, 24]. The maleic anhydride addition to the >C=C< double bonds of 1,4-units of polydiene macromolecules does not take place. As in synthesis of the arylamino derivatives of syndiotactic 1,2-PB, it is connected with steric difficulties preventing the interaction of bulk molecules of the maleic anhydride with inner double bonds of the polymer chain [23].

At syndiotactic 1,2-PB modification by the maleic anhydride, the so-called "neighbor effect" is observed, that is, the introduction of bulk substituents into the polymer chain prevents the functionalization of the neighboring polymer units due to steric difficulties. For this reason the content of anhydride groups in the modified polymer molecules do not exceed ~15 mol. %.

Thus, >C=C< double bonds in 1,2- and 1,4-units of syndiotactic 1,2-PB macromolecules considerably differ in the reactivity due to the polydiene structure. The inductive effect of the alkyl substituents resulting in the increase of the electron density of the inner double bonds of macromolecules determines their high activity in the considered reactions with different electrophilic agents.

At interaction of syndiotactic 1,2-PB with strong electrophiles (ozone, halogens, dichlorocarbene) both inner double bonds and side vinyl groups of polydiene macromolecules are involved in the reaction. It results in polymer products formation with quite a high functionalization degree. In the case when the used reagent does not display enough activity (interaction of syndiotactic 1,2-PB and hydrogen chloride and aliphatic peracids), the process is controlled by electronic factors: more active double bonds in 1,4-units of the polymer chain are subjected to modification whereas the formed polymer products are characterized by a relatively low functionalization degree.

Polymer modification reactions are mostly carried out through vinyl groups at appearance of steric difficulties. They are connected with formation of a bulky intermediate complex or usage of reagents of big-sized molecules (reactions with aromatic amines, maleic anhydride, t-BuOOH / Mo(CO)$_6$). Such reactions are controlled by steric factors. The predominant course of the reaction by the side vinyl groups of polymer macromolecules is determined by their more accessibility to the reagent attack. However, in such reactions high functionalization degree of syndiotactic 1,2-PB cannot be achieved due to steric difficulties arousing through the introduction of bulky substituents into the polymer chain. They limit the reagent approaching to the reactive polydiene bonds.

Thus, a targeted chemical modification of the polydiene accompanied by obtaining polymer products of different content and novel properties can be carried out using differences in the reactivity of $>C=C<$ double bonds of syndiotactic 1,2-PB. Various polymer products with a set complex of properties is possible to obtain on the syndiotactic 1,2-PB basis varying the nature of the modifying agent, a functionalization degree of the polymer and synthesis conditions.

KEYWORDS

- **chemical modification**
- **functionali-zation degree**
- **polymer products**
- **syndiotactic 1,2-polybutadiene**

REFERENCES

1. Kochnev, A. M., Galibeev, S. S., *Khimiya i khimicheskaya tekhnologiya*. 2003, *46(4)*, 3–10.
2. Byrihina, N. N., Aksenov, V. I., Kuznetsov, E. I., Patent, R. F., 2177008. 2001.
3. Ermakova, I., Drozdov, B. D., Gavrilova, L. V., Shmeleva, N. V., Patent, R. F., 2072362. 1998.
4. Luo Steven. Patent US 6284702. 2002.
5. Wong Tang Hong, Cline James Heber. Patent US 5986026. 2000.
6. Ni Shaoru, Zhou Zinan, Tang Xueming. *Chinese Journal of Polymer Sci.*, 1983, *2*, 101–107.
7. Monteil, V., Bastero, A., Mecking, S., *Macromolecules*, 2005, *38,* 5393–5399.
8. Obata, Y., Tosaki Ch., Ikeyama, M., *Polym. J.* 1975. *Vol. 7*, № 2, 207–216.
9. Glazyrin, A. B., Sheludchenko, A. V., Zaboristov, V. N., Abdullin, M. I., *Plasticheskiye massy.* 2005, *8*, 13–15.

10. Abdullin, M. I., Glazyrin, A. B., Sheludchenko, A. V., Samoilov, A. M., Zaboristov, V. N., *Zhurnal prikl.khimii*. 2007, 80, № 11, 1913–1917.
11. Xigao, J., Polymer Sci. 1990. 28, №. 9. 285–288.
12. Kimura, S., Shiraishi, N., Yanagisawa, S., *Polymer-Plastics Technology and Engineering*, 1975, 5, № 1, 83–105.
13. Lawson, G. Patent US 4960834. 2003.
14. Gary, L., Patent US 5278263. 2001.
15. Dontsov, A. A., Lozovik, G. Y., Chlorinated polymers. Khimiya: Moscow. 1979. 232 p.
16. Asfandiyarov, R. N., Synthesis and properties of halogenated 1,2-polybutadienes: Ph.D. Chem. Science.–Ufa, Bashkir State. University, 2008. 135 p.
17. Kayumova, M. A., Synthesis and properties of the oxygen-and aryl-containing derivatives of 1,2-syndiotactic polybutadiene: Ph.D. Chem. Science.–Ufa, Bashkir State. University, 2007. 115 p.
18. Abdullin, M. I., Kukovinets, O. S., Kayumova, M. A., Sigaeva, N. N., Ionova I. A., Musluhov, R. R., Zaboristov, V. N., *Vysokomolek. soyed*. 2004. 46, № 10, 1774–1778.
19. Gainullina, T. V., Kayumova, M. A., Kukovinets, O. S., Sigaeva, N. N., Musluhov, R. R., Zaboristov, V. N., Abdullin, M. I., *Polymer Sci. Ser. B*. 2005. 47, № 9, 248–252.
20. Abdullin, M. I., Kukovinets, O. S., Kayumova, M. A., Sigaeva, N. N., Musluhov, R. R., *Bashkirskiy khimicheskiy zhurnal*. 2006. 13, № 1, 29–30.
21. Gainullina, T. V., Kayumova, M. A., Kukovinets, O. S., Sigaeva, N. N., Musluhov, R. R., Zaboristov, V. N., Abdullin, M. I., *Vysokomolek. soyed*. 2005. 47, № 9, 1739–1744.
22. Abdullin, M. I., Gaynullina, T. V., Kukovinets, O. S., Khalimov, A. R., Sigaeva, N. N., Musluhov, R. R., Kayumova, M. A., *Zhurnal prikl. khimii*. 2006. 79, № 8, 1320–1325.
23. Abdullin, M. I., Glazyrin, A. B., Kukovinets, O. S., Basyrov, A. A., *Khimiya i khimicheskaya tekhnologiya*. 2012. 55, № 5, 71–79.
24. Kukovinets, O. S., Glazyrin, A. B., Basyrov, A. A., Dokichev, V. A., Abdullin, M. I., *Izvestiya ufimskogo nauchnogo tsentra rossiyskoy akademii nauk*. 2013. 1, 29–37.
25. Valekzhanin, I. V., Abdullin, M. I., Glazyrin, A. B., Kukovinets, O. S., Basyrov, A. A., *Aktualnyye problemy gumanitarnykh i yestestvennykh nauk*, 2012. 3, 13–14.
26. Abdullin, M. I., Basyrov, A. A., Kukovinets, O. S., Glazyrin, A. B., Khamidullina, G. I., *Polymer Sci., Ser. B*. 2013. 55, № 5–6, 349–354.
27. Abdullin, M. I., Glazyrin, A. B., Kukovinets, O. S., Basyrov, A. A., *Mezhdunarodnyy nauchno-issledovatelskiy zhurnal*, 2012. 4, 36–39.
28. Abdullin, M. I., Glazyrin, A. B., Kukovinets, O. S., Valekzhanin, I. V., Klysova, G. U., Basyrov, A. A., Patent, R. F., 2465285. 2012.
29. Abdullin, M. I., Glazyrin, A. B., Kukovinets, O. S., Valekzhanin, I. V., Kalimullina, R. A., Basyrov, A. A., Patent, R. F., 2456301. 2012.
30. Kurmakova, I. N., *Vysokomolek. soyed*. 1985. 21, №12, 906–910
31. Hayashi, O., Kurihara, H., Matsumoto, Y., Patent US 4528340. 1985.
32. Blackborow, John, R., Patent US 5034471. 1991.
33. Jacobi, M., M., Viga, A., Schuster, R., H. *Raw materials and application*. 2002. 3, 82–89.

34. Emmons, W. D., Pagano, A. S., *J. Am. Chem. Soc.* 1955. 77, № 1, 89–92.
35. Xigao Jian, Allan, S., Hay. *J. Polym. Sci.: Polym. Chem. Ed.* 1991. 29, 1183–1189.
36. Abdullin, M. I., Glazyrin, A. B., Asfandiyarov, R. N., *Vysokomolek. soyed.* 2009. 51, №8, 1567–1572
37. Glazyrin, A. B., Abdullin, M. I., Muslukhov, R. R., Kraikin, V. A., *Polymer Sci., Ser. A.* 2011. 53, № 2, 110–115.
38. Abdullin, M. I., Glazyrin, A. B., Akhmetova, V. R., Zaboristov, V. N., *Polymer Sci., Ser. B.* 2006. 48, № 4, 104–107.
39. Abdullin, M. I., Glazyrin, A. B., Asfandiyarov, R. N., *Zhurnal prikl. khimii.* 2007. 10, 1699–1702.
40. Abdullin, M. I., Glazyrin, A. B., Asfandiyarov, R. N., Akhmetova, V. R., *Plasticheskiye massy.* 2006. 11, 20–22.
41. Lishanskiy, I.S, Shchitokhtsev, V. A., Vinogradova, N. D., *Vysokomolek. soyed.* 1966. 8, 186–171.
42. Komorski, R. A., Horhe, S. E., Carman, C. J., *J. Polym. Sci.: Polym. Chem. Ed.* 1983. 21, 89–96.
43. Nonetzny, A., Biethan, U., *Angew. makromol. Chem.* 1978. 74, 61–79.
44. Glazyrin, A. B., Abdullin, M. I., Kukovinets, O. S., *Vestnik Bashkirskogo universiteta.* 2009. 14, №3, 1133–1140.
45. Glazyrin, A., B., Abdullin, M., I., Muslukhov, R., R. *Polymer Sci., Ser. B.* 2012. 54, 234–239.
46. Glazyrin, A. B., Abdullin, M. I., Khabirova, D. F., Muslukhov, R. R., Patent RF 2456303. 2012.
47. Glazyrin, A. B., Abdullin, M. I., Sultanova, R. M., Dokichev, V. A., Muslukhov, R. R., Yangirov, T. A., Khabirova, D. F., Patent RF 2443674. 2012.
48. Glazyrin, A. B., Abdullin, M. I., Sultanova, R. M., Dokichev, V. A., Muslukhov, R. R., Yangirov, T. A., Khabirova, D. F., Patent RF 2447055. 2012.
49. Gareyev, V. F., Yangirov, T. A., Kraykin, V. A., Kuznetsov, S. I., Sultanova, R. M., Biglova, R. Z., Dokichev, V. A., *Vestnik Bashkirskogo universiteta.* 2009. *14*, №1, 36–39.
50. Gareyev, V. F., Yangirov, T. A., Volodina, V. P., Sultanova, R. M., Biglova, R. Z., Dokichev, V. A., *Zhurnal prikl. khimii.* 2009. *83*, №7, 1209–1212.
51. Glazyrin, A. B., Abdullin, M. I., Dokichev, V. A., Sultanova, R. M., Muslukhov, R. R., Yangirov, T. A., *Polymer Sci., Ser. B.* 2013. 55, 604–609.
52. Shapiro Ye.A., Dyatkin, A. B., Nefedov, O. M., Diazoether. Nauka: Moscow. 1992. 78 p.
53. Fedtke, M., Chemical reactions of polymers. Chemistry: Moscow, 1990. 152 p.
54. Kuleznev, V. N., Shershnev, V. A., The chemistry and physics of polymers. Kolos: Moscow. 2007. 367 p.
55. Vatsuro, K. V., Mishchenko, G. L., Named Reactions in Organic Chemistry. Khimiya: Moscow. 1976. 528 p.

CHAPTER 2

IMPROVING THE QUALITY OF GAS-FILLED POLYMER MATERIALS BASED ON POLYPROPYLENE BY SONICATION OF THEIR MELT DURING THE FOAMING PROCESS

ANANIEV VLADIMIR VLADIMIROVICH and
SOGRINA DARYA ALEXANDROVN

Moscow State University of Food Production, 125080, Moscow, Volokolamskoye shosse, 11, Russia, E-mail: kaf.vms@rambler.ru

CONTENTS

ABSTRACT

Effect of the ultrasonic treatment during the foaming process on the structure and properties of the gas-filled polypropylene, for example, polypropylene foam, was examined in this article.

2.1   INTRODUCTION

Wide application of polymeric foamed materials, based on the polyolefin's, explained by their mechanical, insulating and operational properties. Foamed polypropylene (PP) is widely used material in numerous applications, but existing foaming technologies of this polymer have some general disadvantages. Foaming of PP requires the use of specific blowing agents, specific types of polymer and special process conditions [1].

From the literature revealed, that significant impact on technological and operational properties of the polymer products renders ultrasonic treatment during their production [2–4]. Based on changes occurring in polymers under sonication, we assume, that using of ultrasonic irradiation, may be promising for the foaming processes. Incidentally, ultrasonic treatment during extrusion of polymer foam can change structure (size of cells, cells distribution in the material volume) of foamed PP and by that change bulk density of foamed materials.

2.2   EXPERIMENTAL PART

Russian polypropylene grade 21020 and Hydrocerol BM 70, as blowing agent, have been chosen for foaming. In order to more clearly demonstrate practical opportunities of using ultrasound for obtaining of the gas-expanded materials were used standard Russian brands of polypropylene for the study (melt flow index 2,26 g/10 min). The blowing agent, Hydrocerol BM 70, is injected in amounts from 0.1 to 1.5% by weight of the composition Hydrocerol BM 70 is a chemical substance that decompose or react by the influence of heat. This is chemical, endothermic foaming agent. To achieve an optimum gas yield, a processing temperature of at least 180 °C is suggested.

Based on the analysis of modern equipment for PP foaming, it was decided to create special laboratory installation. This installation is made on the basis of the single screw extruder, with barrier screw, screw diameter 12 mm. Maximum productivity of installation is 1.5 kg/hr. An installation includes: pressure sensor, unit of ultrasonic processing and a set of capillaries for rheological studies directly during extrusion. The kit of ultrasonic unit includes an ultrasonic generator (oscillation frequency 22 kHz and capacity 300 W), piezoceramic vibration transducer, titanium sonotrode. At the output of the camera is set slit die (width 100 mm). For receiving of flat film is used the system of rolls with an air-cooling. Thus, on the installation with a single screw extruder, a strand can be obtained as well as the film material.

Ultrasonic apparatus for processing the polymer melt represents a complicated system of blocks and elements consisting of transducer of electrical oscillations, system for concentrating of ultrasonic vibrations, instrument for input of ultrasonic vibrations, ultrasonic vibration generator, system of control and automation. Beforehand it was experimentally established that, neither type of generator or type electroacoustic transducer does not affect the results obtained in the process modification polymer melt.

During the experiment it was observed that, PP foaming without ultrasonic treatment, was accompanied by abrupt changes in pressure (both strands and films). While abrupt changes in pressure were not observed during foaming with ultrasound. Samples foamed with ultrasonic treatment of its melt had even surface and uniform thickness over the length. It should be mentioned, that formation of different thickness strands had random nature, while the occurrence of irregular portions of smaller diameter is associated with non-uniform nucleation in polypropylene, foamed without ultrasound. Reducing defects in the treated samples indicates that ultrasound promotes more uniform nucleation and growth of the bubbles in the foam. In previously studies conducted by us, we found, that ultrasonic treatment of the various polymer melts facilitates formation of more uniform and amorphized structure, which shall give rise to favorable conditions for the foaming processes of the melt. Another reason also improving the conditions for foaming, is a better distribution of the components introduced into the polymer melt under the action of ultrasonic vibrations. The result is a more uniform nucleation and growth of bubbles in the foamed material, and reduction of defects in the processed samples.

2.3 RESULTS AND DISCUSSION

At the first stage, were studied the rheological properties of primary material. The test was conducted on the special device intended for laboratory determination of MFI for powdered, granulated and pressed thermoplastics. It was established that PP has a melt flow index 2,26 g/min, that is, as claimed by the manufacturer MFR value for used PP grade.

To confirm the assumption, that ultrasonic treatment caused changes in the structure the foam, we examined the samples by optical microscopy method. Figure 2.1 shows photomicrographs of foamed samples, prepared under the same conditions and with the same magnification. Given photographs were obtained in transmitted light. They showing the general nature of pore distribution in the samples.

Analyzing the micrographs of foamed polypropylene, it is easy to see that for the same parameters of the extrusion process, sonication leads

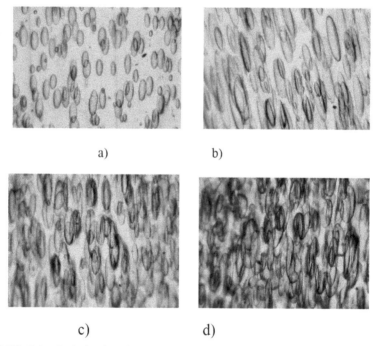

a)   b)

c)   d)

FIGURE 2.1 Optical microphotos of foamed polymeric PP samples. (a) PP sample produced at a temperature of 210°C and containing 0.6% of a blowing agent without the influence of ultrasound; (b) PP sample produced at a temperature of 210°C and containing 0.6% of the blowing agent under the influence ultrasonic; (c) PP sample produced at a temperature of 220°C and containing 0.6% of a blowing agent without the influence of ultrasound; (d) PP sample produced at 220°C and containing 0.6% of the blowing agent under the influence ultrasound.

to increasing of the number gas cells and an increasing in their average size. Increasing extrusion temperature at 10 °C also leads to an increase in the number of pores and increase their sizes The effect from action of ultrasound is stored at elevated temperatures, but sonication promotes the formation of a homogeneous structure with a uniform distribution of cells by volume of the polymer matrix. We plan to get more complete information spending microscopic examination of cross sections of the samples.

Tensile Test Method was used to determine the physico-mechanical characteristics of the samples. We used the method of State Standard 14236–81 for testing. "Polymer films. Tensile test methods." The experiment was conducted on a tensile testing machine type RM−10. Since for test samples was used a heterogeneous material, it was important to obtain accurate data at any given time, so breaking machine with a computer interface was used for testing, allowing to adjust the load up to 10,000 measurements at the second. Limit of allowable measurement error in the direct course does not exceed ± 1% of the measured load. Date physico-mechanical investigations shows an improvement in strength characteristics of irradiated PP samples. Such improvement can be caused by the

TABLE 2.1   The Bulk Density of Foamed Polypropylene Samples

The test material	Sample thickness, micron	Sample weight, g.	The overage bulk density, g/cm³
PP 0.6%HYDROCEROL BM 70 220°C, not sonicated during foaming	0.0308	0.0201	0.69
PP 0.6%HYDROCEROL BM 70 220°C, sonicated during foaming	0.0364	0.0207	0.58
PP 0.6%HYDROCEROL BM 70 210°C, not sonicated during foaming	0.023	0.016	0.72
PP 0.6%HYDROCEROL BM 70 210 °C, sonicated during foaming	0.0339	0.0206	0.65
PP 0.4%HYDROCEROL BM 70 215 °C, not sonicated during foaming	0.0224	0.0187	0.78
PP 0.4%HYDROCEROL BM 70 215 °C, sonicated during foaming	0.0251	0.0108	0.56

formation of more homogeneous structure, foam cells grows and reducing the number of defects.

Further the bulk density of the obtained samples has been investigated. It has been established that the ultrasonic treatment of the polypropylene melt during the foaming process reduces the bulk density of test materials (Table 2.1).

Thus, ultrasound modification of polypropylene melt during foaming promotes the formation of gas-filled material with a more homogeneous structure and fewer defects. Moreover, processed samples have improved physico-mechanical properties and a lower bulk density.

2.4   CONCLUSION

Based on carried out researches it can be concluded, that ultrasonic treatment of the polypropylene melt improves the basic mechanical characteristics of foamed polypropylene. On the results of research on the structure of foamed PP samples by optical microscopy can be concluded, that sonicated PP samples have a more homogeneous structure. It is established that under the action of ultrasound there is a decrease of apparent density of foam that can be associated with the formation of larger pores and a more uniform cells distribution in the volume of material. Thus, we can conclude, that there is a theoretical possibility of changing the structure and properties of foamed materials based on polypropylene using ultrasonic treatment of their melts during the foaming process.

KEYWORDS

- blowing agent
- bulk density
- foaming
- microscopy
- polypropylene
- ultrasound

REFERENCES

1. Klempner, D., Sendzharevich, B. Polymeric foams and foam technology, Profession: St. Petersburg, 2009, 600 pp.
2. Ananev, V. V., Gubanova, M. I., A. Kirsh, I., Semenov, G. V., Kozmin, D. V. Modification of polyethylene initiated by ultrasound. Plastics, №6 (in Rus.): Moscow, 2008, 6–9pp.
3. Semenov, G. V., Ananev, V. V., Kirsch, I. A., Kozmin, D. V., Gubanova, M. I. Recycling of plastic waste under the influence of ultrasound.Plastics, №10(in Rus.): Moscow, 2008, 41–44pp.
4. Ananev, V. V., Filinskaya, Y. A., Kirsh, I. A., Bannikova, O. A., Utkin, A. O. Improving the quality of polymer composite materials and packaging design. Food processing, №1(in Rus.): Moscow, 2012, 16–18 pp.

CHAPTER 3

THE AQUA MELAFEN SOLUTIONS' INFLUENCES ON THE PARAMETERS OF THE MICE'S SOLID *CARCINOMA LUIS* DEVELOPMENT

A. V. KREMENTSOVA, V. N. EROKHIN, V. A. SEMENOV, and O. M. ALEKSEEVA

Emanuel Institute of Biochemical Physics, Russian Academy of Sciences, ul. Kosygina, 4, Moscow, 119334, Russia, Fax: (499) 137 41 01; E-mail: akrementsova@mail.ru

CONTENTS

ABSTRACT

This investigation deals with the actions of aqua Melafen (melamine salt of bis (oximethyl) phosphinic acid) solutions, applied in agriculture, as the plant growth regulator, to the animal tumor cells in vivo. The aqua Melafen solutions influences to the growth kinetics of experimental malignant neo-plasm of mice's solid tumors carcinoma Luis were tested. We measured the changes of tumor sizes, and the duration of animal's life in control and experimental groups of mice. The tested doses of Melafen (10^{-12}, 10^{-9}, 10^{-5} mol/kg) suppressed the growth of carcinoma. Notably, the rate of tumor growth decelerated, and the average tumor volume was decreased at the point of animal death time. The means of lifespan of experimental and control animals were similar. Thereby, we obtained that the Melafen had the appreciable suppressive actions to the tumor development of mice's Luis carcinoma under the all tested dozes, even under the super low dozes (10^{-12} mol/kg). The values of the latent periods of the tumor development in the experimental mice's groups had the same duration, as in control mice groups. The obtained data testify that aqua Melafen solutions slowed the mouse tumor growth, but its does not increase the life expectancy of experimental animals.

3.1 INTRODUCTION

Melafen (melamine salt of bis (oximethyl) phosphinic acid) is the prepara-tion, applied in agriculture, as a plant growth regulator. Subject to concen-tration it can operate, as the stimulator (10^{-18}–10^{-13} M) or as the inhibitor (10^{-9}–10^{-3} M) of the rate of developments of the plant body. We used the aqueous solutions of Melafen from large concentrations up to ultra small ones (10^{-3}–10^{-21} M) when studying its effects on the experimental objects. Taking into account the close interdependence of vegetation and animal bodies in nature, it was necessary to investigate the action of plant growth regulator to the fate of objects by animal's origin.

At the earlier papers [1–5] the data of Melafen influences to the cel-lular components by animal's origin were obtained and were discussed. At that works there were number measurements of Melafen actions to the experimental objects were provided. The model and cellular objects were selected on the principle of the forward complicating of structure and/or function. There were the lipids and proteins compounds of animal cellular membranes: multilammelar liposomes, formed by dimyristoylphosphatidyl-choline or egg lecithin and erythrocytes ghost; and soluble protein – bovine serum albumin; and isolated native cells – erythrocytes and Ehrlich ascetic carcinoma cells. At first, that investigations [1] deals with the influence of Melafen, at the wide concentration range (10^{-3}–10^{-21} M), on the structural properties of lipid membranes with different composition. The lipid-Melafen interactions were tested with differential scanning microcalorymetry and small-angle X-ray diffraction. Authors did not reveal by X-ray diffraction any noticeable structural changes of the egg lecithin membranes at concentrations of Melafen used at crop production. But on the basis of differential scanning microcalorymetry data it was concluded that the microdomains structure of dimyristoylphosphatidylcholine bilayer membrane was changed by Melafen under the polymodal manner.

At the next investigations authors [2, 3] did not reveal any noticeable structural changes of the proteins, which are soluble or bounded by membranes. Bovine serum albumin (soluble protein – the main component of blood serum) and proteins of erythrocyte ghost membranes were tested under the concentrations of Melafen that used at crop production. The Melafen actions to the native isolated erythrocytes were tested by the measurements the value of microviscosity and the rate of spontaneous hemolysis. Melafen had not any destruction actions to the membrane-bounded and soluble proteins under the concentrations that activate the plant growth.

Also the Melafen actions were tested on the protein's functions with aid the activation of purine-depended Ca^{2+}- transduction of transformed ascetic Ehrlich carcinoma (EAC) mouse cells [4, 5]. It was disclosed that over a wide range of concentration (from 10^{-13} M up to 10^{-3} M) the aqueous solution of Melafen oppressively influenced to the purine-depended Ca^{2+}-signal system, witch dealing with the activation of Ca^{2+}-depended K^+- and Cl^--channels, in EAC cells. Melafen exerts the destructive effect on the functioning of transformed EAC cells under the certain region of concentration the Melafen aqueous solution (from 10^{-7} M up to 10^{-3} M).

The next logical stage of research was the study of Melafen aqua solution effects at the level of organism entire. Because the Melafen solution, in low concentrations, is powerful plant growth stimulant, and in big concentrations – the inhibitor, then logically would be suppose, that it is acting on the divided animal's cells too. As the experimental object we chose the animals with transformed cells. It is known that the constant proliferation is the main properties of tumor (malignant) cells. On this basis, the influence of solution Melafen on the development of experimental malignant neoplasm in animals was studied, over a wide range of concentrations. In earlier papers [4, 5] has been demonstrated the life activity oppression of isolated cells of short-lived ascetic Ehrlich carcinoma. As the next object for investigations the long-lived solid Luis carcinosarcoma of mouse was chosen. Due to of long-term monitoring of tumor development and life expectancy of tumor-bearing mice we were able to come to some conclusions about influence of aqueous solutions of Melafen over a wide range of concentrations at the level of organism entire.

The solid carcinoma Luis is the widespread model for the testing of anticancer drugs: nature substances and synthetic ones. Melafen is synthetic hydrophilic agent, derivative melamine and bisphophinic acid. It was applauded as aqueous solutions in doses (10^{-12} mol/kg, 10^{-9} mol/kg, 10^{-5} mol/kg) concerning the mice weights.

3.2   MATERIAL AND METHODS

Carcinoma was transplanted to mice of line C57Bl and their of F1 hybrids (C57BlxDBA). For the watching of development of tumor process in mice there were recorded the change of tumor sizes and the duration of animal life in control (the tumor growth without action) and in experimental (the solution introduction of Melafen) groups.

The animals, which had been transplanted the tumor cells of Luis, were divided on any groups on 7 animals in each group. The first: № 1 group – it was the "clean" control (the tumor developed without any actions). Second № 2 group – was the control animals, which were inputted intraperitoneally on 0.2 mL of distilled water (the water used for the preparation of Melafen solutions). Other groups consisted of animals which were inputted on 0.2 mL of solutions that containing the needed doses of Melafen. So, in third animals group (№ 3) the Melafen solution was inputted to mice body at every day in dose 10^{-5} mol/kg. Over all the 15 injections were done. In fourth animals group (№ 4) the Melafen solutions were inputted daily in dose 10^{-5} mol/kg during the entire period of animal life (up to death of last animal). In № 5-group of animal the Melafen solutions were inputted daily in dose 10^{-9} mol/kg. Over all 15 injections were done. In № 6-group of animal the Melafen solution were inputted daily in dose 10^{-9} mol/kg, during the entire period of animal life (up to death last animal). In № 7-group of animal the Melafen solutions were inputted daily in dose 10^{-12} mol/kg. Over all 15 injections was done. In № 8-group of animal the Melafen solutions were inputted daily in dose 10^{-12} mol/kg, during the entire period of animal life (up to death last animal).

We watched on change in size of tumor growing for registration of tumor process development in mice body. The measurements were made in three of mutually perpendicular directions (that corresponding to the linear dimensions x, y, z) that allowed to estimate the swelling of volume. The volume calculated from the formula:

$$D = \frac{1}{6}\pi xyz$$

The individual kinetic curves were made for everyone animal groups. These curves reflected the dependence of change of tumor sizes from time that past after the carcinoma transplantation to mice (the longitudinal study). The obtained curves of tumor growth were approximated by the Gompertz function [6]:

$$D = D_\infty e^{-be^{-ct}}$$

where, D – the tumor volume, D_∞ – fixed the maximum volume of tumor (it was chosen by the equal to $D_\infty = 12000$, on the assumption of work [6]), b – parameter on kinetic curve, determining the latent period after carcinoma transplantation to mice, c – parameter, rate-controlling of tumor growth.

3.3 RESULTS AND DISCUSSION

In Fig. 3.1, we showed the kinetic curves of tumor growth in mice body, when three different doses of Melafen solution were injected to animals. In practice when all injection schemes of solution mice of Melafen introductions to animal's body, we observed the inhibition of growth of carcinosarcoma the Luis. The curves, obtained for experimental groups of

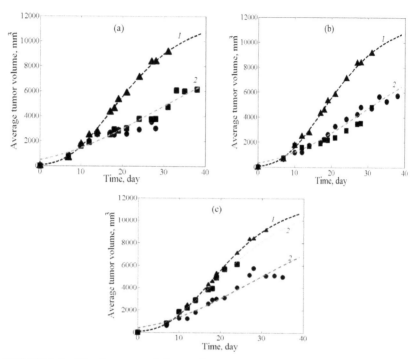

FIGURE 3.1   Kinetic curves of changes in average tumor volume in mice at the control ("triangle", regression – curve 1) and at the administration Melafen solution to mice (regression – curve 2) in different doses: (a) 10^{-5} mol/kg, group № 3 – "points", group № 4 – "squares"; (b) 10^{-9} mol/kg, group № 5 – "points", group № 6 – "triangles"; (c) 10^{-12} mol/kg, group № 7 – "points", group № 8 – "squares."

mouse, are located lower than curve was built for control group of mice (at the figure it was the curve 1). There were not significant differences from control groups. Only group of mice № 8 curve was not significant different from control group.

All individual kinetic curves of tumor growth were fitted by Gompertz function and approximating parameters b and c were calculated. The corresponding values of kinetic parameters of b and c (their median data with 25–75% intervals) are introduced on Fig. 3.2. The significant differences between control group and administration Melafen solution to mice groups was estimated by nonparametric Mann-Whitney U test.

Estimating the parameter of b, characterizing the latent period after the carcinoma transplantation mice (Fig. 3.2), it is possible to note that the

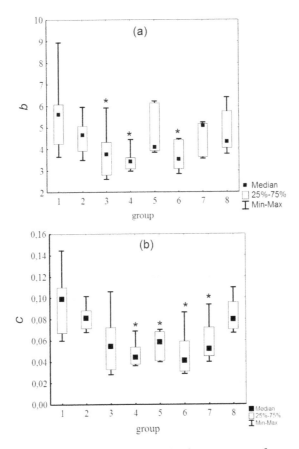

FIGURE 3.2 The Melafen influence on the kinetic parameters of tumor growth. The median, 25–75% intervals, minimum and maximum values of parameters. (a) latent period of tumor development b; (b) rate of tumor growth c. Sing (*) were denoted significant difference by Mann-Whitney U test from control set ($p<0.05$).

animals group № 3 and the group № 4 (the Melafen solution in dose 10^{-5} mol/kg, when different injection circuits) and the group № 6 (the Melafen solution in dose 10^{-9} mol/kg at the introduction up to death of last animal) showed the significant differences from control group. Animals groups № 7 and № 8 (the most small dose of Melafen in solution, 10^{-12} mol/kg) had not shown significant differences from control. On parameter of c, reflecting the rate of tumor growth in mice body, significant difference were found in groups of animals № 4, № 5, № 6 and even in group № 7 (the most small dose of Melafen in solution 10^{-12} mol/kg, when 15-fold intro-duction). The comparing on both kinetic parameters of b and c animals groups № 1 ("clean" the control) and № 2 (the dissolvent introduction – water) demonstrated the absence of differences.

So that, we indicated that the Melafen solution (in concentrations from 10^{-12} M up to 10^{-5} M) slows the rate of the tumor growth at experimental mice compared to checking, that is, influences the kinetics of developmental expression of carcinosarcoma Luis.

If address to results of comparison average tumor volume at death point of animals (Fig. 3.3), then it should be noted that the tumor size in animals, which inputted the Melafen solution in groups № 3 and № 4 (10^{-5} mol/kg) and groups № 5 and № 6 (10^{-9} mol/kg) were significant smaller than in control. The Melafen solution even in most small dose (10^{-12} mol/kg), used for animals in groups № 7 and № 8, is likely also decreased average the tumor size in mice, but the datasets were not significant different.

Most important finding of antitumor activity of preparation – is the increasing of average duration of animal life. In our work we estimated the mean life span of checking and experimental mice (Fig. 3.4).

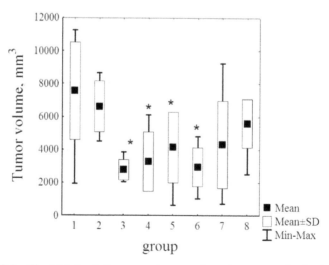

FIGURE 3.3 The Melafen influence on tumor volume in groups of animals. Average tumor volume in mice, standard deviation, minimum and maximum values of tumor volume (mm³) at the time of animal's death. Sing (*) were denoted significant difference by Mann-Whitney U test from control set ($p<0.05$).

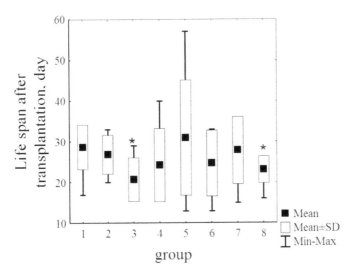

FIGURE 3.4 The Melafen influence on life span of mice in groups of animal after carcinoma transplantation. Mean life span per days (day), standard deviation, and minimum and maximum life span after transplantation of tumor cells to mouse body. Sing (*) were denoted significant difference by Mann-Whitney U test from control set ($p<0.05$).

As it can be seen from Fig. 3.4, the mean life span of animals, which was inputted the Melafen solution, when different dosage schedules of preparation, in practice has not changed. Furthermore, in group № 3 (the Melafen solution 10^{-5} mol/kg that was inputted until to the moment of death of last animal) the toxicity being discussed material was found: the experimental animals were killed significant earlier than checking mouse. Obtained data testify that for Melafen solution the nonlinearity of dose-response relationship has been found, that corresponds famous perceptions of effect dependence from Professor, E.B. Burlakova [7].

3.4 CONCLUSION

So that, the action of aqueous solution of Melafen, over a wide range of concentrations, from 10^{-12} M up to 10^{-5} M, at the level of whole animal organism, as an example of model the carcinosarcoma Luis was studied. When administered to an aqueous solution of the drug at various doses Melafen, the tumor size in animals was significantly lower in all experimental groups compared to tumor size in animals control group. This suggests concluding that the Melafen solution in doses 10^{-12} mol/kg, 10^{-9} mol/kg, 10^{-5} mol/kg slows tumor growth in sick animals.

The values of latent period of tumor development after the carcinoma transplantation to mice body in all experimental groups of animals in practice was not different from one for animals in control groups. But the tumor size reduced when action of solution Melafen. It is possible to suggest that the Melafen solution, without being toxic to cells transformed solid Luis lung carcinoma, to some extent, has a depressing effect on their livelihoods.

The duration of life when different dosage schedules of Melafen solution, over a wide range of concentrations, in experimental groups of mice compared to duration of animal life in control group has not significant changed for most of animals groups. Perhaps this is due to changes in cellular regulation in tumor when exposed to it Melafen.

As its were demonstrated in the Refs. [3–5], the Melafen solution oppressively influence to the purine-dependent calcium signaling in cells with uncontrolled growth – cells of short-lived ascetic Ehrlich carcinoma. In these works was presented the scheme of initiation of Ca^{2+}-signaling when activating of ATP-dependent metabotropic purine receptors P2Y, and also store operated Ca^{2+}-channels (SOC). P2Y and SOC were being as integral proteins at EAC cells plasmalemma. According to set out of scheme and kinetic curves of cell answers to additions of Melafen solution, over a wide range of concentrations, it is possible believe that the Melafen solution, while acting directly or is mediated through the metabolic pathways on Ca^{2+}-channels, changes in cells Ca^{2+}-transitions, that influence the receptors functioning, enzymes activity and channels gates etc. This, are likely, brings about not only to life activity suppressing of individual insulated from cells organism of short-lived tumor of ascetic Ehrlich carcinoma of mice, but, it is possible, and to growth retardation of such growth, as the solid malignancy of carcinosarcoma the Luis in whole organism of mice.

The obtained data provide any evidence rather about regulator role of Melafen solution, used over a wide range of concentrations, when proliferation of the cells with uncontrolled growth in animals, than about antitumor data activity of preparation.

KEYWORDS

- **lifespan**
- **Luis carcinoma**
- **Melafen**
- **tumor development**

REFERENCES

1. Alekseeva, O. M., Krivandin, A. V., Shatalova, O. V., Rykov, V. A., Burlakova, E. B., Fattakhov, S. G., Konovalov, A. I. "The Melafen-Lipid- Interrelationship Determination in phospholipid membranes." "Reports of Russian Academy of Sciences" (in Rus.). 2009. V. 427. № 6. 837–839.
2. Alekseeva, O. M., Krementsova, A. V., Shibriyaeva, L. S., Sofina, S.Yu., Muhamedzanova, E. R., Zaikov, G. E. "Melafen action at objects of animal origin." Vestnic KNITU of Kazan Technological University. 2013. V.16. №8, 195–198.
3. Alekseeva, O. M., Fatkullina, L. D., Kim, Yu.A., Burlakova, E. B., Fattakhov, S. G., Goloshchapov, A. N., Konovalov, A. I. "Melafen influence on structural and functional state of liposome membrane and cells of ascetic Ehrlich carcinoma." Bulletin of experimental biology and medicine. 2009. V.147, № 6, 684–688.
4. Alekseeva, O. M. "Melafen – Plant Growth Regulator and Pathways of cells." Polymers Research Journal, V.7, №1, 2013. 117–128.
5. Alekseeva, O. M. "The influence of Melafen on animal cells." Journal of Information, Intelligence and Knowledge. 2013, V.5, Iss.1, 33–44.
6. Dronova, L. M., Belich, E. I., V. N. Erokhin, Emanuel, N. M. "Kinetic the objective law of the development of rat's erithromielose." "News (Izvestiya) of Russian Academy of Sciences" (in Rus.). series. Biol. 1966. № 5. 743–750.
7. Burlakova, E. B. "Effect of ultrasmall doses." "Herald of Russian Academy of Sciences" (in Rus.). 1994. V. 64. № 5. 425–431.

CHAPTER 4

ANTHROPOGENY BED AND RADIOMETRIC MEASUREMENTS DEPOSITS OF SURFACE WATER SOME COASTAL AREAS IN VALLEY AND ARDON SADON

MARGARITA E. DZODZIKOVA

North Ossetian State Nature Reserve, D.1, Chabahan Basiev St., Alagir, North Ossetia-Alania, 363000, Russia; Tel.: +7-918-822-42-69; E-mail: dzodzikova_m@mail.ru

CONTENTS

ABSTRACT

Conducted morphometric studies and radiometric survey along the riverbanks Ardon and some of its tributaries in the North Ossetian State Natural Reserve, the buffer zone and adjacent areas. Here are conducted works on construction of hydroelectric power plant Zaramag and the gas pipeline in southern portal.

In riverbed Ardon registered 49 loose dump sites, from 21 of them falling down stones until reach the water's edge, artificially narrowing the riverbed Ardon. Located in the surveyed areas as stone crushing company, which also blamed the breed along rivers Sadon and Ardon also narrowing the riverbed above-mentioned rivers. In addition, they will be reset waste process water, increasing the turbidity of the rivers that may be adversely affected hydrobiotite, investigated watercourses. Uncontrolled growth in the number and volume of unfortified dump sites has a special danger in case of flooding.

Analysis of the radiation measurements showed that according to the "Methodology of dosimetric control of industrial waste," approved by the State Standard of the Russian Federation, the background radiation of the investigated areas does not exceed the maximum allowable concentrations and may be considered pure radiation and interventions that do not require.

4.1 INTRODUCTION

The North–Ossetian National Nature Reserve includes 592 streams with a combined length of 831.5 km. The main waterways are the rivers Ardon and Fiagdon. Ardon receives in the sum of 173 left tributaries, which make 225 km long, and 148 right tributaries having to 205 km in length. The basin of river Ardon has 76 glaciers, which cover the area 35.42 km² [1–4].

There are more than 70 unique, in chemical structure, and the richest, by water discharge, mineral springs in the North-Ossetian reserve. Some of which have been well studied by now, with their healing properties having been successfully used in complex therapies such as for the treatment of diseases of gastrointestinal tract, liver, gallbladder, bile ducts and kidneys musculoskeletal as well as cardiovascular and female reproductive systems, respiratory organs, skin diseases and disorders of metabolism [5].

The Reserve is protected and the frontier zones are richly populated with all manner of animals. For instance there are about 1500 turs (*Capra caucasica cylindricornis,* Blyth, 1841) and some 130 chamois (*Rupicarpa rupicarpa,* Linnaeus, 1758), in the upper tributaries of Ardon.

The broad-leaved forests of the North-Ossetian reserve are inhabited by bison (*Bison bonasus*, Linnaeus, 1758), roe deer (*Capreólus capreólus*, Linnaeus, 1758), wild boar (*Sus scrofa*, Linnaeus, 1758), approximately 30–35 bears (*Ursus arctos*, Linnaeus, 1758), jackals (*Canis aureus*, Linnaeus, 1758), wild cats (*Felis silvestris* Schreber, 1777), badgers (*Meles meles*, Linnaeus, 1758), pine martens (*Martes martes*, Linnaeus, 1758), American mink (*Mustela vison*, Schreber, 1777), raccoon dog (*Nyctereutes procyonoides*, Gray, 1834), lepus (*Lepus europaeus*, Pallas, 1778) and a squirrels (*Sciurus*, Linnaeus, 1758) [6].

There are more than 2000 species of invertebrates and insects in the Ardon river valley. Some of them are included in the Russian Federation Red Book: Dybok steppe (*Saga pedo*, Pallas, 1771), Askalaf motley (*Ascalaphus macaronius*, Scopoly, 1763), Carabus caucasicus (*Carabus caucasicus*, Adams, 1817), Carabus hungaricus (*Carabus hungaricus*, Fabricius, 1792), Fiery hunter (*Calosoma sycophanta*, Linnaeus, 1758), butterfly: European swallowtail (*Papilio machaon*, Linnaeus, 1758), Rite swallowtail (*Iphiclides podalirius*, Linnaeus, 1758), Apollo butterfly (*Parnassius apollo*, Linnaeus, 1758), Black Apollo (*Parnassius mnemosyne*, Linnaeus, 1758), Apollo Normans (*Parnassius nordmanni*, Siemaschko, 1850), Underwing moth (*Catocala sponsa*, Linnaeus, 1767), Acherontia atropos (*Acherontia atropos*, Linnaeus, 1758), Daphnis nerii (*Daphnis nerii*, Linnaeus, 1758), Jersey tiger (*Callimorpha quadripunctaria*, Linnaeus, 1758) and scarlet tiger moth (*Callimorpha dominula*, Linnaeus, 1758). Then among the more unusual species of invertebrates for this nature reserve are found Tarantula (*Lycosa*, Latreille, 1804), Solifugae (*Solifugae*, Sundevall, 1833) and Iris polystictica (*Iris polystictica*, Fischer von Waldheim, 1846). There are five different species of bat – with upto 1300 specimens. In the Subi-Nihashi cave: Lesser mouse-eared bat (*Myotis blythi*, Tomes, 1857), Lesser horseshoe bat (*Rhinolophus hipposideros*, Bechstein, 1800), Greater horseshoe-nosed bat (*Rhinolophus ferrumequinum*, Schreber, 1774), Myotis mystacinus (*Myotis mystacinus*, Kuhl, 1817) and Myotis ikonnikovi(*Myotis ikonnikovi*, Ognev, 1912) [7].

207 species of birds live in the nature reserve. Some of them are included in the Red Book of the Russian Federation: Caucasian blackcock (*Lyrurus mlokosiewiczi*, Taczamowski, 1875), Guldenstadt's redstart (*Phoenicurus erythrogaster*, Gould, 1850), Lammergeyer (*Gypaetus

barbatus, Linnaeus, 1758), Erne (*Aquila chrysaetos*, Linnaeus, 1758), Hooded vulture (*Neophron percnopterus*, Linnaeus, 1758) and Peregrine falcon (*Falco peregrinus*, Tunstall, 1771) [8].

There are two species of fish in the rivers of the reserve: Brook trout, *(Salmo trutta*, Linnaeus, 1758) and Tera Barbel (*Barbus ciscaucasicus*, Kessler, 1877) [9].

The Flora of the nature reserve has more than 1500 species. There are species included in the Red Book of the USSR: Ardonic Bell (*Campanula ardonensis*, Rupr., 1867), *Cladochaeta candidissima, Turkish hazel* (*Corylus colurna*, Linnaeus, 1753), Orchis (Órchis militáris, Linnaeus, 1753), *Vavilovia Formosa* (*Vavilovia formosa*, Fed., 1852) among others. There are additionally relics, preserved since ancient times like oriental beech (*Fagus orientalis*, Lipsky, 1898), Common yew (*Táxus baccáta*, Linnaeus, 1753), Oriental spruce (*Picea orientalis*, Peterm, 1845), Nordmann Fir-tree (Ábies nordmanniána, Spash, 1841), Macrophylla thickwalls (*Pachyphragma macrophyllum*, Hoffm.) and others [10].

The moist hollows and the Northern slopes are covered by herb meadows. Here you can find Caucasian rhododendron (*Rhododendron caucasicum*, Pall.), which can be reached up to a height of 3000 m.

The Alpine belt is characterized by the predomination of lichen and moss-lichen wastelands, herb-shrub carpets and low-grass meadow vegetation.

The forests cover over 6000 hectares of the nature reserve. Coniferous forest consist basically of pines and deciduous forests, represented by birches. Juniper Cossacks (Juniperus sabina) is known to be present here. The main forest-forming species are: Koch's pine (*Pinus kochiana*, K.Koch.), Litvinov's birch (*Betula litwinowi*, Trautv., 1887), Oriental beech (*Fugus orientalis*, Lipsky, 1898), Speckled alder (Álnus incána, Moench, 1794). There are areas with prevalence of mountain ash (*Sórbus aucupária*, Linnaeus, 1753), Bird cherry (*Prunus*, Linnaeus, 1753), Common yew (*Táxus baccáta*, Linnaeus, 1753), Trautvetter's maple (*Acep trautwetteri*, Medw.), pine forests with green moss and others [11].

However, through the territory of North-Ossetian national nature reserve runs the Trans-Caucasian highway. Furthermore, in 2009, the construction of the Zaramag hydro power plant (HPP) commenced and since 2010 they have been building the basin. Daily regulation derivation and summing to water diversion tunnels, paved roads, necessary for vast amounts of construction machinery and vehicles causes situation where roads are not moistened as they should be, exhaust gases and dust pollute the environment, and roadside vegetation is dying, so that the normal existence of all life in these territories is disturbed. There are additionally four stone crushing companies operating in the valleys of the rivers Ardon and Sadon, and a tailing damp (complex of special facilities and equipment intended for the storage or disposal of radioactive, toxic and other waste dump mineral called tails. On mountain-concentrating combines from entering mined ore receive a concentrate, and waste recycling move in the tailings pond) of the Mizur's concentration plant in the area of the village Nizhniy Unal, in the immediate vicinity of the river Ardon.

The purpose of this study is to investigate the ecological state and radiation background of the banks of the river Ardon and some of its tributaries, in the natural reserve, protective zone and cross-border region adjacent territories, which are undergoing increasing anthropogenic impact due to active construction works on: the objects of Zaramag hydroelectric power station, the gas pipeline in Judjniy portal, the stone crushing enterprises and unauthorized dumping rocks along the riverbeds.

4.2   MATERIALS AND METHODOLOGY

The conducted morphological research of the state of the banks of the river Ardon and some of its tributaries on the reserve buffer zone and on the adjacent area at different stages of the formation of artificial coastal zones who are filled by waste rock. In addition on these plots were measures of radiation. For radiometric measurements the dosimeter of gamma-radiation, namely Geiger-Muller DKG-03D was used "Grach". Work was performed taking into account the "Norms of radiation safety", "Main sanitary rules of radiation safety" and "safety Rules at operation of electrical installations" (Resolution № 47, on approval of the sanitary rules and norms 2.6.1.2523–09, 7 July 2009, Registered at the Ministry of Justice of the Russian Federation of August 14, 2009 N 14534).

4.3   RESULTS AND DISCUSSION

The practice of dumping rocks along a channel of the river Ardon was formed gradually. At first it was single discharges, which continued for some time and were ended in response to the observations from the management of the Park in 2009 and 2010. But at some stage they took on an uncontrollable character. As a result of this in October 2011, it was decided to measure the morphometry of the coastal zone, of the river Ardon, along the extent of the problem areas. The results of this study are presented in Table 4.1.

The study showed that as of October 2011 along the river Ardon a total of approx. 855,010 m^3 rock dump had been unloaded.

In April 2012, the measurement of heap rocks along the river Ardon were again repeated. The results are presented in Table 4.2.

Thus, it is revealed that if, as of October 2011, along a channel river Ardon about 855,010 m^3 rock dump was unloaded (Table 4.1), then by April 2012, at the aforesaid territories a total of 1,415,597 m^3 was uploaded, species (PL. 2), that is, for 6 months the volume of the dumps increased by 560,587 m^3, and this without taking account of the whole range of small-volume of unauthorized dumps and landfills of solid household and construction waste.

There are 13 tunnels on a relatively small segment, between 62 and 70 km of the gas pipeline along the bed of river Ardon (from the village Buron to Head HPP) and 26 dump zones have been registered (less lengths of the tunnels) (Fig. 4.1).

The water's edge (rarer coastline) – the line of intersection of the water surface of any of the basins (river flows or reservoirs) with the surface of the land. The height above sea level of the watercourse (reservoir) is determined by the elevation of the water's edge.

And beside that, apart from those volumes of the rock dumps, which have been filled on the foundation pit's perimeter of the daily runoff pound (including the Bad gorge's direction), it was found that 15 of 21 dump sites, which have been studied along the river Ardon, are composed of rocks, exported from the construction sites of the pool of the daily run-off pound to the derivation tunnel.

TABLE 4.1 Morphometric Measurements (Length, Width, Height, and the Approximate Amount) Dumps Along the Banks of the River Ardon (Upstream) (17.10.2011)

The location of the dump	Length, m	Width, m	Height, m	Volume, m³ ≈
Right riverside[2] (26 km²)	130	12	11	17 160
Left riverside[1], the village of Mizur (before the entrance to the 1st tunnel)	60	15	18	16,200
Right bank[1], before the river Baddon	400	10	2.5	2,220
River Sadon, the left inflow of the river Ardon	900	3	2.9	7,830
Left bank[1], the village Nusal (all)	580	16	79	46,110
Right bank[1], the village Nusal, below the top of the bridge	85	6	5	2,550
Right bank[1], above the top of the bridge and the settlement of Nusal	80	7	19	10,640
58-th km of the gas pipeline[2]	180	10	23	41,400
Right riverside[1], before the village Buron	380	9	18	61,560
Right riverside[1], above the village Buron	350	95	7	232,750
Right bank[1] before turning in Kosh gorge	240	5	8	9,600
Left bank[1], pipelines tunnel[2]	7	3	10	210
Left bank[1], 50 m below the Head HPP	480	5	18	43,200
Left bank[1], 1 km below the Head HPP	1050	10	3	31,500
Right bank[1], 2 km below the Head HPP	100	40	0,9	3,600
Right bank[1] observation platform over HPP	450	4	28	50,400
Left bank[1], the road to Mamisoni gorge	240	4	32	30,720
Left bank[1], (tunnel "Kosh № 5")	140	4	37	20,720
Left bank[1], the 64-th km of the gas pipeline[2]	95	26	9.5	23,465
Left bank[1], the 62–64-th km of the gas pipeline[2]	1700	2.5	13.5	57,375
Dumps in the construction of roads (10 tunnels), the 68–70-th km of the gas pipeline[2]	1800	3	27	145,800
Sum total				855,010

[1]The river Ardon.

[2]The gas pipeline Dzuarikau – Tskhinval.

TABLE 4.2   Measurements Dumps Along the River Ardon (Upstream) (2.4.2012)

The location of the dump	Length, m	Width, m	Height, m	Volume, $m^3 \approx$
Right bank[2] (26 km[2])	245	17	14	58,310
Left bank[1], the village of Mizur (before the entrance to the 1st tunnel)	100	18	20	36,000
Right bank [1], before the river Baddon	700	13	3	6150
River Sadon, the left inflow of the river Ardon	1200	5	5	30,000
Left bank[1], the village Nusal (all)	800	20	83	79,675
Right bank[1], the village Nusal, below the top of the bridge	100	6	7	4200
Right bank[1], above the top of the bridge and the settlement of Nusal	100	10	25	25,000
58-th km of the gas pipeline[2]	200	12.5	25	62,500
Right bank[1], before the village Buron	450	10	22.5	101,250
Right bank[1], above the village Boron	400	120	9	432,000
Right bank[1] before turning in Kosh gorge	300	4	9.5	11,400
Left bank[1], pipelins tunnel[2]	8	2	12	192
Left bank[1], 50 m below the Head HPP	500	5	20	50,000
Left bank[1], 1 km below the Head HPP	1200	11	3.5	46,200
Right bank[1], 2 km below the Head HPP	150	50	1.5	11,250
Right bank[1] observation platform over HPP	492	3.25	30	47,970
Left bank[1], the road to Mamisoni gorge	300	4	35	42,000
Left bank[1], (tunnel "Kosh № 5")	150	4	40	24,000
Left bank[1], the 64-th km of the gas pipeline[2]	100	30	12.5	37,500
Left bank[1], the 62–64-th km of the gas pipeline[2]	2000	2.5	20	100,000
Dumps in the construction of roads (10 tunnels), the 68–70-th km of the gas pipeline[2]	2000	3.5	30	210,000
Sum total				1,415,597

[1] The river Ardon.

[2] The gas pipeline Dzuarikau – Tskhinval.

FIGURE 4.1   Dumps along the right Bank of river Ardon, reaching to the edge water (Photo by Margarita El. Dzodzikova).

We did not include in the above-represented tables a number of single and small areas of heaped rocks and places of domestic waste discharges, located at some distance from the bed of the river Ardon. Only spoil heaps, reaching up to the water's edge and really narrowing the riverbed, were of interest to us (Fig. 4.2). On the whole, 49 spoil heaps are registered in the bed of the river Ardon, Heaped rocks reach up to the water's edge on the 21 spoil heaps.

Thus, comparative analysis of morphometric data of the state of the banks along the river Ardon shows that the uncontrolled discharge of loose rocks, which are not strengthened, has continued despite repeated warnings by the nature reserve's administration.

Twice a day the head hydroelectric power station makes monitoring discharges of water, and these flows wash away the bottom edge of the rock dumps. Spoil heaps regularly creep down into the river in these

FIGURE 4.2 Dumps (right) and dumping of household and building rubbish on the left Bank of river Ardon, near the village of Mizur (Photo by Margarita El. Dzodzikova).

places. Territories in the vicinity of the villages Mizur and Buron and the tail-water of the head hydroelectric power station are of particular concern in the case of river floods. We can see clearly in Figs. 4.2 and 4.3 that firstly, rock dumps reach up to the water's edges, and secondly, they are in no way strengthened.

There are also four stone crushing companies in the surveyed territories, which discharge without control dump rocks and screenings along the bed of the rivers Sadon and Ardon, narrowing the beds of these rivers. Moreover, they have produced the draining of technical water that has been amply illustrated in Fig. 4.4. These discharges increase the turbidity of rivers, which affects negatively the state of their hydrobiota.

Radiometric measurements were made with the use of the gamma radiation dosimeter – DKG-OZD "Grach" on all available radiometric research, the above-stated surveyed areas of the river Ardon's bank zone within maximum proximity to the water's edge. The examples of measurement

FIGURE 4.3   Moldboard ground North head hydroelectric power station up to the water edge (Photo by Margarita El. Dzodzikova).

are shown in Fig. 4.5. These figures show most closely the location of the dosimeter to the water's edge by taking down indicators.

The results of radiometric measurements are presented in Table 4.3.

In all cases, the measurement error ranged from 8% to 13%, the common average error indexes for all surveyed areas amounted to 10.7% (at the permissible limit of 15%). Indicators of radiation power γ μ^3в/s (millisieverts per sec) were converted to ICRI/hour (microroentgen per hour) for convenience.

On average the capacity of gamma radiation for the surveyed areas is 0,27 SV/s (26,9 ICRI/hour), in six places, approaching the maximum permissible concentration and only in one place – the Right Bank above the village Boron (Table 4.3) is somewhat exceeding the maximum allowable concentration.

The all-round apparent lack of living creatures (even tadpoles) in turbid waters below stone crushing enterprises has been drawn to attention.

FIGURE 4.4 Stone-crushing plant on the river Sadon – the left tributary river Ardon (dumps along the river; the turbidity of river discharge waste technical water) (Photo by Margarita El. Dzodzikova).

FIGURE 4.5 Places of measurements of gamma-radiation – DKG-03D "rook" near the spring (left) and by the river Ardon (right) (Photo by Margarita El. Dzodzikova).

TABLE 4.3   Radiometric Measurements Dumps Along the River Ardon (Upstream)

The location of the dump	Error, %	Capacity of gamma bat. -Dπ, SV/s	Capacity of gamma bat, ICRI/ hour	Maximum perfor- mance, ICRI/h
Right bank[2] (26 km[2])	9	0.27	27	30
Left bank[1], the village of Mizur (before the entrance to the 1st tunnel)	12	0.25	25	30
Right bank[1], before the river Baddon	11	0.28	28	30
River Sadon, the left inflow of the river Ardon	10	0.29	29	30
Left bank[1], the village Nusal (all)	9	0.25	25	30
Right bank[1], the village Nusal, below the top of the bridge	13	0.25	25	30
Right bank[1], above the top of the bridge and the settlement of Nusal	9	0.24	24	30
58-th km of the gas pipeline[2]	-	-	-	-
Right bank[1], before the village Buron	10	0.27	27	30
Right bank[1], above the village Boron	8	0.32	32	30
Right bank[1] before turning in Kosh gorge	11	0.27	27	30
Left bank[1], pipelins tunnel[2]	-	-	-	-
Left bank[1], 50 m below the Head HPP	10	0.29	29	30
Left bank[1], 1 km below the Head HPP	12	0.28	28	30
Right bank[1], 2 km below the Head HPP	11	0.25	25	30
Right bank[1] observation platform over HPP	-	-	-	-
Left bank[1], the road to Mamisoni gorge	10	0.26	26	30
Left bank[1], (tunnel "Kosh № 5")	9	0.29	29	30
Left bank[1], the 64-th km of the gas pipeline[2]	-	-	-	-
Left bank[1], the 62–64-th km of the gas pipeline[2]	-	-	-	-
Dumps in the construction of roads (10 tunnels), the 68–70-th km of the gas pipeline[2]	-	-	-	-
Average	10.7	0.27	26.9	30

[1] The river Ardon.
[2] The pipeline Dzuarikau – Tskhinval.

Figure 4.6 shows full visual lack of living creatures in the water and depleted, feeble vegetation cover. But a large number of fry Terek barbel (Barbus ciscaucasicus Kessler, 1877) by the value of 0.5–7 cm, and sometimes met and larger individuals to 20–30 cm (Fig. 4.6, b) have been observed in the upper reaches of the rivers in the shallows, for example, of the right tributary river Ardon near the village of Nizhny Biragsang.

Analysis of the obtained data of radiation measurements showed that according to the Methodology of radiation control of industrial waste, approved by the State Standard of the Russian Federation "System of accreditation of laboratories of radiation control," approved by the Director of the Metrology Centre of ionizing radiation of the State Enterprise "All-Russian Scientific Research Institute of Physical-Technical and Radiotechnical Measurements," Jarinoi, V.R. (14.07.2000), the radiation background of the surveyed areas (the State Standard of the Russian Federation) does not exceed maximum permissible concentrations and can be recognized as free from radiation zone and not demanding intervention measures [12].

Thus, the results of measurement of the level of gamma radiation of the dump discharges along the bed of the river Ardon presented in Table 4.3, allow us to draw the conclusion that identified indexes of the radiation background of the surveyed areas can be recommended as targets.

Specially protected natural territories are a national treasure, and their value is absolute. As shown previously conducted work, anthropogenic

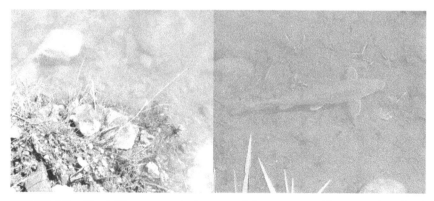

FIGURE 4.6.   No visible creatures in the waters below stone crushing enterprises. Terek barbel (Barbus ciscaucasicus, Kessler, 1877) in the shallow waters of the right tributary river Ardon near the village of Nizhny Biragsang (right) (Photo by Margarita, E. Dzodzikova).

impact disturbs the Reserve's water's chemistry [13, 14], affects the state of the Reserve's air basin and the health of children and adults in popu-lated places of the surveyed territories [15, 16]. If the radiation level in the valley of the river Ardon is relatively stable, the uncontrolled growth of the number and volume of dump sites has a particular danger in case of flooding.

For the frequency of the area of distribution and average annual dam-age floods are first among natural disasters in Russia. The peculiarity of floods and several other emergency situations of natural character is that it is impossible to prevent [17].

Floods in the river Ardon valley occur not every year. Mostly they are caused by the coincidence of heavy rainfall and active snow melting, and as a result water level in the channel and the flow rate increase sig-nificantly. These phenomena reach the maximum by the end of July – the beginning of August.

From 1849 to 2011, 103 catastrophic floods happened in the Ardon River valley, 342 mudflows came down, different parts of the Transcau-casian highway (transkam) were destroyed 32 times and the destruction length of the road was from 100 m to 8 km. In the last years, the floods and mudflows frequency have significantly increased. There were rock slides and landslides, mostly in the Kosh, Tsey and Mamison gorges [18]. If the rise of the water level in the period of summer floods up to 2.5 m is con-sidered as a relative norm and the banks of the river have been prepared to the load with this water body (catastrophic level is considered from 3 to 6 meters and more), the presence of rock dumps increase significantly the destructive force of the stream, because the rise of the water level even for a meter will wash away the spoil heaps and help them to slip away into the river and increase the water body in addition.

In this situation a separate threat comes from the Mizur concentrat-ing factory's tailing dump, which is located near the left bank of the river Ardon (the distance from the fortified region of tailings to the edge of the river Ardon over the slope is about 5 m, the relative height above the back-line is about 3 m). If the rise of the water level in the river exceeds more than 3 meters, there is a danger of wash out of the contents of the tailings into the bed of the river. All these factors greatly increase the anthropo-genic impact on the flora and fauna of the reserve, protected areas and adjacent residential areas, lead to the destruction of fauna and flora, and disruption of the balance of the ecosystem.

4.4 CONCLUSIONS

Research has shown that:
1. The revealed growth of the number and the volume of dump sites along the river Ardon and its tributaries and the activities of the stone crushing enterprises increase the turbidity of waters and narrow the beds of the rivers, disrupt normal life-sustaining activity of the ecosystems of these areas, and unreinforced banks are a special threat in case of flooding.
2. Indexes of background radiation of the surveyed territories in planning future radiometric studies don't exceed the maximum permissible concentration and can be recommended as targets.

KEYWORDS

- gas pipeline
- morphometry
- radiation measurements
- riverbanks
- riverbanks
- stone crushing companies
- tailing dump

REFERENCES

1. Panov, E. Glaciers pool river Terek. Leningrad. Gidrometeoizdat., 1971. 296p. (In Russian).
2. Catalogue of glaciers of the USSR. Volume 8. North Caucasus. Part 8. Pool & Oruch. Part 9. Pool & Ardon [E. Panov, A. S. Borovik]. Leningrad. Gidrometeoizdat. 1976. 76p. (In Russian).

3. Dontsov, V. I., Tsogoev, V. B. Natural resources of the Republic of North Ossetia-Alania. Water resources. Vladikavkaz. Project-press. 2001. 367p. (In Russian).

4. Dzodzikova, M. E., Pogosyan, A. A. Rivers and glaciers of the North-Ossetian nature reserve. Mountain regions: XXI century: collection of scientific papers dedicated to the 75th anniversary PhD geography, B. M. Beroev. Vladikavkaz, 2011. P. 175–179. (In Russian).

5. Chubetzova, R. D., Gubanova, J. C. Therapeutic action of mineral waters of North Ossetia. Ordzhonikidze. IR. 1996. 158p. (In Russian).

6. Amirkhanov, A. M., Lipkovich, A. D., Popov, C. P., Weinberg, P. I., Alekseev, S. K. North Ossetian reserve. Reserves of the USSR: the nature Reserves of the Caucasus. 1990. P. 50–69. http://oopt.info/sosetin/fauna.html (In Russian).

7. Badtiev, U. S., Dzodzikova, M. E., Alagov, A. A. Ecological status of specially protected natural territories of North Ossetia-Alania of Federal significance. Vladikavkaz. Publishing house, V. I. Abaev North Ossetian Institute of Humanitarian and Social Research of Vladikavkaz Scientific Center of the Russian Academy of Sciences and the Government of the Republic of North Ossetia-Alania. 2012. 142p. (In Russian).

8. Komarov, J. E., Khokhlov, A. N., Iluh, B. T. Ecology of some species of birds of North Ossetia-Alania. Stavropol. Stavropol state University. 2006. 257p. (In Russian).

9. Weinberg, P. I. An annotated checklist of the fishes, amphibians and reptiles of North Ossetian reserve. Proceedings of the North-Ossetian state natural reserve. 2006 Vol. 1. P. 142–144.

10. Popov, C. P. Alagir gorge. Vladikavkaz. IR. 2008. 415 p. (In Russian).

11. North-Ossetian state nature reserve. Ordzhonikidze. IR. 1989. 106p. (In Russian).

12. Normative documents for radiation monitoring. http://betagamma.ru/product_info. php?products_id=451 (In Russian).

13. Dzodzikova, M. E., Pavlova, I. G., Gabaraeva, V. M. The impact of waters of different origin on the frequency of occurrence of mammary tumors in rats induced by CMA. Thesises of VII International Conference "Sustainable development of mountain areas in the context of global change". Vladikavkaz.- IR-2010. 124–125. (In Russian).

14. Dzodzikova, M. E., Gridnev, E. A., Pogosyan, A. Chemistry of waters of the North Ossetian reserve. Mountain regions in XXI century: collected scientific articles to the 75th anniversary of, B. M. Beroev. –Vladikavkaz. IR. 2011. P.173–175. (In Russian).

15. Dzodzikova, M. E., Butaeva, F. M. Health Status of the population living on the territories of North-Ossetian reserve and its buffer zone in 2006 to 2011// Materials of Conference "White nights." The International Scientific and Practical Use. St. Petersburg. International Academy of Sciences of ecology and life safety. 2013. 84–86. (In Russian).

16. Dzodzikova, M. E., Ephieva, M. K., Turiyeva, D. V. Environmental problem as one of global problems of the present (on a material of an ecological condition of the air pool and incidence of children in A mountain bush" Alagirsky area RSO-Alania). 5-th International Scientific Conference "Applied Sciences and Technologies in the United States and Europe: common challenges and scientific findings", February 12, 2014. New York. Hosted by the CIBUNET Publishing Conference papers. P. 62–64.

17. The method of predicting flood floods. Training materials and literature on life safety for students Academy of EMERCOM of Russia. http://www.agps-mipb.ru/index.php/2011–01–08–07–37–51/395–3-6-profilakticheskie-i-prog-nosticheskie-meropriyatiya-opolznevyx-processov.html (In Russian).
18. Popov, C. P. Floods as outstanding hydrological phenomena in the North Ossetia-Alania and the Caucasus and their impact on the biota and human economic activities. Bulletin of the North Ossetian Branch of the Russian Geographical Society. 2006. № 11. 50–55. (In Russian).

CHAPTER 5

HYDROSILYLATION ON HYDROPHOBIC MATERIAL-SUPPORTED PLATINUM CATALYSTS[*]

AGATA WAWRZYŃCZAK,[1] HIERONIM MACIEJEWSKI,[1,2] and RYSZARD FIEDOROW[1]

[1]Adam Mickiewicz University, Faculty of Chemistry, Grunwaldzka 6, Poznań, Poland

[2]Poznań Science and Technology Park, A. Mickiewicz University Foundation, Rubież 46, Poznań, Poland

CONTENTS

ABSTRACT

Catalytic activity of platinum supported on styrene-divinylbenzene copolymer and fluorinated carbons was tested in reactions of hydrosilylation

---

[*]Dedicated to Professor Gennady, E. Zaikov on the occasion of his 80th birthday in appreciation of his outstanding contributions to chemistry and related sciences.

of several olefins and the way of platinum binding to the support surface as well as platinum oxidation states were determined. Catalysts based on the above hydrophobic materials make it possible to obtain high yields of desirable reaction products and in most cases the highest activity was shown by polymer-supported platinum catalyst. In addition to hydrophobicity, another factor appeared to influence catalytic activity, namely the kind of the catalyst precursor. It was established that some amount of unpolymerized vinyl groups that were present on the polymeric support surface was involved in the interaction with platinum. XPS spectra enabled to determine that platinum was present on the catalyst surface in 0 and +2 oxidation states, however, Pt° clearly predominated.

5.1   INTRODUCTION

Hydrosilylation, that is, the addition of Si-H bond to multiple bonds, is an important reaction from the viewpoint of both laboratory and commercial scale applications. It was discovered in 1947 [1] and since then it arouses an unceasing interest of researchers that is reflected by thousands of papers published on this topic. However, a straight majority of them was devoted to hydrosilylation in homogeneous systems, where transition metal complexes in solutions were employed as catalysts, or to hydrosilylation with the use of the mentioned complexes anchored to surfaces of inorganic or organic solids (hydrosilylation by heterogenized complexes) [2, 3]. Much less publications concerned hydrosilylation in typical heterogeneous systems, although the latter found an application to the commercial process of trichlorosilane addition to allyl chloride in the presence of active carbon-supported platinum catalyst [4]. However, in recent years an increase in the interest in heterogeneously catalyzed hydrosilylation has been observed [5, 6] because the latter offers a possibility of easier separation of catalysts from postreaction mixtures thus facilitating catalyst reuse. The above advantage acquires a particular significance when the price of a catalyst is high and this is the case of catalysts for hydrosilylation processes since they are realized in the presence of costly transition metals. If hydrosilylation proceeds in a homogeneous system, the separation of a transition metal complex from reaction products and unreacted parent substances, for example, by distillation, can be associated with its decomposition because many of these complexes undergo thermal degradation even below 150 °C and other methods, e.g. extraction, usually result in a considerable loss of a precious metal [7]. Such a situation does not occur if the process is carried in heterogeneous system, because even in the case of a slight leaching of a noble metal, its loss is significantly smaller compared to that observed during attempts of catalyst recovery from homogeneous systems.

The choice of an appropriate support is of no less importance than that of active phase of a catalyst. We have focused our attention on the application of hydrophobic supports to prepare effective platinum catalysts for hydrosilylation since our preliminary experiments have shown that in a number of hydrosilylation reactions hydrophobic material-supported catalysts appeared to be superior to those based on hydrophilic supports such as alumina and silica. We have also aimed at selecting such supports which, in addition to their hydrophobicity, do not have acid centers on their surfaces and, due to this, they do not catalyze undesirable side reactions of isomerization. The supports selected for our study were styrene-divinylbenzene copolymer (SDB) and fluorinated carbon (FC), because nonfunctionalized SDB is free of acid sites and surface acidity of FC is extremely weak (H_0 » 9) [8]. The performance of SDB- and FC-supported platinum catalysts was studied in several reactions of hydrosilylation.

5.2 MATERIALS AND METHODS

5.2.1 PREPARATION OF CATALYSTS

Supports employed for the preparation of catalysts were styrene-divinylbenzene copolymer (SDB) purchased from Aldrich and fluorinated carbons FC10, FC28 and FC65 (fluorine content: 10, 28 and 65 wt.%, respectively) obtained from Advanced Research Chemicals, Inc. (U.S.A.). Moreover, a commercial catalyst from Degussa – 1% Pt supported on active carbon (Pt/C) was used for the sake of comparison. Surface areas of the supports, measured by low-temperature nitrogen adsorption on an ASAP 2010 sorptometer (Micromeritics) were: 1172 m²/g for SDB and 178, 173 and 366 m²/g for FC10, FC28 and FC65, respectively. The supports were impregnated with chloroformic solution of platinum(II) acetylacetonate (from ABCR GmbH) or with a solution of hexachloroplatinic acid (from Aldrich) in 2-propanol to result in 1 wt.% Pt in a catalyst. Reduction of platinum catalyst precursors introduced onto surfaces of the above supports was performed at 160 °C in hydrogen flow. The polymer-supported catalyst prepared by using $Pt(acac)_2$ was labeled Pt/SDB, whereas that obtained by impregnation with H_2PtCl_6 was denoted by asterisk: Pt/SDB*.

5.2.2   REACTIONS STUDIED

Reactions of hydrosilylation were carried out in the liquid phase in glass vials of 10 mL capacity. The vials were loaded with reacting substances and a catalyst taken in such amounts that the following ratio was met: 2×10^{-4} mole Pt: 1 mole of a compound with Si-H bond: 1 mole of a compound with C=C bond. In the case of allyl polyether hydrosilylation with poly(hydromethyl-codimethyl)siloxane a greater amount of platinum (5×10^{-3} mole Pt) was used as well. Before the start of a reaction, each vial was sealed with a headspace aluminum cap and a teflon-lined septum followed by immersing into an oil thermostated bath. Mixtures placed in the vials were stirred during reactions. All reactions were carried out at 100 °C, except for that between polyether and polysiloxane, where the reaction temperature was 125–130 °C due to problems with homogenization of the mixture. Post-reaction mixtures were analyzed on a gas chromatograph SRI 8610C equipped with a CP-Sil CB column (30 m). In the case of polyether – polysiloxane mixture, another method of evaluating conversion of reacting substance was applied, namely the determination of the loss of Si-H bonds on the basis of FT-IR spectra recorded on a *Bruker Tensor* 27 Fourier transform spectrometer equipped with a SPE-CAC *Golden Gate* diamond ATR unit. The reason for replacing the GC analysis with infrared spectroscopic analysis was low vapor pressure of the polysiloxane, molecular weight of which was 5500.

5.2.3   CP/MAS 13C NMR ANALYSIS

The measurements were performed on a 9.4 T Bruker DMX NMR spectrometer. Powdered sample was placed in a zirconia rotor of 4 mm outer diameter followed by spinning at the magic angle (54.44°) at 5 kHz. All spectra were recorded at 21 °C.

5.2.4   X-RAY PHOTOELECTRON SPECTROSCOPIC (XPS) ANALYSIS

XPS spectra were recorded on a VSW spectrometer (Vacuum Systems Workshop Ltd., England) using a nonmonochromatized Al Kα radiation (1486.6 eV). The X-ray gun was operated at 10 kV and 20 mA. The working pressure was 3×10^{-8} mbar.

5.2.5   DETERMINATION OF PLATINUM CONTENT IN CATALYSTS

To determine if leaching of platinum occurred during experiments with multiple use of catalyst, analyzes for platinum content before and after using a catalyst in a hydrosilylation reaction were performed on an ICP-OES (inductively coupled plasma – optical emission) spectrometer (Vista MPX, Australia).

5.3   RESULTS AND DISCUSSION

5.3.1   CATALYTIC ACTIVITY FOR HYDROSILYLATION REACTIONS

The choice of reactions to be carried out in the presence of hydrophobic material-supported platinum catalysts was based on the practical importance of reaction products. Epoxy functional silanes are among the most important adhesion promoters applied to create bonds between filler and polymer matrix thus improving physicochemical and strength parameters of composites. Of practical importance are also epoxy functional siloxanes that are applied to the modification of epoxy resins thus making them more flexible, less susceptible to water sorption and more resistant to heat. Moreover, epoxy functional siloxanes find application to the manufacture of ionic silicone surfactants.

Octyl- and hexadecylsilanes are commonly applied as effective agents for hydrophobization and consolidation of building materials, architectural elements, monuments, etc. Alkylsiloxanes with long alkyl chain (at least 8 carbon atoms in alkyl group) are called silicone waxes which due to their moistening, softening, lubricating and spreading properties are widely applied in cosmetics, household chemistry as well as lubricating agents.

Silicone polyethers are important nonionic surfactants that are also used in cosmetics and household chemistry. However, their most important application is the manufacture of polyurethane foams, both rigid and flexible ones. There are no substitutes for them and their role consists in the facilitation of mixing of foam components. They prevent from the formation of large bubbles, facilitate the control of fluidity of liquid mixture (that expands due to the bubble growth) and they enable accurate control of time and degree of foam opening.

5.3.2.1 Addition of 1,1,1,3,5,5,5-heptamethyltrisiloxane To Allyl Glycidyl Ether

Measurements of catalytic activity for hydrosilylation of allyl glycidyl ether with heptamethyltrisiloxane were conducted at 100 °C for 30, 60 and 180 min. Catalytic activity expressed as percent yield of desirable product, that is, (3-glycidoxypropyl)bis(trimethylsiloxy)methylsilane, was presented in Fig. 5.1.

The yield of (3-glycidoxypropyl)bis(trimethylsiloxy)methylsilane after 3 h from the beginning of the reaction carried out in the presence of polymer-supported platinum catalysts was well over 80%, whereas in the case fluorinated carbon-supported ones it was considerably lower. The best of the latter catalysts (Pt/FC28) has reached the yield over 70% only after 3 h, while in the case of SDB-supported catalysts such an activity level was obtained already after 30 min (Fig. 5.1). The desirable product yield appeared to depend not only on the kind of support, but also on platinum precursor. The impregnation with platinum(II) acetylacetonate resulted

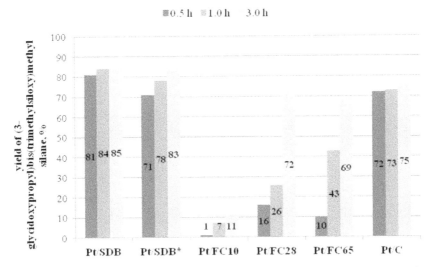

FIGURE 5.1 Activity of polymer- and carbon-supported platinum catalysts for hydrosilylation of allyl glycidyl ether with heptamethyltrisiloxane at 100 °C.

in a more active catalyst than that prepared with the use of hexachloroplatinic acid. The activity of polymer-supported catalysts appeared to be clearly higher than that of the commercial catalyst. The selectivity to (3-glycidoxypropyl)bis(trimethylsiloxy)methylsilane was high (96–98%) in the presence of all catalysts.

5.3.2.2 Addition of Triethoxysilane To Allyl Glycidyl Ether

Catalytic performance of polymer- and fluorinated carbon-supported platinum catalysts was determined at the same time intervals as in the case of hydrosilylation of the above ether with heptamethyltrisiloxane. Results of the measurements (shown in Fig. 5.2) point to high activity of both kinds of investigated catalysts which after 3 h reaches the level of 87% in the presence of all catalysts, except for that supported on fluorinated carbon with the lowest fluorine content (FC10). However, even in the latter

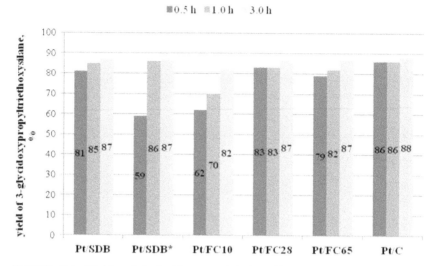

FIGURE 5.2 Yields of 3-glycidoxypropyltriethoxysilane obtained in the reaction between triethoxysilane and allyl glycidyl ether proceeding in the presence of SDB- and FC-supported platinum catalysts.

case the yield of desirable product, that is, 3-glycidoxypropyltriethoxysilane, exceeded 80%.

The activity of the reference (Pt/C) catalyst after 3 h was similar to that of catalysts prepared by us, but it reached the level of 86% as early as after 30 min. All the catalysts studied were characterized by a high selectivity to the desirable product ($\geq$ 95%).

5.3.2.3 Addition of Triethoxysilane To 1-Hexadecene and 1-Octene

where $n = 13$ or 5 in the case of the addition to 1-hexadecene and 1-octene, respectively.

Both SDB-supported catalysts made it possible to obtain good yields of desirable reaction product (1-hexadecyltriethoxysilane), but similarly as it was in the case of allyl glycidyl ether hydrosilylation with heptamethyltrisiloxane, results shown in Table 5.1 point to platinum(II) acetylacetonate as a better platinum catalyst precursor than H_2PtCl_6. Also FC-supported catalysts have shown a very good performance in the discussed reaction that was particularly impressive (91%) in the presence of Pt/FC28 (Fig. 5.3).

Hydrosilylation of 1-octene with triethoxysilane on polymer-supported catalysts (Table 5.2) proceeded, to a considerable extent, in the way similar to that of hexadecene. Platinum(II) acetylacetonate as a platinum catalyst precursor appeared to be again a more advantageous choice.

TABLE 5.1 Yield and Selectivity to the Desirable Reaction Product and Conversion of Parent Substances in the Process of 1-Hexadecene Hydrosilylation With Triethoxysilane Carried Out at 100 °C on Polymer-Supported Platinum Catalysts

Reaction time, h	Yield of desirable reaction product, %	Selectivity to desirable reaction product, %	Conversion degree of triethoxysilane, %	Conversion degree of hexadecene, %
Pt/SDB				
0.5	84	96	86	87
1	84	96	86	87
3	86	96	88	87
Pt/SDB*				
3	74	97	78	79

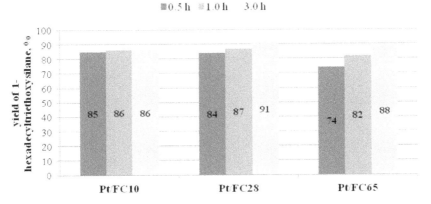

FIGURE 5.3 Catalytic performance of the catalysts studied for the addition of triethoxysilane to 1-hexadecene at 100 °C.

However, somewhat unexpected results were obtained while using FC-supported catalysts (Fig. 5.4). The Pt/FC10 catalyst, the performance of which in the reactions described previously was poorer than that of other fluorinated carbon-supported catalysts, this time appeared to be the most active, surpassing even polymer-supported catalysts. Water contact angle for hydrophobic materials should be greater than 90° and in the case of FC10 it is a bit below the above value, namely it equals to 85°, whereas for FC28 to 120° [9] (for SDB copolymer it is in the range of 109–117° [10])

TABLE 5.2   Yield and Selectivity to 1-Octyltriethoxysilane and Conversion of Parent Substances in the Process of Triethoxysilane Addition to 1-Octene Carried Out at 100 °C on Polymer-Supported Platinum Catalysts

Reaction time, h	Yield of desirable reaction product, %	Selectivity to desirable reaction product, %	Conversion degree of triethoxysilane, %	Conversion degree of octene, %
Pt/SDB				
0.5	69	95	75	73
1	83	95	85	86
3	86	95	87	86
Pt/SDB*				
3	67	94	83	82

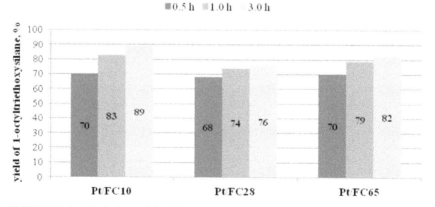

FIGURE 5.4   Catalytic activity for 1-octene hydrosilylation with triethoxysilane at 100 °C in the presence of Pt supported on fluorinated carbon of different fluorine content.

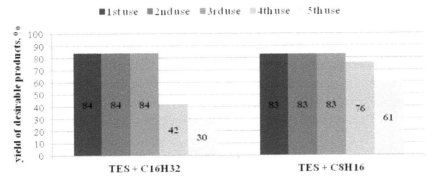

FIGURE 5.5 Yields of desirable products (hexadecyltriethoxysilane and octyltriethoxysilane, respectively) obtained after 1h in tests for multiple use of Pt/SDB catalyst in hexadecene and octene hydrosilylation with triethoxysilane (TES) at 100 °C.

and this fact brings into conclusion that support hydrophobicity, although it plays an important role in hydrosilylation reactions, is not the only factor affecting the activity of catalysts for hydrosilylation processes. It is worth of mentioning that the catalytic performance of platinum supported on fluorinated carbon containing 65% F, which is characterized by the highest water contact angle (125° [9]) from among supports employed in our study, is not the best catalyst among them.

As it was already mentioned, hydrosilylation in heterogeneous systems facilitates catalyst reuse. This is why we have undertaken tests for multiple use of catalysts in the reactions of hydrosilylation of 1-hexadecene and 1-octene with triethoxysilane (Fig. 5.5).

It results from Fig. 5.5 that Pt/SDB catalyst is characterized by a high activity that remains on a constant level for three reaction runs and then declines, most likely due to platinum leaching from the catalyst as can be concluded from ICP-OES analyzes for platinum content in catalysts after their use in the fifth run of reactions between TES and hexadecene (0.40% Pt) as well as TES and octene (0.44% Pt).

5.3.2.4 Addition of Poly(hydromethyl–codimethyl)siloxane To Allyl Polyether

First measurements of catalytic activity for the discussed reaction were performed at the mole ratio of platinum to parent substances (polysiloxane and polyether) equal to 2×10^{-4} and resulted in Si-H conversion of 73% and 77% after 2.5 and 5 h, respectively, of conducting the reaction. Then the ratio was increased to 5×10^{-3} in hope for raising the conversion. Si-H conversion observed at the latter ratio after 5h of the reaction was 80% and after 10 and 15h it was 82 and 86%, respectively. The obtained results show that in the presence of Pt/SDB catalyst it is possible to reach a high Si-H conversion already at the ratio of Pt to polyether and polysiloxane equal to 2×10^{-4} and a further increase in the ratio seems pointless taking into account the cost of platinum-containing catalyst.

5.3.3   PLATINUM SPECIES ON CATALYST SURFACES

In the previous section we have shown that platinum supported on styrene-divinylbenzene copolymer makes an effective catalyst for hydrosilylation. Now we should answer the question: how platinum is bound to the support surface? X-ray diffraction and hydrogen chemisorption measurements [11, 12] showed the presence of a considerable number of large Pt crystallites which are weakly bound to the support surface and thus are vulnerable to leaching. However, on SDB surface, there are also sites capable of inter-acting with platinum in a stronger way. Potential centers for the interaction between the metal and the support are vinyl groups, certain number of which (rather small one) could remain after the polymerization process. Such a hypothesis was put forward in Ref. [11] and in the present study we verified this conjecture by recording solid-state ^{13}C NMR spectra of SDB support and Pt/SDB catalyst (Fig. 5.6). Such measurements enabled to determine a possible loss of signal originated from carbon atoms present in $-CH=CH_2$ group.

The studied system is simple for analyzing because the signal coming from double-bonded carbon atoms of vinyl groups appears at about 112 ppm and the signal ascribed to carbon atoms of benzene ring is located at about 127 ppm. Areas under each peak were measured and ratios of peak areas originating from carbon atoms of vinyl groups and those of aromatic rings were calculated. Results of the calculations are shown in Table 5.3.

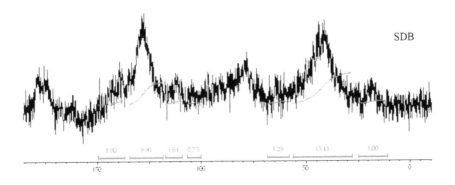

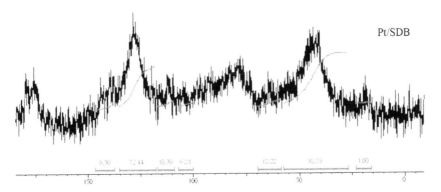

FIGURE 5.6 ^{13}C NMR spectra of SDB support and Pt/SDB catalyst.

TABLE 5.3 Changes in the Number Unpolymerized Vinyl Groups in SDB Occurring As A Result of Introducing Platinum onto the Support

Sample	–CH=CH$_2$ (1)	–CH=CH– (2)	(1)/(2)
SDB	1.61	9.90	0.16
Pt/SDB	10.39	72.44	0.14

Data presented in Table 5.3 show that the introduction of platinum onto the support causes a small decrease in the number of double bonds originating from free vinyl groups. This result supports the hypothesis presented in Ref. [11] that a surface complex was formed by the interaction between platinum and vinyl ligands.

Next question concerns the oxidation state of platinum in Pt/SDB catalysts. Admittedly, X-ray diffraction pattern presented in Ref. [12] clearly indicated the presence of Pt°, but small amounts of other platinum species (undetected by X-ray technique) possibly can also exist on the catalyst surface. Temperature-programed reduction profile of SDB-supported Pt(acac)$_2$ contained two peaks, the first of which corresponded to the reaction of hydrogen with of platinum(II) acetylacetonate and the second one to the reaction between hydrogen and products of partial decomposition of Pt(acac)$_2$ [12]. X-ray photoelectron spectroscopy (XPS) measurements carried out in the present study have shown the presence of platinum species in the +2 oxidation state, in addition to those in zero oxidation state (Table 5.4).

The spectrum of Pt/SDB catalyst was deconvoluted into two components at 72.2 and 75.1 eV (Fig. 5.7). Peak corresponding to binding energy of 72.2 eV can be ascribed to Pt°, whereas that at 75.1 eV to Pt^{+2} [13, 14]. The deconvolution of Pt/SDB* spectrum resulted in two components at 73.3 and 75.3 eV ascribed to Pt° and Pt^{+2}, respectively (Table 5.4).

Results of XPS analysis bring into conclusion that reduction of platinum precursors at 160 °C leads to a clear predominance of Pt°, although some amount of Pt^{+2} remains on the catalyst surface. This is in agreement with data obtained earlier from XRD and TPR H$_2$ analyzes. It is possible that both Pt° and Pt^{2+} take part in catalyzing hydrosilylation reactions, however, the main contribution to catalytic process seems to come from Pt° which predominates on surfaces studied.

TABLE 5.4 Results of XPS Analysis of Pt/SDB and Pt/SDB* Catalysts

Band	Peak maximum, eV	Concentration in the sample, %	Quantitative ratio	Ascription
Pt/SDB				
Pt 4f$_{7/2}$	72.2	0.60	83	Pt°
	75.1	0.10	17	Pt^{2+}
Pt/SDB*				
Pt 4f$_{7/2}$	73.0	0.70	73	Pt°
	75.3	0.25	27	Pt^{2+}

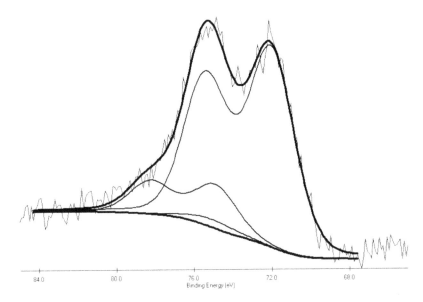

FIGURE 5.7   The Pt 4f XPS spectrum of Pt/SDB catalyst.

5.4   CONCLUSIONS

Hydrophobicity of catalyst support, although very important to many hydrosilylation reactions carried out in heterogeneous systems, is not the only factor deciding of the performance of supported platinum catalysts for hydrosilylation. Results obtained by using different platinum compounds for the preparation of catalysts for hydrosilylation show that the kind of catalyst precursor also belongs to the factors influencing catalytic activity.

In addition to high activity, the studied catalysts are highly selective. One of reasons for their high selectivity is the absence of acid centers on surfaces of styrene-divinylbenzene copolymer and fluorinated carbons, therefore undesirable side reactions of isomerization do not proceed.

Tests for multiple use of catalysts for hydrosilylation of hexadecene and octene with triethoxysilane have shown that yields of desirable products are maintained on a high constant level for three runs and then they decline due to platinum leaching from the catalysts.

The presence of a certain amount of unpolymerized vinyl groups on polymeric support surface results in their interaction with platinum and the complex formed between them plays a role in binding platinum to the support surface.

Platinum is present on the polymeric support surface in 0 and +2 oxidation states and the former of the oxidation states clearly predominates.

KEYWORDS

- fluorinated carbon
- platinum(II) acetylacetonate
- polyether
- polysiloxane
- styrene-divinylbenzene copolymer
- triethoxysilane

REFERENCES

1. Sommer, L. H., Pietrusza, E. W., Whitmore, F. C., *"Peroxide-catalyzed Addition of Trichlorosilane to 1-Octene"*, J. Am. Chem. Soc. 69, 188 (1947).
2. Marciniec, B., Maciejewski, H., Pietraszuk, C., Pawluć, P., *"Hydrosilylation. A Comprehensive Review on Recent Advances"*, Springer, Dordrecht, 2009.
3. Marciniec, B., Maciejewski, H., Pietraszuk, C., Pawluć, P., J. Guliński "Hydrosilylation" in "Encyclopedia of Catalysis", Wiley 2010, *onlinelibrary.wiley.com/* DOI: 10.1002/0471227617.eoc117.
4. Marciniec, B., Maciejewski, H., Duczmal, W., Fiedorow, R., Kityński, D., *"Kinetics and Mechanism of the Reaction of Allyl Chloride with Trichlorosilane Catalyzed by Carbon-supported Platinum"*, Appl. Organometal. Chem. 17, 127–134 (2003).
5. Fiedorow, R., Wawrzyńczak, A., in *"Catalysts for Hydrosilylation in Heterogeneous Systems"* in "Education in Advanced Chemistry" (Marciniec, B., ed.), Wydawnictwo Poznańskie, Poznań – Wrocław 2006, vol. 10, pp. 327–344.
6. Pagliaro, M., Ciriminna, R., Pandarus, V., Béland, F., *"Platinum-Based Heterogeneously Catalyzed Hydrosilylation"*, Eur. J. Org. Chem. 6227–6235 (2013).
7. Cole-Hamilton, D. J. *"Homogeneous Catalysis – New Approaches to Catalyst Separation, Recovery, and Recycling"* Science 299, 1702–1706 (2003).
8. Fiedorow, P., Krawczyk, A., Fiedorow, R., Chuang, K. T., *"Studies of the Surface of Fluorinated Carbon in the Aspect of Its Catalytic Properties"* Mol. Cryst. Liq. Cryst. 354, 435–442 (2000).
9. Spagnolo, D. A., Maham, Y., Chuang, K. T., *"Calculation of Contact Angle for Hydrophobic Powders Using Heat of Immersion Data"*, J. Phys. Chem. 100, 6626–6630 (1996).

10. Wu, J., C. S., Chang, T.-Y., *"VOC Deep Oxidation Over Pt Catalysts Using Hydrophobic Supports"*, Catal. Today 44, 111–118 (1998).
11. Maciejewski, H., Wawrzyńczak, A., Dutkiewicz, M., Fiedorow, R., *"Silicone waxes – synthesis via hydrosilylation in homo- and heterogeneous systems"*, J. Molec. Catal. A 257, 141–148 (2006).
12. Wawrzyńczak, A., Dutkiewicz, M., Guliński, J., Maciejewski, H., Marciniec, B., Fiedorow, R., *"Hydrosilylation of n-alkenes and allyl chloride over platinum supported on styrene-divinylbenzene copolymer"*, Catal. Today 169, 69–74 (2011).
13. Wu, J., C.-S., Chang, T.-Y., *"VOC deep oxidation over Pt catalysts using hydrophobic supports"*, Catal. Today 44, 111–118 (1998).
14. Moulder, J. F., Stickle, W. F., Sobol, P. E., Bomben, K. D., *"Handbook of X-Ray Photoelectron Spectroscopy: A Reference Book of Standard Spectra for Identification and Interpretation of XPS Data"*, Physical Electronics Inc., Eden Prairie 1995.

CHAPTER 6

SOME PECULIARITIES OF ACRYLATE LATEX FORMATION IN EMULSIFIER-FREE EMULSION POLYMERIZATION

N. V. KOZHEVNIKOV, M. D. GOLDFEIN, and
N. I. KOZHEVNIKOVA

*Saratov State University, 410012 Saratov, Russian Federation,
E-mail: KozevnikovNV@info.sgu.ru*

CONTENTS

ABSTRACT

Emulsion polymerization of alkyl acrylates and their copolymerization with some water-soluble monomers (metacrylic acid, acrylonitrile) in the absence of emulsifier were investigated by means of dilatometry and the turbidity spectrum method. The feasibility of synthesis of stable polymeric dispersions under specially selected conditions by temperature, the concentration of the initiator (ammonium persulfate), the presence and nature of comonomers is shown. Several mechanisms of the formation and stabilization of latex particles are discussed. Particles can be stabilized by the end-groups of macromolecules, surfactant oligomers (capable of acting as emulsifiers), and the polar groups of comonomers.

6.1   AIM AND BACKGROUND

Polymer dispersions are widely used in industry, agriculture, building, and home. They are typically prepared by emulsion polymerization in the presence of stabilizers for polymer particles (emulsifiers) [1]. However, there is a possibility of producing some synthetic latexes by emulsion polymerization without specially added surfactants, which allows one to obtain an environmentally friendly product [2]. The synthesis of polymer dispersions is complicated by the need to ensure their colloidal stability, which is achieved by introducing an emulsifier in conventional emulsion polymerization [3].

This chapter presents the results of our study of the emulsifier-free polymerization of the alkyl esters of acrylic acid and their copolymerization with methacrylic acid (MAA) and acrylonitrile (AN). These monomers are widely used for producing industrially important acrylic polymer dispersions synthesized by conventional emulsion polymerization. The kinetics and mechanism of this reaction initiated by ammonium persulfate (APS) were investigated by us earlier [4–6], when sulfated oxyethylated alkylphenols or sodium lauryl sulfate acted as emulsifiers. In these studies it is indicated that, unlike the classical notions of emulsion polymerization [7], not only the micellar mechanism but also the homogeneous mechanism of nucleation [8], the interaction of radicals in the aqueous phase with the water-dissolved monomer (chain propagation) and other oligomeric radicals (bimolecular termination) should be considered in the polymerization system under discussion. Besides, a conclusion was made of the presence of several growing radicals in the polymer-monomer particles (PMP), which causes the gel effect, and of PMP flocculation proceeding at various stages of the process.

The effect of these reactions should increase in the absence of emulsifier. When emulsifier-free polymerization with persulfate-type initiators, the polymer particles are stabilized by the initiator's charged end groups. The water-dissolved oligomeric radicals $\cdot M_nSO_4^-$ grow until some critical chain length is reached, whereby they lose their solubility and are extracted from solution to form nucleus PMP, leading to homogeneous nucleation [9]. Stabilization of the particles can be enhanced by copolymerization of monomers with ionizing or highly hydrophilic monomers.

Moreover, the "micellar" nucleation mechanism can be realized in the emulsifier-free conditions, which, with visible similarity, substantially differs from the particle formation occurring during conventional emulsion polymerization. In the absence of emulsifier micelles, the primary charged radicals resulting in the aqueous phase, when decomposition of a water-soluble persulfate-type initiator, after several acts of growth (interacting with the water-dissolved monomer) react with each other to form oligomeric molecules with surface activity and being able to create micelle-like structures, which play the role of an "own" emulsifier. Subsequently, the monomer and oligomer radicals are absorbed by these "micelles", where chains can grow [3].

Therefore, preconditions are created for PMP stabilizing and the implementation of polymerization by the emulsion mechanism and without the participation of a specially introduced emulsifier.

6.2 EXPERIMENTAL PART

A study was made of emulsifier-free polymerization in a biphasic water-monomer reaction system (a monomer fraction of 20%) in a helium atmosphere under constant stirring (a magnetic stirrer), nearly under the same conditions as for the conventional emulsion polymerization process [4–6] but in the absence of emulsifier. The reaction solutions were freed from air oxygen (the monomer and aqueous phases separately) by multiple repetition of the process of their freezing, high-vacuum pumping, and thawing in vacuum. The reaction rate was measured by means of dilatometry [10] (in glass devices of our original design), allowing continuous monitoring of the process progress from its very initial stage until the end. The size and number of polymer particles in the resulting dispersion was estimated by the turbidity spectrum method [11] (an SP-26 spectrophotometer). The monomer composition, initiator concentration and temperature values in these experiments are indicated in the Figs. 6.1–6.7. All reagents were preliminary purified by standard methods.

6.3   RESULTS AND DISCUSSION

In the absence of emulsifier, the polymerization of the monomer systems discussed proceeds by the emulsion mechanism, primarily in monomer-polymer particles. The basic regularities typical for the reaction in the presence of emulsifier are observed. This is indicated by the formation of an emulsion just at the initial stage of polymerization and relatively stable polymer dispersion upon its completion, the similarity of the kinetic curves (Fig. 6.1) and the kinetic regularities of both emulsifier-free and

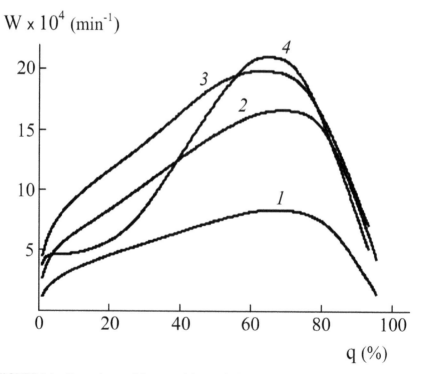

FIGURE 6.1   Dependence of the rate of the emulsifier-free emulsion copolymerization of MA with AN (4%) and MAA (14%) on the monomer conversion degree. [APS] $\times$ 10^3 = 1 (*1*), 4 (*2*), 10 (*3*), and 20 mol/L (*4*). 70°C.

traditional emulsion polymerization in the presence of specially added emulsifiers [6]. The most important differences between these two processes are the ways of particle generation and stabilization.

The kinetics of classical emulsion polymerization is characterized by three main stages, namely: a rapid increase in the rate (due to the formation of primary latex particles), the stationary reaction (while monomer droplets are present in the aqueous phase and its equilibrium concentration is set in the particles), and the completion of the process (as the monomer in the particles is exhausted) [4–6]. The same stages are also peculiar to emulsifier-free polymerization [3] but their course and duration depend on individual monomers and reaction conditions.

The relatively low stability of the polymer dispersions synthesized in the absence of a special emulsifier is expressed in the fact that, under certain conditions, a coagulum could be formed during synthesis, as well as a precipitate and a clarified upper layer in the dispersion during its storage. This indicates the existence of particles of different sizes: the larger ones form a precipitate while the smaller ones remain in a dispersed state. The considerable variation of the latex particles by size is possible if the step of their formation has a long duration. In the presence of emulsifier, more uniform dispersions are formed since PMP are formed with a high rate when radicals entering from the aqueous phase into the monomer-filled emulsifier micelles, where polymerization then proceeds with an equal probability (the polymerization in the presence of certain emulsifiers soluble in the monomer is an exception, which the formation of new particles during the reaction is also characteristic of Ref. [5]). The appearance of a clarified layer without precipitate formation indicates that in the absence of a specially added emulsifier, though large but size-uniform particles may occur which, under insufficiently effective stabilization, gradually settle to form two layers with a high and low content of latex particles.

Emulsifier-free polymerization proceeds at a slower rate than in the presence of emulsifier, and an increase in the rate, longer by time and deeper by conversion degree, is observed, that is, the first stage of the process (latex particle formation) lasts longer. The further variation of the reaction rate (W) with increasing the degree of conversion (q) (Fig. 6.1), as well as in the case of conventional emulsion polymerization, is caused by changes in the number of particles in the emulsion because of their flocculation at high conversions or additional nucleation in the course of the reaction, and peculiarities of the reaction proceeding inside PMP: the formation of coarse particles during emulsifier-free polymerization contributes to the gel effect and an increased rate at high conversions, despite of the lower monomer concentration therein.

It is shown that an increased temperature leads to an increased polymerization rate and an increased number of latex particles in the final product (N), their reduced size (r), and a lower amount of coagulum formed (P) with the dispersion stability improved. Increasing temperature accelerates the decomposition of the initiator and leads to more frequent penetration of oligomeric radicals into particles to increase their charge and stabilization. The amount of radicals in the aqueous phase increases and the probability of their bimolecular interaction to form associates of water-soluble surfactant oligomers having micelle-forming properties and promoting nucleation rises. The probability of oligomeric radicals to reach a critical chain length at which they would lose their solubility and form new particles increases as well. Therefore, changing temperature affects not only the rate of generation of primary radicals and their growth in PMP but also other processes involved in nucleation. Besides, increasing temperature influences the value of gel effect, which also affects the effective activation energy of polymerization. The ratio of the maximum rate to that at 10% conversion (corresponding to the second stationary stage of the reaction) can be regarded as a measure of the gel effect. In the copolymerization of butyl acrylate (BA) with AN and MAA in the absence of emulsifier, it increases with temperature up to 70–75 °C, and decreases at higher values (Fig. 6.2), reflecting opposite tendencies in the effect of the reaction temperature on the probability of coexistence of several radicals in PMP. This probability increases with the initiation rate but decreases when the particle size reduces, which occurs under these conditions. As a result, the maximum rate of polymerization (W_{max}) does not obey the Arrhenius dependence, and the experimental data in the lg W_{max} vs. 1/T coordinates do not fit a straight line (Fig. 6.2), indicating a changed reaction mechanism in various temperature ranges and the role of individual factors which determine the rate of the process.

The initiator concentration also renders significant impact on emulsifier-free polymerization. With its increasing, the amount of latex particles

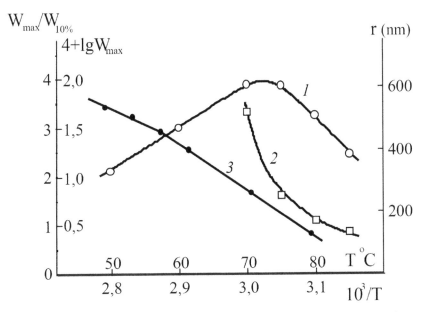

FIGURE 6.2 Temperature dependence of the size of latex particles (*2*) and the maximum rate–steady rate differences (*1*) of the emulsion copolymerization of BA with AN (4%) and MAA (14%). *3* – Arrhenius' dependence for W_{max}. [APS] = 40 × 10^{-3} mol/L.

in the final dispersion increases but decreases at higher concentrations, passing through a maximum (Fig. 6.3). When BA is copolymerized with AN and MAA, this maximum is observed at [APS] = 15 × 10^{-3} mol/L (85 °C). The reaction rate also changes, to various degrees, depending on the monomer-to-polymer conversion degree. At low conversions (10%) the rate, like the number of particles, has an extreme dependence on the APS concentration.

When the initiator concentration increases, the depth of polymerization, at which the maximum rate is achieved for given conditions of the process, shifts towards higher values, and the rate itself rises. But its logarithmic dependence on the initiator concentration is nonlinear. In a BA-based monomer system the order by initiator ($n_и$) decreases from 0.6 when [APS] < 10^{-2} mol/L down to 0.3 at higher concentrations. (According to the classical theory, $n_и$ = 0.4 [7]). In the case of the copolymerization of methyl acrylate (MA) at a sufficiently high content of the initiator (>10^{-2} mol/L) W_{max} almost no longer depends on the initiation rate as well.

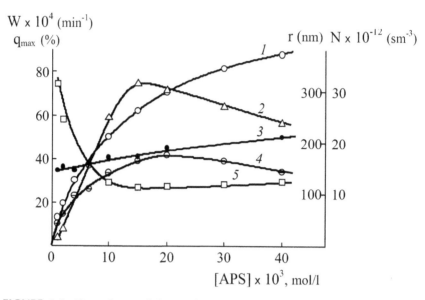

FIGURE 6.3 Dependence of the maximum rate (*1*) of the emulsifier-free emulsion copolymerization of BA with AN (4%) and MAA (14%), the corresponding degree of conversion (*3*), the number of particles in the resulting polymer dispersion (*2*), their average radius (*5*), and the rate at a 10% conversion (*4*) on the initiator concentration. 85°C.

Dilatometric measurements in these systems were carried out at temperatures not exceeding 70 °C, because the studied monomers are low-boiling (80 °C for MA, 77 °C for AN). However, the synthesis of dispersions without registering rate (in sealed ampoules) was performed at higher temperatures as well.

The initiation rate also affects the stability of the resulting product. BA-based dispersions which form neither precipitate nor clarified layer during storage were obtained at high temperatures (80–85 °C) only in a narrow range of APS concentration (10–20) × 10⁻³ mol/L and simultaneous copolymerization with MAA and AN. In the presence of only one of the said comonomers within a concentration range of 0–14% we have failed to avoid the appearance of a clarified layer, but its volume and rate of formation decrease when the comonomer concentration increases.

Delamination to form a clarified layer (even more pronounced) occurs during storage of MA-based dispersions as well. This process is more probable in the case of the synthesis of dispersions with low initiation rates, and the amount of the coagulum produced has a minimum value in the range of APS concentrations $(1-10) \times 10^{-3}$ mol/L (90 °C) (Fig. 6.4).

Therefore, stable polymer dispersions were obtained only at a high temperature of synthesis, a relatively high content of hydrophilic comonomers, within a narrow concentration range of the initiator.

It should be noted that in the emulsifier-free copolymerization of alkyl acrylates the stability of emulsions and final dispersions decreases in the row butylacrylate > ethylacrylate > methylacrylate, that is, the stability deteriorates with an increased water solubility (polarity) of the primary monomer. BA-based dispersions consist of a much larger number of smaller particles than those on the basis of MA. Their better stabilization, with surfactant oligomers involved, may be associated with the fact that the surface activity of the "own" surfactants increases with an increased hydrophobicity of the particle surface. This is confirmed by the increase in the interaction energy of the emulsifier with the organic phase as the polarity of alkyl acrylate decreases [1].

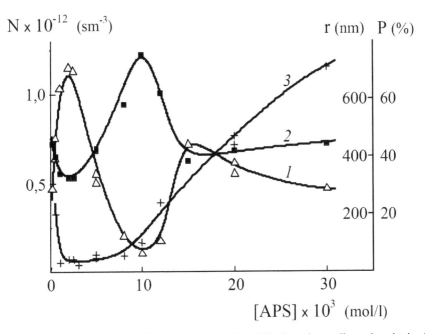

FIGURE 6.4 Dependence of the number of particles (*1*) in the polymer dispersion obtained by the emulsifier-free polymerization of MA with AN (4%) and MAA (14%), their average radius (*2*), and the amount of the coagulum formed (*3*) on the initiator concentration. 90°C.

The extreme dependences on the APS concentration observed for the number of particles in the final dispersion and the reaction rate at low conversions (Figs. 6.3 and 6.4), just at polymerization in the presence of emulsifier [4–6], are associated with insufficient stabilization of latex particles and their flocculation which occurs at various degrees of conversion, and not only at the initial stage of polymerization, as follows from the classical theory [7]. The reduction of the number of particles in the polymer emulsion when increasing the conversion at the third stage of emulsifier-free polymerization has been found experimentally. In the range of APS concentration corresponding to the maximum number of particles in the dispersion and their minimum size, the lowest number of the coagulum formed and the highest stability of the dispersion during storage are observed. Moreover, the content of the initiator affects the balance among various nucleation mechanisms.

At polymerization with low initiation rates, the aqueous phase of the emulsion system contains a low concentration of oligomeric radicals and their recombination probability is low. Under these conditions, many of them would have time to reach a critical size and to form primary particles by the homogeneous mechanism. Their stabilization occurs due to the polar groups of the monomers and the surface charge arising due to the presence of end groups in the persulfate initiator. To create a sufficient charge density, the primary particles flocculate at the stage of their formation. During subsequent polymerization, together with an increase in the volume (and surface) of the particles, their charge increases due to the periodic introduction of charged oligomeric radicals $\cdot M_n SO_3^-$ to PMP. An increased initiation rate leads to an increased number of radicals turning into primary particles, and the total number of PMP increases. More frequent penetration of oligomer radicals to the particles occurs as well, which improves their stabilization and prevents flocculation at later stages of the reaction.

At higher initiator concentrations, the role of the bimolecular chain breakage increases due to recombination of oligomeric radicals in the aqueous phase. As a result, water-soluble surfactant oligomers appear to play the role of an emulsifier. If their concentration in water exceeds the critical micelle concentration (which is possible at a sufficiently high initiation rate) then the associates of these oligomers form micelles which the monomer diffuses into and wherein polymerization can proceed. Under these conditions, the contribution of the micellar mechanism of PMP nucleation rises. Their surface is protected from the very beginning with the "own" emulsifier, which inhibits flocculation at the first stage of the reaction and promotes increasing, N. However, if the "own" emulsifier is formed and adsorbed more slowly than the surface requiring stabilization grows (large rate constants of chain propagation and the concentration of the monomer in particles but a relatively small initiation rate), flocculation takes place at the second stage of polymerization.

At the third step of the reaction corresponding to the disappearance of monomer droplets and the reduction of the monomer concentration in PMP, the protective effect of the "own" emulsifier deteriorates due to an increased surface charge of the particles due to the introduction of new charged oligomeric radicals therein. As a result, flocculation of the particles occurs until the effect of another stabilization mechanism (due to the surface charge as in the case of homogeneous nucleation) gets sufficiently effective. The transition from stabilization by surfactant oligomers to stabilization due to electrostatic forces occurring at the third stage of polymerization is accompanied by reduction in the PMP number. As a result, N passes through a maximum with increasing APS concentration.

In the case of MA-based dispersions, the appearance of another peak on the N vs. [APS] dependence was detected at high initiator concentrations (Fig. 6.4, curve 1). Under these conditions the rate of formation of the "own" emulsifier by recombination of oligomeric radicals in water may turn out to be so high that flocculation at the second step of polymerization begins to play a less important role, which leads to an increased number of particles. However, flocculation at the third stage of the reaction enhances with increasing the APS concentration, because oligomeric radicals enter the particle more often and the surface charge grows faster to worsen the operating conditions for the emulsifier. Therefore, the flocculation increase with an increased content of the initiator reduces N again.

The surface activity of the "own" emulsifier depends on the structure and composition of the oligomers, and, consequently, on the properties of the comonomers. In the case of BA no second peak has been detected, but it may, like the first maximum, occur at higher initiation rates than that of methyl acrylate latexes, outside of the range investigated. It can be assumed that the transition to the micellar nucleation mechanism as well as the termination of flocculation at the second stage of the reaction, at BA copolymerization occurs at higher rate initiation than in the case of MA., that is, the surfactant oligomers resulting from radical recombination in the aqueous phase are produced slowly. These oligomers must have a different composition at copolymerization of MA or BA due to their different reactivity and the solubility in water (5 and 0.2%, respectively [1]).

Taking into account that the MAA distribution coefficient between the aqueous and organic phases is less than unity, we can assume that within the investigated concentration range of MAA in the monomer system (£14%) the content of MA in water is higher than for MAA. However, MAA is more actively involved in the copolymerization reaction [12]. Therefore, the oligomeric radicals formed from MA and MAA copolymerization apparently consist of units of both monomers. In contrast, the concentration of BA in water is less than that of MAA, and the activity of MAA in copolymerization with BA is also high ($r_{MAA} = 1.3$, $r_{BA} = 0.3$ [13]). In this case, the oligomeric radicals are mainly composed of MAA. It is known that at copolymerization of monomers of different polarity a tendency to cross breakage is observed due to the effects of electron transfer [14]. The rate constant of cross break is higher than at homopolymerization of the corresponding monomers. In this connection, breakage in the aqueous phase to form surfactant oligomers at copolymerization of MA is more probable, which explains the lower (than for BA) APS concentrations, which the maximum number of particles in the dispersion is achieved at.

At copolymerization in the MA-based ternary system, all the monomers are (partially) dissolved in the aqueous phase, namely: MA, MAA, and AN. The copolymerization of AN with MA is characterized by the constants: $r_{AN} = 1.5$, $r_{MA} = 0.84$ [15]. Involving AN to oligomeric radicals leads to acceleration of their bimolecular interaction since the rate constant k_o of chain termination of AN is almost two orders of magnitude higher than that of MA [14]. This promotes strengthening of the role of the micellar mechanism of particle formation.

In the case of ternary BA-based systems, copolymerization of MAA and AN should mainly occur in the aqueous phase. And AN less actively participates in the reaction, as evidenced by its much slower consumption when the copolymerization reaction (by chromatographic data) and the resulting oligomeric radicals are mainly composed of MAA units. The breakage constant at polymerization of this monomer in water [16] is close to k_o of acrylic esters.

Flocculation of particles at different stages of the polymerization process and its dependence on the concentration of the initiator influence not only the number of particles in the final dispersion but, together with the gel effect, also the value of the reaction rate at different conversions and, consequently, the shape of the kinetic curves. Apparently, just the flocculation at the second stage, growing within some range of the APS concentration, causes an end of the rate growth with the polymerization depth (or even its slight decrease) at relatively small depths of conversion observed under these conditions (Fig. 6.1, curve 4).

Unlike conventional emulsion polymerization with a specially introduced emulsifier which is spent for the formation and stabilization of the resulting latex particles, at emulsifier-free polymerization during the reaction an "own" emulsifier is formed, which participates in the origination of new particles. The gradual emergence of new surfactant oligomers increases the nucleation stage duration. Therefore, with the growth of the initiation rate, the instant of reaching the maximum rate of polymerization shifts towards higher conversion degrees (Fig. 6.3). Besides, this increases the heterogeneity degree of particles by size, promotes enhancing flocculation, coagulum and precipitate formation during storage of the dispersion. On the other hand, an increased APS concentration promotes a better stabilization of the particles and the clarified layer decreases.

The formation of latex particles during emulsifier-free polymerization occurs by various mechanisms whose relative contributions depend on the reaction conditions. It has turned out that changes in some of them can influence the nature of the effects of other factors. E.g. an increased MAA concentration at relatively low initiation rates ([APS] = 5×10^{-3} mol/L, 70 °C) leads to a decreased rate of the emulsifier-free polymerization of methyl acrylate, ethyl acrylate, and butyl acrylate and the number of latex particles, an increased conversion corresponding to the maximum rate. The particle size and the amount of coagulum formed increase as well (Fig. 6.5). On the contrary, at high initiation rates (85–90 °C) MAA improves the stability of the dispersion, increases the number of particles and the reaction rate (Fig. 6.6).

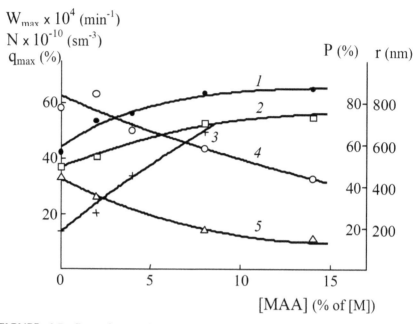

FIGURE 6.5 Dependence of the maximum rate of the emulsifier-free emulsion copolymerization of MA with MAA (4), the corresponding depth of polymerization (1), the number of particles in the polymer dispersion (5) and their average radius (2), and the amount of the coagulum formed (3) on the MAA concentration. [APS] = 5 × 10⁻³ mol/L; 70°C.

In the presence of MAA, the probability of recombination of oligomeric radicals in the aqueous phase increases due to their hindered entry into PMP due to the poor solubility in the monomers. In the presence of a specifically introduced emulsifier this leads to reduction in the number of PMP and the polymerization rate and to strengthening of the gel effect in the resulting larger particles [6]. The same phenomena are also observed while emulsifier-free polymerization at relatively low initiation rates (low temperatures) when the homogeneous nucleation mechanism predominates. However, at higher temperatures (higher initiation rates), PMP are mainly formed by the micelle mechanism as a result of the formation of their "own" emulsifier. Strengthening of bimolecular breakage in water under the influence of MAA promotes the appearance of water-soluble surfactant oligomers, playing the role of an emulsifier, the improved particle stabilization, an increase in their number and polymerization rate.

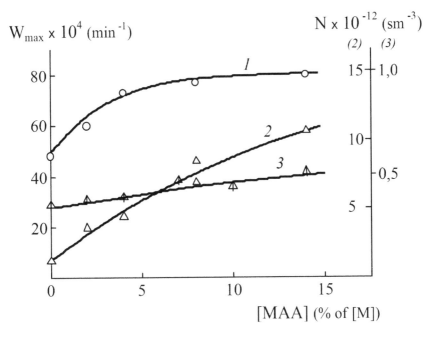

FIGURE 6.6 Effect of the MAA concentration on the maximum rate of reaction (*1*) and the number of particles in the dispersion (*2, 3*) formed by the emulsifier-free emulsion copolymerization of BA (*1, 2*) and MA (*3*). [APS] × 10³ = 5 (*3*) and 15 mol/L (*1, 2*); T = 85 (*1, 2*) and 90 °C (*3*).

Therefore, some of the processes proceeding in the reaction system may result in various effects at conventional and emulsifier-free emulsion polymerizations.

Similar differences in the effect on emulsifier-free emulsion polymerization at different initiation rates have been detected for AN as well, which retards the polymerization of MA and reduces N at 70 °C (the predominance of the homogeneous nucleation mechanism) but increases it at 90 °C, that is, under those conditions in which the micellar nucleation mechanism involving surfactant oligomers is better expressed. Moreover, it turns out that the effect of one of the comonomers on the nucleation process may depend on the availability of other comonomers. E.g., an increased AN concentration in the ternary monomer system with a high MAA content (14%), when the micellar nucleation mechanism seems more probable, leads to an increased number of particles at relatively low temperatures (70 °C) as well, although in the absence of MAA or at its low concentrations, a decrease in N is observed (Fig. 6.7).

An increased AN content in BA-based systems leads to an increased number of particles in the final polymer dispersion but the polymerization rate lowers. Increasing N in emulsifier-free conditions is associated with the improved particle stabilization during copolymerization with a hydrophilic monomer having polar CN- groups. However, just as in the presence of a specific emulsifier [6], AN weakens the gel effect, thereby reducing the reaction rate in the particle. Moreover, it suppresses flocculation at the third polymerization stage, whereby the number of particles reduces by the end of the reaction to a lesser extent than in the absence of AN. This leads to the observed differences in the effect of comonomer on the number of particles and the rate of emulsion polymerization.

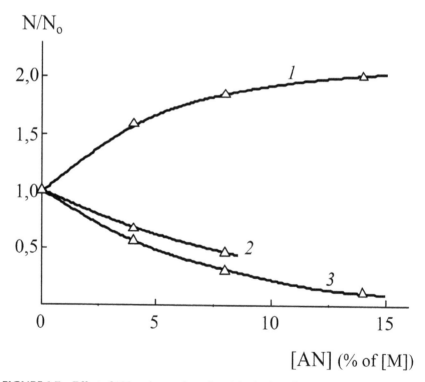

FIGURE 6.7 Effect of AN on the number of particles in the polymer dispersions produced during the emulsifier-free emulsion copolymerization of MA with AN (3) with MAA (4%) and AN (2), and with MAA (14%) and AN (1). [APS] = 4×10^{-3} mol/L; 70 °C.

6.4 CONCLUSION

Thus, this study shows the possibility of the synthesis of polymer dispersions in the absence of any emulsifying agent, but only under specially selected conditions (temperature, initiator concentration, the presence and nature of comonomers). The influence of these factors on the rate of emulsifier-free emulsion polymerization and the stability of the resulting latexes is discussed.

KEYWORDS

- acrylonitrile
- alkyl acrylates
- ammonium persulfate
- emulsifier-free emulsion polymerization
- kinetics
- mechanism
- metacrylic acid
- stabilization of dispersions

REFERENCES

1. Eliseeva, V. I., Polymer Dispersions. Moscow, *Khimiya*, 1980.
2. Aslamazova, T. R., Bogdanova, S. V., Movchan, T. G., Bases of Creation of Ecologically Safe Technology of Synthesis of Latex without Use of Emulsifiers: *Ross. Chemical Journal*, 37 (4), 112–114 (1993).
3. Eliseeva, V. I., Aslamazova, T. R., Emulsion Polymerization without Emulsifier and Latexes on its Basis: *Uspekhi Khimii*, 60 (2), 398–429 (1991).
4. Kozhevnikov, N. V., Goldfein, M. D., Trubnikov, A. V., Kozhevnikova, N. I., Emulsion Polymerization of (Meth) Acrylates: Characteristics of Kinetics and Mechanism: *J. Balkan Tribological Association,* 13 (3), 379–386 (2007).
5. Kozhevnikov, N. V., Kozhevnikova, N. I., Goldfein, M. D., Some Features of Kinetics and Mechanism of Emulsion Polymerization of Methyl Acrylate: *Proceeding of Higher Educational Establishments. Chemistry and Chemical Technology,* 53 (2), 64–68 (2010).
6. Kozhevnikov, N. V., Goldfein, M. D., Kozhevnikova, N. I., Kinetics and Mechanism of Emulsion Copolymerization of Methyl Acrylate with Hydrophilic Monomers: *J. Characterization and Development of Novel Materials*, 2 (1), 53–62 (2011).

7. Smith, W. V., Ewart, R. H., Kinetics of Emulsion Polymerization: *J. Chem. Phys.*, 16 (6), 592–599 (1948).
8. Hansen, F. K., Ugelstad, J., Particle Nucleation in Emulsion Polymerization. A Theory for Homogeneous Nucleation: *J. Polymer Sci.: Polymer Chem. Ed.*, 16 (8), 1953–1979 (1978).
9. Fitch, R. M., The Homogeneous Nucleation of Polymer Colloids: *Brit. Polymer, J.*, 5, 467–483 (1973).
10. Gladyshev, G. P., Popov, V. A., Radical Polymerization at High Conversion Degrees. Moscow, *Nauka.* 1974.
11. Kozhevnikov, N. V., Kozhevnikova, N. I., Goldfein, M. D., Turbidity Spectra of Polymer-Monomer Particles Generated During Emulsion Polymerization of Acrylates: *Journal of Applied Spectroscopy,* 72 (3), 313–316 (2005).
12. Nazarova, I. V., Eliseeva, V. I., Determination of the Methyl Acrylate – Metacrylic Acid Copolymerization Constant in an Aqueous Medium and in an Organic Solvent: *Mendeleyev Soc. J.*, 12 (4), 567–588 (1967).
13. Kulikov, S. A., Yablokova, N. V., Nikolaeva, T. V., Emulsion Copolymerization of Butyl Acrylate with Metacrylic Acid: *Vysokomol. Soed. Series, A.* 31 (11), 2322–2326 (1989).
14. Bagdasariyan, H. S., Theory of Radical Polymerization. Moscow, *Nauka,* 1966.
15. Encyclopaedia of Polymers. Vol. 1. Moscow, *Sovetskaya Encyclopaedia*, 1972.
16. Kabanov, V. A., Zubov, V. P., Yu. Semchikov, D., Complex-Radical Polymerization. Moscow, *Khimiya,* 1987.

CHAPTER 7

CONCEPTS OF THE PHYSICAL CHEMISTRY OF POLYMERS IN TECHNOLOGIES AND ENVIRONMENT PROTECTION

M. D. GOLDFEIN and N. V. KOZHEVNIKOV

Saratov State University named after, N.G. Chernyshevskiy, Russia, E-mail: goldfeinmd@mail.ru

CONTENTS

ABSTRACT

In article results of researches of kinetics and the mechanism of the radical polymerization proceeding on air and in oxygen-free conditions, in mass, solution and an emulsion, in the presence of components of salts of metals of variable valence, aromatic amines, cupferronates and stable radicals are provided. The specified substances can be used as effective stabilizers of monomers. Scientific bases of technology of synthesis high-molecular flocculent for cleaning natural and waste water, technology of receiving rigid polyurethane foam heat insulation and technology of synthesis of the polymeric latex which isn't containing the surface active agents are developed. The key kinetic parameters of the studied responses are determined.

7.1 INTRODUCTION

Physical chemistry is a most important fundamental science in the modern natural sciences. The concepts of physical chemistry qualitatively and quantitatively explain the mechanisms of various processes, such as, the reactions of oxidation, burning and explosion, obtaining food products and drugs, oil hydrocarbon cracking, making polymeric materials, biochemical reactions underlying metabolism and genetic information transfer, etc. Physical chemistry comprises notions of the structure, properties, and reactivity of various substances, free radicals as active centers of chain processes proceeding in both mineral and organic nature, the scientific foundations of low-waste and resource-saving technologies. The design and wide usage of synthetic polymeric materials is a lead in chemistry. This results in the emergence of new environmental problems due to pollution of the environment with these materials and the wastes of their production and monomer synthesis. The presented below results of our studies of the kinetics and mechanism of polymerization of vinyl monomers which can proceed in quite various conditions point to their usability in the scientific justification of optimization of technological modes of monomer and polymer synthesis, and at solving both local and global environmental problems.

To obtain reliable experimental data and to correctly interpret them, we used such physicochemical and analytical techniques as dilatometry, viscometry, UV and IR spectroscopy, electronic paramagnetic resonance, light scattering spectroscopy, electron microscopy, and gas-liquid chromatography. To analyze the properties of polymeric dispersions, the turbidity spectrum method was used, and the efficiency of flocculants was estimated gravimetrically and by the sedimentation speed of special suspended imitators (e.g. copper oxide).

7.2 RESULTS AND DISCUSSION

7.2.1 EFFECT OF SALTS OF METALS OF VARYING VALENCY

Additives of the stearates of iron (IS), copper (CpS), cobalt (CbS), zinc (ZS), and lead (LS) within a certain concentration range were found to increase the polymerization rate of styrene and methylmethacrylate (MMA) in comparison with their thermopolymerization [1, 2]. By initiating activity, they can be arranged as LS < CbS < ZS < IS < CpS. The decrease in the values of the effective activation energy, the activation energy of reaction initiation, and the kinetic reaction order by monomer point to the active participation of the monomer in chain initiation. IR spectroscopy shows that an intermediate monomer–stearate complex is formed which then decomposes into active radicals to initiate polymerization. The benzoyl peroxide (BP)–IS (or CpS) systems can be used for effective polymerization initiation. A concentration inversion of the catalytic properties of stearates has been found, which depends on salt concentration and conversion degree. The efficiency of the accelerating influence decreases with increasing temperature, the BP–IS system possesses the highest initiating activity. The initiating mechanism for these systems is principally different that the redox one. It follows from experimental data (color changes before and in the course of polymerization, the absorption and IR spectra of reactive mixtures, electron microscopy observations, etc.) that initiation occurs due to stearate radicals formed at decomposition of the complex consisting of one BP molecule and two stearate ones.

The phenomenon of concentration and temperature inversion of the catalytic properties of gold, platinum, osmium and palladium chlorides at thermal and initiated polymerization of styrene and MMA has been discovered. The mechanism of ambiguous action of noble metal salts is caused by the competition of the initiating influence of monomer complexes with colloidal metal particles and the inhibition reaction proceeding by ligand transfer.

7.2.2 HOMOPOLYMERIZATION AND COPOLYMERIZATION OF ACRYLONITRILE IN AN AQUEOUS SOLUTION OF SODIUM SULFOCYANIDE

When acrylonitrile (AN)-based fiber-forming polymers are obtained, a spinning solution ready for fiber formation appears as a result of polymerization in some solvent [1, 3–6]. Organic solvents or solutions of inorganic salts are used as solvents in these cases. First, a comparative study was made of the kinetics and mechanism of polymerization of AN in DMFA and in an aqueous solution of sodium sulfocyanide (ASSSC) initiated with AIBN and some newly synthesized azo nitriles (azobiscyanopentanol, azobiscyanovalerial acid, azobisdimethylethyl-amidoxime). The polymerization rate in ASSSC turns out to be significantly higher in comparison with the reaction in DMFA, in spite of the initiation rate in the presence of the said azo nitriles being more than 1.5 times lower than that in DMFA. The lower initiation rate in ASSSC is associated with stronger manifestation of the "cell effect" due to the higher viscosity of the water–salt solvent and the ability of water to form H-bonds (which hinders initiation). The ratio of the rate constants of chain propagation and termination ($K_p/K_t^{0.5}$) was found to be ca. ten-fold higher in ASSSC than in DMFA. The molecular mass of the polymer formed is correspondingly higher. These differences are caused by the influence of medium viscosity on K_t and formation of H-bonds with the nitrile groups of the end chain of a macroradical and the added monomer. But sodium sulfocyanide (which forms charge-transfer complexes with a molecule of AN or its radical to activate them) mainly contributes into increasing K_p.

The kinetics of AN copolymerization with methylacrylate (MA) or vinylacetate (VA) in ASSSC is qualitatively analogous to AN homopolymerization in identical conditions. At the same time, the initial reaction rate, copolymer molecular mass, effective activation energy (E_{ef}), orders by initiator and total monomer concentration differ. For example, the decrease in E_{ef} is due to the presence of a more reactive monomer (MA), and in the AN–VA system the non-end monomer chains in macroradicals influence the rate constant of cross chain termination. For these binary systems, the copolymerization constants were estimated, whose values point to a certain mechanism of chain growth, which leads to the MA concentration in the copolymer being significantly higher than in the source mixture. The same is observed in the case of AN with VA polymerization (naturally, the absolute amount of AN in the final product is much higher than MA or VA due to its higher initial concentration in the mixture).

Obtaining synthetic PAN fiber of a nitrone type is preceded by the formation of a spinning solution by means of copolymerization of AN with MA (or VA) and itaconic acid (IA) or acrylic acid (AA) or methacrylic acid (MAA) or methallyl sulfonate (MAS). Usually, the mixtures contain 15% of AN, 5–6% of the second monomer, 1–2% of the third one, and ca. 80 wt. % of 51.5% ASSSC. When a third monomer is introduced into the reaction mixture, the rate of the process and the molecular mass of the copolymer decrease, which makes these comonomers be peculiar low-effective inhibitors. In such a case, it becomes possible to estimate the inhibition (retardation) constant which is, in essence, the rate constant of one of the reactions of chain propagation. At copolymerization of ternary monomeric systems based on AN in ASSSC, the chain origination rate increases with the total monomer concentration. The initiation reaction order relative to the total monomer concentration varies from 0.5 to 0.8 depending on the degree of the retarding effect of the third monomer. Besides, AN in the three-component system is shown to participate less actively in the chain origination reaction than at its homopolymerization and copolymerization with MA or VA.

Thus, the obtained results enable regulating the copolymerization kinetics and the structure of the copolymer formed, which, finally, is a way of chemical modification of synthetic fibers.

7.2.3 STABLE RADICALS IN THE POLYMERIZATION KINETICS OF VINYL MONOMERS

Free radicals are neutral or charged particles with one or more uncoupled electrons [7–15]. Unlike usual (short-living) radicals, stable ones (long-living) are characteristic of paramagnetic substances whose chemical particles possess strong delocalized uncoupled electrons and sterically screened reactivity centers. This is the very cause of the high stability of many classes of nitroxyl radicals of aromatic, fatty-aromatic and heterocyclic series, and ion radicals and their complexes.

Peculiarities of thermal and initiated polymerization of vinyl monomers in the presence of anion radicals of tetracyanoquinodimethane (TCQM) were investigated. TCQM anion radicals are shown to effectively inhibit both thermal and AIBN-initiated polymerization of styrene, MMA, and methylacrylate in acetonitrile and dimethylformamide solutions. Inhibition is accomplished by the recombination mechanism and by electron transfer to the primary (relative to the initiator) or polymeric radical. The electron-transfer reaction leads to the appearance of a neutral TCQM, which regenerates the inhibitor in the medium of electron-donor solvents. Our calculation of the corresponding radical-chain scheme has allowed us to derive an equation to describe the dependence of the induction period duration on the initiator and inhibitor concentrations, and how the polymerization rate changes with time. The mechanism of the initiating effect of the peroxide—TCQM system has been ascertained, according to which a single-electron transfer reaction proceeds between an anion radical and a BP molecule, with subsequent reactions between the formed neutral TCQM and benzoate anion, and a benzoate radical and one more anion radical TCQM. Free radicals initiating polymerization are formed at the redox interaction between the products of the said processes and peroxide molecules. The TCQM anion radical interacts with peroxide only at a rather high affinity of this peroxide to electron (BP, lauryl peroxide); in the presence of cumyl peroxide, the anion radical inhibits polymerization only.

Iminoxyl radicals, which are stable in air and are easily synthesized in chemically pure state (mainly, crystalline brightly colored substances) present a principally new type of nitroxyl paramagnets. Organic paramagnets are used to intensify chemical processes, to increase the selectivity of catalytic systems, to improve the quality of production (anaerobic hermetics, epoxy resins, and polyolefins). They have found application in biophysical and molecular-biological studies as spin labels and probes, in forensic medical diagnostics, analytical chemistry, to improve the adhesion of polymeric coatings, at making cinema and photo materials, in device building, in oil-extracting geophysics and defectoscopy of solids, as effective inhibitors of polymerization, thermal and light oxidation of various materials, including polymers.

In this connection, systematic studies were made of the inhibiting effect of many stable mono- and polyradicals on the kinetics and mechanism of vinyl monomer polymerization. The efficiency of nitroxyls as free radical acceptors has promoted their usage to explore the mechanism of polymerization by inhibition. Usually, nitroxyls have time to react only with a part of the radicals formed at azonitrile decomposition, and they do not react at all with primary radicals at peroxide initiation. Iminoxyls have been found to terminate chains by both recombination and disproportionation in the presence of azo nitriles. Inhibitor regeneration proceeds as a result of detachment of a hydrogen atom from an iminoxyl by an active radical to form the corresponding nitroso compound. The mechanism of inhibition by a nitroso compound is addition of a growing chain to a $-N=O$ fragment to form a stable radical again. The interaction of iminoxyls with peroxides depends on the solvent type. For example, in vinyl monomers, induced decomposition of benzoyl peroxide occurs to form a heterocyclic oxide (nitrone) and benzoic acid. In contrast to iminoxyls, aromatic nitroxyls in a monomeric medium interact with peroxides to form nonradical products. Imidazoline-based nitroxyl radicals possess advantages over common azotoxides, which are their stability in acidic media (owing to the presence of an imin or nitrone functional groups) and the possibility of complex formation and cyclometalling with no radical center involved.

7.2.4 STABILIZATION OF MONOMERS

The practical importance of inhibitors is often associated with their usage for monomer stabilization and preventing various spontaneous and undesirable polymerization processes [16–24]. In industrial conditions, polymerization may proceed in the presence of air oxygen and, hence, peroxide radicals MOO· serve active centers of this chain reaction. In such cases, compounds with mobile hydrogen atoms, e.g. phenols and aromatic amines, are used for monomer stabilization. They inhibit polymerization in the presence of oxygen only, that is, they are antioxidants. As inhibitors of polymerization of (met)acrylates proceeding in the atmosphere of air, some aromatic amines known as polymer stabilizers were studied, namely, dimethyldi-(n-phenyl-aminophenoxy)silane, dimethyldi-(n-β-naphthylaminophenoxy)silane, 2-oxy-1,3-di-(n-phenylaminophenoxy) propane, 2-oxy-1,3-di-(n-β-naphthylaminophenoxy)propane. These compounds have proven to be much more effective stabilizers in comparison with the widely used hydroquinone (HQ), which is evidenced by high values of the stoichiometric inhibition coefficients (by 3–5 times higher than that of HQ). It has been found that inhibition of thermal polymerization of the esters of acrylic and methacrylic acids at relatively high temperatures (100°C and higher) is characterized by a sharp increase in the induction periods when some critical concentration of the inhibitor $[X]_{cr}$ is exceeded. This is caused by that at polymerization in air, the formation of polymeric peroxides as a result of copolymerization of the monomers with oxygen should be taken into account. Decomposition of polyperoxides occurs during the induction period as well and can be regarded as degenerated branching. The presence of critical phenomena is characteristic of chain branched reactions. However, in early works describing inhibition of thermooxidative polymerization, no degenerated chain branching on polymeric peroxides was taken into account. It follows from the results obtained that the value of critical inhibitor concentration $[X]_{cr}$ can be one of the basic characteristics of its efficacy.

Inhibition of spontaneous polymerization of (meth)acrylates is necessary not only at their storage but also in the conditions of their synthesis which proceeds in the presence of sulfuric acid. In this case, monomer stabilization is more urgent, since sulfuric acid not only deactivates many inhibitors but is also capable of intensifying the process of polymer formation. The concentration dependence of induction periods in these conditions has a brightly expressed nonlinear character. Unlike polymerization in bulk, in the presence of sulfuric acid, decomposition of polymeric peroxides is observed at relatively low temperatures, and the values $[X]_{cr}$ for the amines studied are by ca. 10 times lower than $[HQ]_{cr}$.

Synthesis of MMA from acetonecyanhydrine is a widely spread technique of its industrial synthesis. The process proceeds in the presence of sulfuric acid in several stages, when various monomers are formed and interconverted. Separate stages of this synthesis were modeled with reaction systems containing, along with MMA, methacrylamide and methacrylic acid, and water and sulfuric acid in various ratios. As heterogeneous and homogeneous systems appeared at this, inhibition was studied in both static and dynamic conditions. The aforesaid aromatic amines appear to effectively suppress polymerization at different stages of the synthesis and purification of MMA. Their advantages over hydroquinone are strongly exhibited in the presence of sulfuric acid in homogeneous conditions, or under stirring in biphasic reaction systems.

Besides, application of polymerization inhibitors is highly needed in dynamic conditions at the stage of esterification.

The usage of monomer stabilizers to prevent various spontaneous polymerization processes implies further release of the monomer from the inhibitor prior to its processing into a polymer. It is usually achieved by monomer rectification, often with preliminary extraction or chemical deactivation of the inhibitor, which requires high-energy expenditures and entails large monomer losses and extra pollution of the environment. It would be optimal to develop such a way of stabilization where the inhibitor would effectively suppress polymerization at monomer storage but would almost not affect it at polymer synthesis. The usage of inhibitors low-soluble in the monomer is one of possible variants. When the monomer is stored and the rate of polymerization initiation is low, the quantity of the inhibitor dissolved could be enough for stabilization. Besides, as the inhibitor is spent, its permanent replenishment is possible due to additional dissolution of the earlier unsolved substance. The ammonium salt of N-nitroso-N-phenylhydroxylamine (cupferon), and some cupferonates were studied as such low-soluble inhibitors. The solubility of these compounds in acrylates, its dependence on the monomer moisture degree, the influence of the quantity of the inhibitor and the duration of its dissolution on subsequent polymerization were studied. Differences in the action of cupferonates are due to their solubility in monomers, their various stability in solution and the ability of deactivation; all this results in poorer influence of the inhibitor on monomer polymerization at producing polymer.

7.2.5 SOME PECULIARITIES OF THE KINETICS AND MECHANISM OF EMULSION POLYMERIZATION

Emulsion polymerization, being one of the methods of polymer synthesis, enables the process to proceed with a high rate to form a polymer with a high molecular mass, high-concentrated latexes with a relatively low viscosity to be obtained [25–34]. Polymeric dispersions to be used at their processing without separation of the polymer from the reaction mixture, and the fire-resistance of the product to be significantly raised. At the same time, the kinetics and mechanism of polymerization in emulsion feature ambiguity, which is caused by such specific factors as the multiphasity of the reaction system and the variety of kinetic parameters whose values depend not so much on the reagent reactivity as on the character of their distribution over phases, reaction topochemistry, the way and mechanism of nucleation and stabilization of particles. The obtained results pointing to the discrepancy with classical concepts, can be characterized by the following effects: (i) recombination of radicals in an aqueous phase leading to a reduction of the number of particles and to the formation of surfactant oligomers capable of acting as emulsifiers; (ii) the presence of several growing radicals in polymer-monomer particles, which causes the appearance of gel effect and the increase in the polymerization rate at high conversion degrees; (iii) a decrease in the number of latex particles with the growth of conversion degree, which is associated with their flocculation at various polymerization stages; (iv) an increase in the number of particles in the course of reaction when using monomer-soluble emulsifiers, and also due to the formation of an "own" emulsifier (oligomers).

Surfactants (emulsifiers of various chemical nature) are usually applied as stabilizers of disperse systems, they are rather stable, poorly destructed under the influence of natural factors, and contaminate the environment. The principal possibility to synthesize emulsifier-free latexes was shown. In the absence of emulsifier (but in emulsion polymerization conditions) with the usage of persulfate-type initiators (e.g., ammonium persulfate), the particles of acrylate latexes can be stabilized with ionized endgroups of macromolecules. The $M_nSO_4^-$ ion radicals appearing in the aqueous phase of the reaction medium, having reached a critical chain length, precipitate to form primary particles, which flocculate up to the formation of aggregates with a charge density providing their stability. Besides, due to recombination of radicals, oligomeric molecules are formed in the aqueous phase, which possess properties of surfactants and are able to form micelle-like structures. Then, the monomer and oligomeric radicals are absorbed by these "micelles," where chains grow. In the absence of a specially introduced emulsifier, all basic kinetic regularities of emulsion polymerization are observed, and differences are concerned only with the stage of particle generation and the mechanism of their stabilization, which can be strengthened at copolymerization of hydrophobic monomers with highly hydrophilic co-monomers. Increasing temperature results in the growth of the polymerization rate and the number of latex particles in the dispersion formed, decreasing their sizes and the quantity of the formed coagulum, and in improved stability of the dispersion. At emulsifier-free polymerization of alkyl acrylates, the stability of emulsions and obtained dispersions rises in the monomer row: methylacrylate < ethylacrylate < butylacrylate, that is, the stability growth at lowering the polarity of the main monomer.

Our account of the aforesaid factors influencing the kinetics and mechanism of emulsion polymerization (in both presence and absence of an emulsifier) has enabled the influence of comonomers on the processes of formation of polymeric dispersions based on (meth)acrylates to be explained. Changes of some conditions of reaction have turned out to affect the character of influence of other ones. E.g., increasing the concentration of methacrylic acid (MAA) at its copolymerization with methylacrylate (MA) at a relatively low initiation rate leads to a decrease in the rate and particle number and to an increase in the coagulum amount. But at high initiation rates, the number of particles in the dispersion in the presence of MAA rises and their stability improves. The same effects were revealed for emulsifier-free polymerization of butylacrylate as well, when at high temperatures its partial replacement by MAA results in better stabilization of the dispersion, an increase in the reaction rate and the number of particles (whereas their decrease was observed in the presence of an emulsifier). Similar effects were found for AN as well, which worsens the stability of dispersion at relatively low temperatures but improves it at high ones. Increasing in the AN concentration in the ternary monomeric system with a high MAA content leads to a higher number of particles and better stability of dispersion at relatively low temperatures as well.

With the aim to explore the possibility to synthesize dispersions whose particles would contain reactive polymeric molecules with free multiple C=C bonds, emulsion copolymerization of acrylic monomers and unconjugated dienes was studied. The usage of such latexes to finish fabrics and some other materials promotes getting strong indelible coatings. The kinetics and mechanism of emulsion copolymerization of ethylacrylate (EA) and butylacrylate (BA) with allylacrylate (AlA) (with ammonium persulfate (APS) or the APS—sodium thiosulfate system as initiators) were studied. The found constants of copolymerization of AlA with EA

($r_{AlA} = 1.05$, $r_{EA} = 0.8$) and AlA with BA ($r_{AlA} = 1.1$, $r_{BA} = 0.4$) point to different degrees of the copolymer's unsaturation with AlA units. Emulsion copolymerization of multicomponent monomeric BA-based systems (with AN, MAA, and the unconjugated diene acryloxyethyl maleate (AOEM) as comonomers) was also studied. The influence of AN and MAA is described above. Copolymerization with AOEM depends on reaction conditions. E.g., in MAA-free systems, AOEM reduces the polymerization rate. In the presence of MAA, the rate of the process at high conversion degrees increases with the AOEM concentration, this monomer promoting the gel effect due to partial chain linking in polymer–monomer particles by the side groups with C=C bonds. The degree of unsaturation of diene units in the copolymer is subject to the composition of monomers, AOEM concentration, and temperature. In the BA–AOEM system, an increase of the diene concentration results in an increase in the unsaturation degree. This means that the diene radical is added to "its own" monomer with a higher rate than to BA ($r_{AOEM} = 7.7$). The higher unsaturation of diene units in MAA-containing systems points to that the AOEM· radical interacts with MAA with a higher rate than with BA, and the probability of cyclization reduces, the degree of polymer unsaturation increases, and the diene's retarding action upon polymerization weakens.

7.2.6 COMPOSITIONS FOR PRODUCTION OF RIGID FOAMED POLYURETHANE

In the field of polymer physicochemistry, studies were made according to the requirements of the Montreal Protocol (1987), which demands drastically reduce the production and consumption of chlorofluorocarbons (CFC, Freon™) and even replace oxone-dangerous substances by ozone-safe ones [24, 35–37]. Our investigations dealt with the replacement of trichlorofluoromethane (Freon-11), which had been used as a foam-maker in the synthesis of rigid foamed polyurethane (FPU) over a long period of time, which was thermal insulator in freezing chambers and building constructions. On the basis of our experimental dependences of the kinetic parameters of the foaming process (the instant of start, the time of structurization, and the instant of foam ending rising), the values of density and heat conductivity of pilot foamed plastics samples on the concentration of the reactive mixture components and physicochemical conditions of FPU synthesis, optimal compositions (recipes) of mixtures with Freon-11 replaced by a ozone-safe (with an ozone-destruction potentials by an order of magnitude lower in comparison with Freon-11) azeotropic mixture of dichlorotrifluoroethane and dichlorofluoroethane were found. Their practical implementation requires no principal changes of known technological procedures and usage of new chemical reagents, which is an important merit of our developments.

7.2.7 SCIENTIFIC BASICS OF THE SYNTHESIS OF A HIGH-MOLECULAR-WEIGHT FLOCCULENT

There exists a problem of purification of natural water and industrial sewage from various pollutants, including suspended and colloid-disperse particles, associated with the growing consumption of water and its deteriorating quality (owing to anthropogenic influence) [23, 24, 35, 37]. It is known that flocculants can be used for these purposes, which are high-molecular-weight compounds capable of adsorption on disperse particles to form quickly sedimenting aggregates. Polyacrylamide (PAA) is most active. In connection with that many countries (including Russian Federation) suffer from acrylamide (AA) deficit, we have developed modifications of the synthesis of PAA flocculent by means of the usage of AN and sulfuric acid to implement the reactions of hydrolysis and polymerization. It has turned out that in the presence of a radical initiator of polymerization and sulfuric acid, AN participates in both processes simultaneously, and, as AA is formed from AN, their joint polymerization begins. Desired polymer properties were achieved at AN polymerization in an aqueous solution of sulfuric acid up to a certain conversion degree with subsequent hydrolysis of the polymerizate (a two-stage synthetic scheme) or at achieving on optimal ratio of the rates of these reactions proceeding in one stage. The influence of the nature and concentration of initiator, the content of sulfuric acid, temperature and reaction duration on the quantity and molecular mass of the polymer contained in the final product, its solubility in water and flocculating properties were studied. The required conversion degree at the first stage of synthesis by the two-stage scheme is determined by the concentrations of AN and the aqueous solution of sulfuric acid.

At one-stage synthesis, changes in temperature and monomer amount almost equally affect the reaction rates of hydrolysis and polymerization and do not strongly affect the copolymer composition. But changes in the concentration of either acid or initiator rather strongly influence the molecular mass and composition of macromolecules, which causes external dependences of the flocculating activity on these factors.

7.3   CONCLUSIONS

Thus, on the basis of our above results, the following conclusions can be made.

1. New kinetic regularities have been revealed at polymerization of vinyl monomers in homophasic and heterophasic conditions in the presence of additives of transition metal salts, azo nitriles, peroxides, stable nitroxyl radicals and anion radicals (and their complexes), aromatic amines and their derivatives, emulsifiers and solvents of various nature.
2. The mechanisms of the studied processes have been established in the whole and as elementary stages, their basic kinetic characteristics have been determined.
3. Equations to describe the behavior of the studied chemical systems in the reactions of polymerization proceeding in various physico-chemical conditions have been derived.
4. Scientific principles of regulating polymer synthesis processes have been elaborated to allow optimization of some industrial technologies and solving most important problems of environment protection.

KEYWORDS

- environment
- flocculent
- inhibitor
- initiator
- kinetics
- polymerization

- **radical**
- **research**
- **technology**

REFERENCES

1. Goldfein, M. D., Kozhevnikov, N. V., Trubnikov, A. V., Kinetics and Mechanism of Regulation Processes of Synthesis Polymers. Saratov State University. 1989. 178 p. (in Rus.).
2. Goldfein, M. D., Kinetics and Mechanism of Radical Polymerization of Vinyl Monomers. Saratov State University. 1986. 139 p. (in Rus.).
3. Goldfein, M. D., Kozhevnikov, N. V., Rafikov, E. A., Polymer Science. Series A (in Rus.). 1975. Vol. 17. No 10, 2282–2287.
4. Goldfein, M. D., Rafikov, E. A., Kozhevnikov, N. V., Polymer Science. SeriesA (in Rus.). 1977. Vol. 19. No 2, 275–280.
5. Goldfein, M. D., Rafikov, E. A., Kozhevnikov, N. V., Polymer Science. Series A (in Rus.). 1977. Vol. 19. No 11, 2557–2562.
6. Goldfein, M. D., Zyubin, B. A., Polymer Science. Series A (in Rus.). 1990. Vol. 32. No 11, 2243–2263.
7. Stepukhovich, A. D., Kozhevnikov, N. V., Leonteva, L. T., Polymer Science. Series A (in Rus.). 1974. Vol. 16. No 7, 1522–1529.
8. Kozhevnikov, N. V., Polymer Science. Series A (in Rus.). 1986. Vol. 28. No 4, 675–683.
9. Kozhevnikov, N. V., Higher of Educational Institute. Series – Chemistry and Chemical Technologies (in Rus.). 1987. Vol. 30. No 4, 103–106.
10. Kozhevnikov, N. V., Stepukhovich, A. D., Polymer Science. Series A (in Rus.). 1980. Vol. 22. No 5, 963–971.
11. Trubnikov, A. V., Goldfein, M. D., Kozhevnikov, N. V., Polymer Science. Series A (in Rus.). 1983. Vol. 25. No 10, 2150–2156.
12. Rozantsev, E. G., Goldfein, M. D., Trubnikov, A. V., Success of Chemistry (in Rus.). 1986. Vol. 55. No 11, 1881–1897.
13. Rozanrsev, E. G., Goldfein, M. D., Pulin, V. F., Organic Paramagnets. Saratov State University (in Rus.). 2000. 340 p.
14. Rozantsev, E. G., Goldfein, M. D., Oxidation Communications. Sofia. 2008. Vol. 31. No 2, 241–263.
15. Rozantsev, E. G., Goldfein, M. D., Polymers Research Journal. New York. Nova Science Publishers, Inc. 2008. Vol. 2. No 1, 5–28.
16. Goldfein, M. D., Kozhevnikov, N. V., Trubnikov, A. V., Polymer Science. Series B (in Rus.). 1983. Vol. 25. No 4, 268–271.
17. Kozhevnikov, N. V., Polymer Science. Series B (in Rus.). 1988. Vol. 30. No 8, 613–616.

18. Goldfein, M. D., Gladyshev, G. P., Success of Chemistry (in Rus.). 1988. Vol. 57. No 11, 1888–1912.
19. Goldfein, M. D., Kozhevnikov, N. V., Trubnikov, A. V., Chemical Industry (in Rus.). 1989. No 1, 20–22.
20. Kozhevnikov, N. V., Zyubin, B. A., Tsyganova, T. V., Aizenberg, L. V., Higher of Educational Institute. Series – Chemistry and Chemical Technologies (in Rus.). 1989. Vol. 32. No 11, 98–102.
21. Goldfein, M. D., Gladyshev, G. P., Trubnikov, A. V., Polymer Yearbook. 1996. No 13, 183–190.
22. Kozhevnikov, N. V., Goldfein, M. D., Kozhevnikova, N. I., Journal of the Balkan Tribological Association. Sofia. 2008. Vol. 14. No 4, 560–571.
23. Kozevnikov, N. V., Kozhevnikova, N. I., Goldfein, M. D., Higher of Saratov State University. Seria – Chemistry, Biology, ecology (in Rus.). 2010. Vol. 10. No 2, 34–42.
24. Goldfein, M. D., ivanov, F. V., Kozhevnikov, N. V., Fundamentals of General Ecology, Life Safety and Environment Protection. New York. Nova Science Publishers, Inc. 2010. 251 p.
25. Kozhevnikov, N. V., Goldfein, M. D., Zyubin, B. A., Trubnikov, A. V., Polymer Science. Series A (in Rus.). 1991. Vol. 33. No 6. 1272–1280.
26. Goldfein, M. D., Kozhevnikov, N. V., Trubnikov, A. V., Polymer Science. Series A (in Rus.). 1991. Vol. 33. No 10. 2035–2049.
27. Kozhevnikov, N. V., Goldfein, M. D., Polymer Science. Series A (in Rus.). 1991. Vol. 33. No 11, 2398–2404.
28. Goldfein, M. D., Kozhevnikov, N. V., Trubnikov, A. V., Polymer Yearbook. 1995. No 12, 89–104.
29. Kozhevnikov, N. V., Goldfein, M. D., Terekhina, N. V., Chemical Physics (in Rus). 1997. Vol. 16. No 12, 97–102
30. Kozhevnikov, N. V., Goldfein, M. D., Terekhina, N. V., Higher of Educational Institute. Series – Chemistry and Chemical Technologies (in Rus.). 1997. Vol. 40. No 3, 78–83.
31. Kozhevnikov, N. V., Goldfein, M. D., Trubnikov, A. V., Inter. Journal Polymer Mater. 2000. Vol. 46, 95–105.
32. Kozhevnikov, N. V., Goldfein, M. D., Trubnikov, A. V., Kozhevnikova, N. I., Journal of the Balkan Tribological Association. Sofia. 2007. Vol. 13. No 3, 379–386.
33. Kozhevnikov, N. V., Kozhevnikova, N. I., Goldfein, M. D., Higher of Educational Institute. Series – Chemistry and Chemical Technologies (in Rus.). 2010. Vol. 53. No 2, 64–68.
34. Kozhevnikov, N. V., Goldfein, M. D., Kozgevnikova, N. I., Higher of Saratov State University. Series – Chemistry. Biology. Ecology (in Rus.). 2012. Vol. 12. No 2, 3–8.
35. Goldfein, M. D., Kozhevnikov, N. V., Problems of local ecology (in Rus.). 2005. No 4, 92–85.
36. Goldfein, M. D., Higher of Saratov State University. Series – Chemistry. Biology. Ecology (in Rus.). 2009. No 2, 79–83.
37. Kozhevnikov, N. V., Goldfein, M. D., Kozhevnikova, N. I., Journal of the Balkan Tribological Association. 2007. Vol. 13. No 4, 536–541.

CHAPTER 8

APPLICATION ABLE RADICALS FOR STUDY OF BEHAVIOR OF BIOLOGICAL SYSTEMS

M. D. GOLDFEIN and E. G. ROZANTSEV

Saratov State University named after, N.G. Chernyshevsky,
E-mail: goldfeinmd@mail.ru

CONTENTS

ABSTRACT

Research of the condensed phases containing stable radicals, by means of radiospectroscopy represents a method of paramagnetic sounding. This method (using iminoxyl radicals) allowed to set the mechanism of interaction of antigenes with antibodies in case of study of immune gamma globulins, particularly structural transitions in biological membranes, to study the structure of some model systems. It is shown that certain derivatives of iminoxyl radicals have low toxicity and exhibit a relatively high antileukemic activity, greatest quantities were observed for the inhibition ratios hemocytoblasts in peripheral blood, bone marrow, and during chemotherapy of certain cancers.

8.1   INTRODUCTION

The presence of paramagnetic particles in liquid or solid objects opens new opportunities of their studying by the EPR technique. Ready free radicals and substances forming paramagnetic solutions due to spontaneous homolization of their molecules in liquid and solid media (such as triphenylmethyl dimer, Frémy's salt or 4,8-diazaadamantan-4,8-dioxide) can act as sources of paramagnetic particles.

The experimental technique of radio spectroscopic examination of condensed phases with the aid of paramagnetic impurities is usually called the paramagnetic probe method. Though iminoxyl radicals have found broadest applications for probing of biomolecules, nevertheless, the first application of the paramagnetic probe technique to study a biological system is associated with a quite unstable aminazine radical cation:

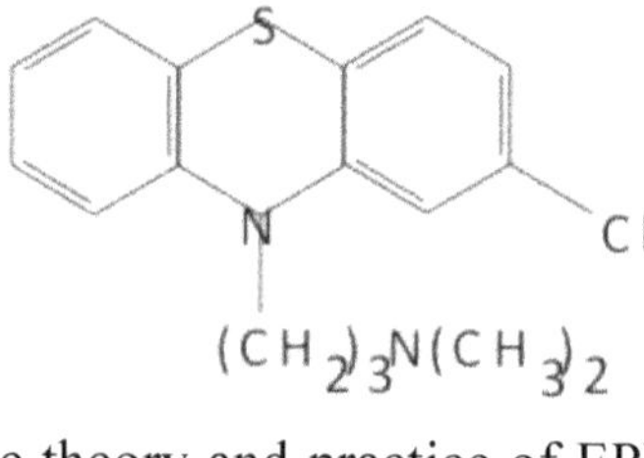

The progress in the theory and practice of EPR usage in biological research is restrained by the narrow framework of chemical reactivity of nonfunctionalized stable radicals with a localized paramagnetic center like:

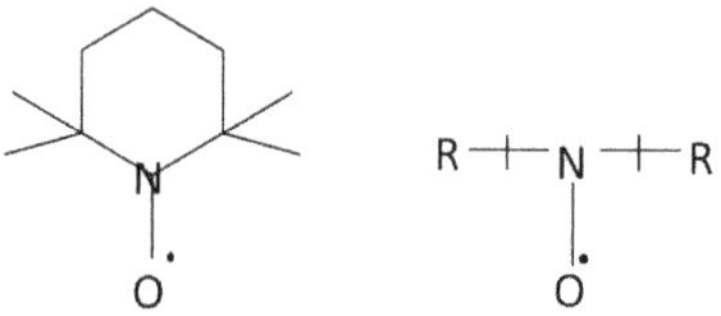

The substances of this class only enter into common, well-known free radical reactions, namely: recombination, disproportionation, addition to multiple bonds, isomerization, and β-splitting [1]. All these reactions proceed with the indispensable participation of a radical center and steadily lead to full paramagnetism loss. And still the synthesis of nonfunctionalized stable radicals plays a very important role. No expressed delocalization of an uncoupled electron over a multiple bond system has been shown to be obligatory for a paramagnetic to be stable.

Despite of the basic importance of the discovery of stable radicals of a nonaromatic type [2], this event has not changed contemporary ideas on the reactivity of stable radicals.

In the early 1960s, one of the authors of this book laid the foundation of a new lead in the chemistry of free radicals, namely, the synthesis and reactivity of functionalized stable radicals with an expressed localized paramagnetic center [3].

The opportunity to obtain and study a wide range of such compounds with various functional substituents arose in connection with the discovery of free radical reactions with their paramagnetic center unaffected.

Functionalized free radicals have found broad applications as paramagnetic probes for exploring molecular motion in condensed phases of various natures. The introduction of a spin label technique (covalently bound paramagnetic probe) is associated with their usage; its idea is not new and based on the dependence of the EPR spectrum shape of the free radical on the properties of its immediate atomic environment and the way of interaction of the paramagnetic fragment with the medium. The reactions of free radicals with their paramagnetic center unaffected (Neumann–Rozantsev's reactions) [4] have became the chemical basis of obtaining spin-labeled compounds.

The concept of the usage of nonradical reactions of radicals to study macromolecules was formulated at the Institute of Chemical Physics (USSR Academy of Sciences) by Lichtenstein [5] in 1961, and the theoretical bases of this method, calculation algorithms for the correlation times of rotary mobility of a paramagnetic particle from their EPR spectrum shape were developed by McConnell [6], Freed and Fraenkel [7], Kivelson [8], and Stryukov [9]. Have investigated the behavior of iminoxyl radicals in various systems and obtained important and interesting results [10].

Let us cite several important aspects of the application of organic paramagnets to researching biological systems.

8.2 APPLICATION OF IMINOXYL FREE RADICALS FOR STUDYING OF IMMUNE GAMMA GLOBULINS

From the physicochemical viewpoint, the mechanism of various immunological reactions is determined by changes in the phase state of a system. Despite of the wide use of these reactions in medical practice, the nature of interaction of antigens with antibodies is not quite clear. To study this process, gamma globulins labeled with iminoxyl radicals were used [11]. The molecule of gamma globulin is known to consist of four polypeptide chains bound with each other with disulfide bridges. When the interchain disulfide bonds are split, the polypeptide chains continue to keep together. A spin label (an iminoxyl derivative of maleimide)

was attached to the sulfhydryl groups obtained by restoration of disulfide bonds with β-mercaptoethanol. Experiments were made on the rabbit and human gamma globulins. The EPR spectra in both cases corresponded to rather high mobility of free radicals (correlation times $\tau = 1.1 \times 10^{-9}$ s for human gamma globulin and 7.43×10^{-9} s for rabbit's one). By comparing the correlation times in these proteins with the values obtained in experiments with serum albumin labeled with sulfhydryl groups, treated with urea ($\tau = 1.09 \times 10^{-9}$ s) and dioxane (2.04×10^{-9} s), it is possible to conclude that the fragments of polypeptide chains bearing free radicals possess no ordered secondary structure. This is in accord with data on very low contents of α-helical structures in gamma globulins. Such a character of the EPR spectrum of immune gamma globulin with preservation of its specific activity opens the possibility to explore conformational and phase transitions at specific antigen–antibody reactions. Sharp distinctions in the mobility of spin labels were revealed at sedimentation of the rabbit antibodies by salting-out with ammonium sulfate and precipitation with a specific antigen (egg albumin). Precipitation of antibodies with a specific antigen led only to a small reduction of the paramagnetic label mobility whereas sedimentation with ammonium sulfate caused strong retardation of the rotary mobility of free radicals (Fig. 8.1).

These results can be considered as direct confirmation of the alternative theory [12] according to which the precipitate formation is associated with the immunological polyvalency of the antigen and antibody relative to each other. Really, excepting the location of spin labels inside the antibody's active center, it is possible to conclude that the rather mobile condition of spin labels in the antigen–antibody precipitate may remain if only there is no strong dehydration of the antibodies due to intermolecular interactions. Unlike gamma globulin precipitated with ammonium sulfate, the specific precipitate, according to the lattice theory, has a microcellular structure. At long storage of the antigen–antibody precipitate with no addition of stabilizers, the mobility degree of iminoxyl radicals sharply decreased. This is apparently a result of secondary dehydration of the antibodies owing to protein molecule interactions in the precipitate.

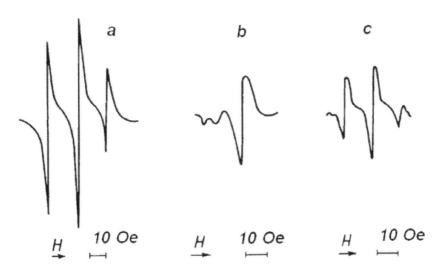

FIGURE 8.1 EPR spectra of gamma globulin labeled with an iminoxyl radical with its free valency unaffected: (a) in solution; (b) in the precipitate obtained by salting-out with ammonium sulfate; (c) in the specific precipitate.

8.3   EXPLORING STRUCTURAL TRANSITIONS IN BIOLOGICAL MEMBRANES

Biological membranes, in particular, mitochondrial membranes, are known to play a huge role in redox processes in the cell, being the very place of respiratory chain enzyme localization. Baum and Riske [13] have found essential distinctions in the properties of one of the mitochondrial membrane fragments (Complex III of Electron Transport Chain) upon transition from the oxidized form to the reduced one: the sulfhydryl groups, easily titrated in the oxidized form of this fragment, become inaccessible in the reduced form. The nature of trypsin digestion of this complex also strongly changes.

Earlier, changes in the repeating structural units of mitochondria in conditions leading to the formation of macroergic intermediate products or their provision, for example, active ion transfer, were revealed by electron microscopy [14]. All this has allowed us to assume that any redox reaction catalyzed by an enzymatic chain of electron transfer is accompanied by some kind of "conformational wave" probably covering not only the protein component of the membrane but also a higher level of its organization, namely: fragments of the membrane of more or less complexity degree, including its lipidic part. To verify this assumption, a modification of the spin label method (a method of noncovalent bound paramagnetic probe) was used. The radical is kept by the matrix (membrane) involving only weak hydrophobic bonds. Such an approach allows studying of weak interactions in the system without essential disturbance of the biochemical functions of the biomembrane and its structure. The paramagnetic probe was 2,2,6,6-tetramethylpiperidine-1-oxyl caprylic ester:

$$H\text{-}C_7H_{15}-C \overset{O}{\underset{O}{\diagup\!\!\!\!\diagdown}}$$

This compound was prepared from caprylic acid chloranhydride and 2,2,6,6-tetramethyl-4-oxpiperidin-1-oxyl in a triethylamine medium by a radical reaction with free valency unaffected. The paramagnetic probe was introduced into a suspension of electron transport particles (ETP) isolated from the bull heart mitochondria by the technique described in Ref. [14] at the Laboratory of Bioorganic Chemistry, Moscow State University. These fragments of the mitochondrial membrane are characterized by a rather full set of enzymes of the respiratory chain with the same molar ratio as in the intact mitochondria [14]. The ability of oxidizing phosphorylation, however, is lost under the used way of isolation.

The paramagnetic probe is insoluble in water but solubilized by ETP suspended in a buffer solution. Owing to this, the observed EPR spectrum is free of any background due to the radicals not attached with the object under study. The presence of a voluminous hydrocarbonic chain provides "embedding" of a molecule of the probe into the lipidic part of ETP. Therefore, the EPR spectrum reflects the condition of exactly this fraction of the membrane. To detect conformational transitions, EPR spectra were recorded before and after the introduction of oxidation substrata (succinate and NAD-N), and after oxidation of the earlier reduced respiratory chain with potassium ferricyanide. Typical results are shown in Fig. 8.2 [14].

The enhanced anisotropy of the EPR spectrum of iminoxyl after substratum introduction is clearly seen. The spectrum in the ferricyanide-oxidized ETP almost does not differ from that in the intact ETP.

ETP inactivation by long storage at room temperature or cyanide inhibition eliminated this effect. Comparison of the shape of signals and correlation times shows that the EPR spectrum of iminoxyl in the intact ETP consists of two signals differing by anisotropy. The radical localized in that part of the membrane where the effective free volume available for radical motion is rather large gives a weakly anisotropic signal. The strongly anisotropic (retarded) spectrum belongs to the radicals localized

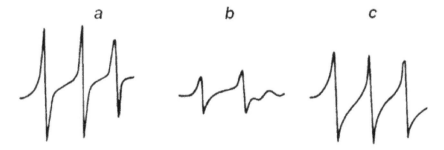

FIGURE 8.2 Changes in the EPR spectrum anisotropy of the hydrophobic iminoxyl radical in a suspension of electron transport particles from bovine heart mitochondria with oxidation substrates added.

in other sites of the system with a smaller effective free volume. Reduction of the respiratory chain with substrata leads, owing to cooperative-type conformational transitions, to a reduced fraction of sites with large free volume (i.e. to an increased microviscosity of the immediate environment of the radical). The correlation time of the whole spectrum changes from 20×10^{-10} s in the oxidized ETP to 4×10^{-10} s in the reduced ETP.

Concurrently with enhancing the anisotropy of the signal, its intensity decreases as well: iminoxyl reduces, apparently, to hydroxylamine derivatives. Potassium ferricyanide inverts this process. It is necessary to consider that oxidation substrata, themselves, do not interact considerably with iminoxyls. Obviously, the conformational transition not only leads to a changed microviscosity but eliminates any steric obstacles complicating reduction of the radical. In principle, this circumstance points to possibly a new, actually chemical, aspect of application of the paramagnetic probe technique.

8.4 EXPLORING THE STRUCTURE OF SOME MODEL SYSTEMS

The application of the paramagnetic probe technique in systems like biological membranes poses a number of questions concerning the behavior of hydrophobic labels in media with an ordered arrangement of hydrophobic chains. As a first stage, mixtures of a nonionic detergent (Tween 80) and water were studied. Tween 80 originates from polyethoxylated sorbitan and oleic acid (polyoxyethylene (20) sorbitan monooleate) and classifies as a nonionic detergent based on polyethylene oxide [14]. Our choice of this object was determined by some methodical conveniences, and also some literature data on the structure of aqueous solutions of Tween, obtained by classical methods (viscometry, refractometry, etc.) The properties of this detergent are also interesting in themselves since it finds quite broad applications for biological membrane fragmentation.

The esters of 2,2,6,6-tetramethyl-4-oxypiperidin-1-oxyl and saturated acids of the normal structure with a hydrocarbonic chain length of 4, 7 or 17 carbon atoms or the corresponding amides were used as a paramagnetic probe. For comparison, the behavior of hydrophobic labels IV and V in these systems was also studied.

I

II

III

IV

V

The course of changes in the correlation time of radical rotation as a function of the Tween concentration is shown in Fig. 8.3.

It is possible to resolve several areas, apparently, corresponding to various structure types. The initial fragment, only distinguishable for the easiest radicals, corresponds to an unsaturated Tween solution in water. This region is better revealed for the detergents with a higher critical micelle concentration (CMC), for example, for sodium dodecyl sulfate (Fig. 8.4) [15, 16]. Then micelle formation occurs. The correlation time of water-insoluble labels increases and comes to a plateau; at further increasing detergent concentration it passes through a maximum, then monotonously increases up to its value in pure Tween. It is useful to compare these data with the results of viscosity measurements by usual macromethods. Figure 8.5 shows that viscosity has one extremum about 60% of Tween. Therefore, at a high Tween concentration, the effective volume available for probe molecule rotation and macroviscosity do not correlate.

These results suggest the following interpretation. In pure Tween the lamellar structure provides easy layer-by-layer sliding, hence, the macroviscosity of the system is low. However, in the absence of water, the interaction

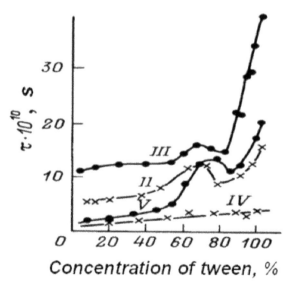

FIGURE 8.3 Changes of the correlation time in the detergent–water system, for nitroxyl radicals with a strongly localized paramagnetic center.

of the polar groups is strong, the hydrocarbonic chains are ordered, and the effective free volume in the field of radical localization is small. Small water amounts lead to the formation of defects in the layered structure. Sliding is hindered, and the viscosity increases. However, moistening breaks the close

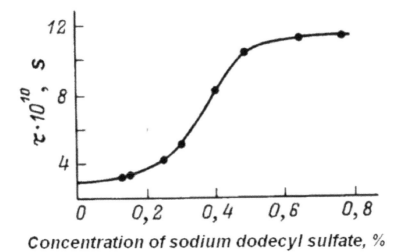

FIGURE 8.4 Estimation of the critical micelle concentration of sodium dodecylsulfate with the aid of 2,2,6,6-tertamethyl-4-hydroxypiperidyl-1-oxyl varerate.

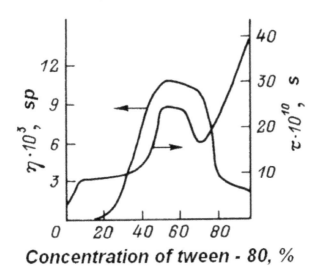

FIGURE 8.5 Changes in the microviscosity and macroviscosity of the water–Tween 80 system.

interaction of the polar groups in Tween. These groups are deformed, at the same time, the hydrocarbonic chains are disordered. The microviscosity of the hydrocarbonic layer so decreases. Upon termination of hydration of the polar groups, water-filled cavities are formed. They are a structural element (micelle) of which the system is built, for example, a hexagonal P-lattice is formed, by Luzzatti. Structure formation manifests itself as increasing microbiscosity and macroviscosity. In the field of the maximum, phase inversion is possible. Structural units of a new type (Tween micelles in water) are formed, passing into colloidal solution upon further dilution. The course of microviscosity changes at high Tween concentrations amazingly resembles the change in correlation time when some lyophilized cellular organelles are moistened. This similarity confirms that in the field of τ maximum, where restoration of the biochemical activity of chloroplasts begins, a phase transition occurs of the same type as in the LC "detergent–water" systems.

Some information on the behavior of radical particles in colloidal systems is provided by the results of temperature measurements. Figure 8.6 illustrates the temperature dependence of the correlation time in Arrhenius' coordinates for several iminoxyl radicals of various hydrophobicity degrees.

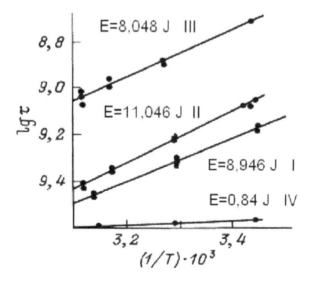

FIGURE 8.6 Activatyion energy of rotational diffusion of free iminoxyl (nitroxyl) radicals in the Tween–water (30% of Tween 80) system.

The strong dependence of the preexponential factor on chain length, apparently, confirms that the observed correlation time really reflects rotation of the radical molecule as a whole. However, this result also allows another interpretation, namely: depending on the hydrocarbonic chain length, the radical introduces into a detergent micelle more or less deeply. This may lead to a changed rotation frequency of iminoxyl groups round the ordinary bonds in the molecule, depending on the environment of the polar end of the radical. If the polar group of the radical is on the micelle–solvent interface, the measured frequency should depend on the surface charge (potential) of the micelle. In our opinion, this circle of colloid chemical problems, closely connected with questions of transmembrane transfer in biological systems, will provide one more application field of the paramagnetic probe technique.

8.5   SOLVING OTHER BIOLOGICAL PROBLEMS WITH STABLE RADICALS

Further progress in the field of the usage of stable paramagnets to solve various biological problems is reflected in numerous reviews and monographs [17–20].

For dynamic biochemistry, of undoubted interest are local conformational changes of protein molecules in solutions. The distances between certain loci of biomacromolecules, in principle, can be estimated quantitatively by means of stable paramagnets. Upon introduction of iminoxyl fragments into certain sites of native protein (NRR-method) the distance between the neighboring paramagnetic centers can be calculated from the efficiency of their dipole–dipole interaction in vitrified solutions of a spin-labeled preparation (the EPR method).

The first attempt to estimate the distances between paramagnetic centers in spin-labeled mesozyme and hemoglobin was undertaken by Liechtenstein [21], who indicated prospects of such an approach. In this regard, there appeared a need of identification of a simple empirical parameter in EPR spectra for quantitative assessment of the dipole–dipole interaction of paramagnetic centers.

A convenient empirical parameter was found when studying vitrified solutions of iminoxyl radicals. It was the ratio of the total intensity of the extreme components of a spectrum to the intensity of the central one (Fig. 8.7). To establish a correlation of the d_1/d value with the average distance between the localized paramagnetic centers, the corresponding calibration plots are drawn (Fig. 8.8). Calculations have shown that the d_1/d parameter depends on the value of dipole–dipole broadening, being in fair agreement with independently obtained experimental results.

Subsequently, the methods of quantitative assessment of the distances between paramagnetic centers in biradicals and spin-labeled biomolecules became reliable tools for structural research.

When determining relaxation rate constants of various paramagnetic centers in solution, the values of these constants have appeared to significantly depend on the chemical nature of the functional groups in iminoxyl radicals.

Sign inversion of the electrostatic charge of the substituent and its distance from the paramagnetic iminoxyl group have the strongest impact on the values of the constants. The electrostatic effect caused by the value and sign of the charge of paramagnetic particles interacting in solution is

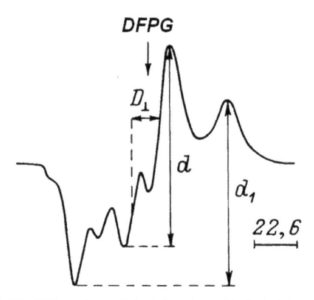

FIGURE 8.7   EPR spectrum of the iminoxyl free biradical (2,2,6,6-tetramethyl-4-hydroxypiperidine-1-oxyl phthalate) vitrified in toluene at 77 K.

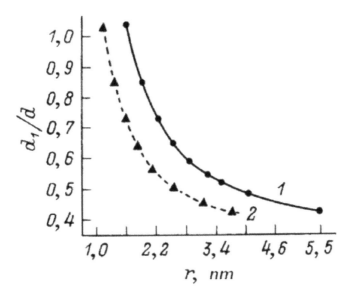

FIGURE 8.8 Dependence of the d_1/d paremeter of an EPR spectrum on the mean distance r between interaction paramagnetic centers of iminoxyl radicals (1) and biradicals (2) at 77 K.

a more essential factor. An increased ionic strength leads to a change in the k value in qualitative agreement with Debye's theory. The substituent's mass is another factor considerably influencing the value of the constant. If the substituent is a protein macromolecule, the value of k decreases twice. The possibilities of application of the paramagnetic probe technique for detection of anion and cation groups and estimation of distances to explore the microstructure of protein were analyzed. The dependences obtained in experiment show that Debye's equation with $D = 80$ can be successfully applied to experimental data analysis, and, in particular, to estimating the distance from the iminoxyl group of a spin label on protein to the nearest charged group if this distance does not exceed 1.0–1.2 nm.

Considering the role of free radical processes in radiation cancer therapy, it was offered to investigate the influence of iminoxyl radicals upon the organism of laboratory animals. Pharmacokinetic studies [22] have shown that the elementary functionalized derivatives of 2,2,6,6-tetramethylpiperidin-1-oxyl possess rather low toxicity and show an expressed antileukemic activity. The highest values of retardation coefficients were observed for hematopoietic stem cells in peripheral blood and marrow. This circumstance stimulated further structural and synthetic research directed to obtaining more effective and less toxic cancerolytics and sensibilizers for radiation cancer therapy.

After the first publication, for a rather short time, many potential cancerolytics were synthesized in the Laboratory of Stable Radicals, Institute of Chemical Physics, USSR Academy of Sciences, among which of greatest interest from biologists were so-called paramagnetic analogs of some known antitumor preparations, for example, a paramagnetic analog of Thiotepa [22]:

A brief review on the usage of stable paramagnets of the iminoxyl series in tumor chemotherapy is presented by Suskina [23].

KEYWORDS

- tumor chemotherapy
- radiation cancer therapy
- organic paramagnetic method
- radio spectroscopy
- sensing
- study

REFERENCES

1. Rozantsev, E. G. Chem. Encyclopedic Dictionary. M. SE.1983. 489 p. [Russ].
2. Rozantsev, E. G., Lebedev, O. L., Kazarnovskii, S. N. Diploma for the opening number 248 of 05.10.1983. TS. 1982. No 6. p. 6. [Russ].
3. Rozantsev, E. G. Doctor dissertation. Moscow: Institute of Chemical Physics, USSR Academy of Sciences. 1965. [Russ].
4. Zhdanov, R. I. Nitroxyl Radicals and Non-Radical Reactions of Free Radicals Bioactive Spin labels. Edited by Zhdanov, R. I. Springer. Berlin. Heidelberg. N-Y. 1991. pp. 24.
5. Liechtenstein, G. I. Method of spin labels in molecular biology, M: Nauka. 1974. Pp. 255. [Russ].
6. McConnell, N. J. Chem. Phys. 1956. Vol.25. p. 709.
7. Freed, J., Fraenkel, G. J. Chem. Phys. 1963.vol. 39. p. 326.
8. Kivelson, D. J. Chem. Phys. 1960.Vol. 33. p. 4094.
9. Stryukov, V. B. Stable radicals in chemical physics. M.: Znanie, 1971. [Russ].
10. Goldfeld, M. G., Grigoryan, G. L., Rozantsev, E. G. Polymer (Gath. of preprints. M. ICP AS SSSR.1979. p. 269. [Russ].
11. Grigoryan, G. L., Tatarinov, S. G., Cullberg, L. Y., Kalmanson, A. E., Rozantsev, E. G., Suskina, V. I. Abst. of USSR Academy of Sciences. 1968. Vol. 178. Pp. 230. Vol. 31. pp. 768. [Russ].
12. Pressman, D. Mol. Structure and Biological Specificity. Washington. D. C. 1957.
13. Baum, H., Riske, J., Silman, H., Lipton, S. Froc. Nature Acad. Sci. US. 1967. Vol. 57. pp. 798.
14. Rozantsev, E. G. Biochemistry of meat and meat products (General Part). Ed. Manual for students. M. Depi. print. 2006. p. 240. [Russ].
15. Goldfeld, M. G., Koltover, V. K., Rozantsev, E. G., Suskina, V. I. Kolloid.Zs. 1970.
16. Schenfeld, H. Nonionic detergents. 1963.
17. Smith, J., Shrier – Muchillo Sh., Murch, D. Method of spin labels. Free radicals in biology. M: Mir. 1979. Vol. 1. pp. 179.
18. Method of spin labels and probes, problems and prospects. M: Nauka. 1986. [Russ].
19. Nitroxyl radicals. Synthesis, chemistry, applications. M: Nauka. 1987. [Russ].
20. Rozantsev, E. G., Goldfein, M. D., Pulin, V. F. Organic paramagnetic. Saratov: Saratov Gov. Univ. 2000. pp. 340. [Russ].
21. Liechtenstein, G. I. J. Molecular Biology. 1968. Vol. 2. P. 234.
22. Konovalova, N. P., Bogdanov, G. N., Miller, V. B. Abst. of USSR Academy of Sciences. T. 1964. Vol. 157. # 3. pp. 707. [Russ].
23. Suskina, V. I. Candidate dissertation. M: Academy of Sciences ICP USSR. 1970. [Russ].

CHAPTER 9

LOCAL STEREOCHEMICAL MICROSTRUCTURES' DEPENDENCE OF β RELAXATION PROCESS IN SOME VINYL POLYMERS: FUNDAMENTALS AND CORRELATIONS

N. GUARROTXENA

Instituto de Ciencia y Tecnología de Polímeros (ICTP), Consejo Superior de Investigaciones Científicas (CSIC). C/ Juan de la Cierva 3, 28006 Madrin, Spain. E-mail: nekane@ictp.csic.es

CONTENTS

ABSTRACT

Dynamic mechanical thermal analysis (DMTA) is an accessible and versatile analytical technique where a sample is submitted to an oscillating stress or strain as a function of oscillatory frequency and temperature. Through this nondestructive measurement, a comprehensive understanding of the dynamic mechanical properties of polymers may be obtained. In this context, two series of polymers having different tacticity-dependent stereochemical microstructures have been submitted to controlled dynamo-mechanical strength. One series consists of three different PVC samples prior to and after substitution reaction to various extents in solution and in melt state. And one series consists of three commercial PMMA samples.

In this review, we provide a straight correlation between β-relaxation and the content of mmr tetrads termini of isotactic sequences of at least one heptad in length, especially with those mmr structures taking the GTTG⁻ TT conformation. Basically it is demonstrated that β-relaxation both, temperature and intensity, decrease with the content of the aforementioned microstructures, giving further support to the implication of magnitudes such as free volume and of coupling degree of the local motions in the mechanisms of the physicochemical processes responsible of the PVC and PMMA mechanical behaviors.

9.1 INTRODUCTION

The structure property relationships are greatly significant industrially and scientifically. Most of abundant research work in this field for the last years has dealt with approaches to the whole simulation of the physical behavior and, thence, to a macroscopic view of the related physical properties of polymers. In contrast, no attempts to correlate the physical properties with the molecular microstructure of polymers have been made. The implication is that little information as to the nature and the mechanisms of the molecular nature involvement in the physical behavior of these materials is available. In fact, the studies to correlate every physical property with any specific molecular structure, in particular the local or sequential chain microstructure regardless of the type of vinyl polymer, are rather scarce if any. The difficulty lies in the fact that both the controlled formation of those microstructures along the polymer chain and their accurate determination are far from being successful.

Within this framework a rather abundant work in our laboratory has revealed that there exists a straight relation between some physical and chemical behaviors and the occurrence of a few repeating steresequences along the chain, especially the mmmr and the mmmmrx (x=m or r) which occur necessarily whenever an isotactic sequence breaks-off. The prominent role of them in a list of physical properties of poly(vinyl chloride) (PVC) including glass transitions, physical aging, dielectrical relaxations and electrical space charge nature and distribution, has been extensively conveyed [1–5]. Most of the correlations found for PVC have been proved to hold true for poly(propylene) (PP) [6,7]. They have been also extended to dynamic mechanical properties of PVC and PMMA. Actually, this paper deals with approaches to secondary relaxation knowledge as inferred from found correlations with the stereochemical microstructures mentioned below.

In general the stereochemical microstructures that have been considered are: (i) the isotactic and syndiotactic sequences, that relate to the extent to which a polymer exhibits a nonBernoullian tacticity distribution; (ii) the mmr-based and the rrm-based local structures which occur necessarily whenever an isotactic sequence or a syndiotactic sequence breaks-off respectively; and (iii) the atactic parts and the pure mrmr moieties.

It is well established that the rheological (mechanical) properties of polymeric materials directly affect their technical performance. Therefore, in order to ensure optimization of their design and application, it is important to accurately characterize their rheological properties and deeply understand the molecular implications on the physical processes involved in their dynamo-mechanical properties (b-relaxation). Among the several techniques used to mechanically characterize polymers, DMTA provides a rapid and nondestructive way to study and quantify the viscoelastic properties of polymeric materials [8–11]. Basically, DMTA involves the application of a sinusoidal stress (or strain) to a sample while measuring the mechanical response as a function of both oscillatory frequency and temperature. The dynamic storage modulus, E', and loss modulus, E", are among the most basic of all mechanical properties, and their importance in any structural property relationship is well applied.

The main purpose of this paper is to consider how far some local repeating stereochemical sequences of PVC and PMMA play a key role on impact strength and dynamic modulus, and to present some evidence that relaxation to stem from the enhanced free volume and the capability to local motions of mmmr structures relative to the remaining stereochemical microstructures in the polymer.

To do the above attempt the authors measured, by dynamic mechanical thermal analysis (DMTA), the β relaxation of: (i) PVC after chemical modification to various extents in solution and in melt state in the presence of some moieties of plasticizer (dioctyl phthalate, DOP), and (ii) PMMA with different stereochemical compositions.

9.2   EXPERIMENTAL PART

9.2.1   MATERIALS

Poly(vinyl chloride). The PVC samples used were an-additive free industrial PVC prepared in bulk polymerization at temperatures of 68 °C (sample X), 60 °C (sample Y) and 45 °C (sample Z). The samples were characterized as described elsewhere [12, 13]. Cyclohexanone (CH) was purified as reported earlier [12]. Dioctyl phthalate (DOP) was used as received. Samples X, Y and Z were modified by nucleophilic substitution reaction with sodium benzenethiolate (NaBT) in CH and in the melt in the presence of DOP moieties following the methods described in previous papers [12, 13]. The tacticity of modified samples was measured by means of ^{13}C NMR decoupled spectra as described elsewhere [12, 13].

Poly(methyl methacrylate). Commercial PMMA (labeled as U, V and W) samples, obtained from Atochem, were purified using tetrahydrofuran (THF, Scharlau) as solvent and water as precipitating agent, washed in methanol and dried under vacuum at 40 °C for 48 h. THF was distilled under nitrogen with aluminum lithium hydride (Aldrich) to remove peroxides immediately before use.

9.2.2   CHARACTERIZATION OF SAMPLES

9.2.2.1   Poly(vinyl chloride)

The tacticities of both the starting (samples X, Y, and Z) and the modified polymers were measured by means of ^{13}C NMR decoupled spectra obtained at 80 °C on an Varian UNITY-500 instrument, operating at 125 MHz using 1,4-dioxane-d$_8$ as solvent under conditions described previously [12, 13]. The resonances used were those of methane carbons of the backbone ranging from 57 to 61 ppm. The 40,000–50,000 scans gave a very satisfactory signal-to-noise and the respective peak intensities were measured from the integrated areas, as calculated by means of an electronic integrator.

9.2.2.2 Poly(methyl methacrylate)

^{1}H NMR spectra were recorded on a Varian UNITY-500 spectrometer operating at 499.88 MHz in CDCl$_3$ as solvent at 50 °C. The parameters of 8000 Hz spectral width and a 1.9 s pulse repetition rate were used. The delay time was set to 0.5 s. The spectra were obtaining after accumulating 64 scans with a sample concentration of 10 wt% solutions. The relative peak intensities were measured from the integrated peak areas, which were calculated with an electronic integrator. The content of the different triads were calculated by the electronic integration of the a-methyl signals, which appear at (1.3–1.1ppm), at (1.1–0.9ppm) and at (0.9–0.7ppm) corresponding to isotactic (mm), heterotactic (mr) and syndiotactic (rr) triads, respectively.

9.2.3 *DYNAMIC MECHANICAL MEASUREMENTS*

9.2.3.1 Poly(vinyl chloride)

Dynamic mechanical measurements were carried out with a Polymer laboratories MkII DMTA working in the tensile modulus at frequencies of 1, 3, 10 and 30 Hz. The temperature was varied from −130 to 90 °C at a heating rate of 2 °C/min. The complex modulus and the loss tangent of each sample were determined at 10 Hz. All specimens were cut into a rectangular strip 2.2 mm wide.

9.2.3.2 Poly(methyl methacrylate)

Mechanical measurements were carried out on rectangular strips of 2.2 mm wide with a Polymer Laboratories MkII DMTA working in the tensile modulus at a frequency of 10 Hz. The measurements were performed in the temperature range -100 to $140\ °C$ at a heating rate of $2\ °C/min$. The complex-modulus and the loss tangent of each sample were determined.

The PVC and PMMA samples were obtained by compression molding of 0.1 g of materials at temperature about $120\ °C$ and 1 MPa of pressure for PVC and around $140\ °C$ and 50 bars of pressure for PMMA The cooling process, under the same pressure, was carried out by quenching the molten polymer with water.

9.3   RESULTS AND DISCUSSION

9.3.1   THE PVC AND PMMA STEREOCHEMICAL MICROSTRUCTURES

The NMR spectroscopies allow one to state and to measure the above-quoted microstructures [14]. ^{13}C NMR and ^1H NMR spectroscopies were used to analyze the sorts of mmr- and rrm- based stereochemical sequences and their surroundings for one series of PVC samples and one series of PMMA samples, of different tacticities. Basically, the spectra allow to determine the content of isotactic (mm), heterotactic (mr & rm) and syndiotactic (rr) triads. In fact, what is directly measured on the spectra are the integrated areas of signals belonging to the corresponding pentads. In the case of the isotactic pentad of PVC the spectra resolution is good enough for the content of the three sorts of pentad, mmmm, mmmr and rmmr to be determined accurately. From the triad content it is possible to calculate on the one hand the content of isotactic (m), heterotactic (h) and syndiotactic (r) dyads and on the other hand, the probability that one placement r follows one m placement or one r placement, P_{mr} and P_{rr} respectively. The corresponding equations have been given previously [14]. Obviously, if Bernoullian statistics applies then $P_{mr}=P_{rr}$. Therefore, $P_{mr}-P_{rr}$ is an accurate measure of the sample departure from Bernoullian behavior (Note that the same is valid for P_{mm} and P_{rm} denoting the probability that one m placement follows one m or one r placement, respectively). In addition in case of Bernoullian behavior the probability of occurrence of any repeating sequence may be obtained through universal relations [14]. The results so obtained for the samples of this work are presented in Table 9.1. As can be

TABLE 9.1   Results Obtained For the Samples of PVC and PMMA Work

	Samples	Mn	Triads			Tetrads	Pentads			Bernoullian Probabilities			h
			mm	mr	rr	mmr	mmmm	mmmr	rmmr	P_{mm}	P_{rm}	$P_{mm}-P_{rm}$	
PVC	X	42000	20.1	49.65	30.25	22.65	3.97	9.61	6.52	0.447	0.451	(0)0.004	0.8
	Y	53500	18.89	49.38	31.72	20.31	4.21	9.06	5.63	0.443	0.438	0.005	0.6
	Z	64000	18.66	49.36	31.99	20.68	3.77	9.11	5.79	0.421	0.431	(-)0.01	0.5

	Samples	Mn	Triads			Diads		Bernoullian Probabilities		
			mm	mr	rr	m	r	P_{mm}	P_{rm}	$P_{mm}-P_{rm}$
PMMA	U	45503	8.56	41.81	49.63	29.46	70.53	0.291	0.296	(-0)0.005
	V	43131	12.8	39.92	47.28	32.27	67.24	0.391	0.297	0.094
	W	44868	20.97	36.5	42.53	39.22	60.78	0.535	0.535	0.235

seen PVC samples, prior to modification (Table 9.1) are all Bernoullian. Thus there appear to be not so different in tacticity arrangement. Nevertheless the whole isotacticity clearly decreases from sample X to sample, Z. As widely argued [14] this means that the isotactic sequences of at least one heptad in length decreases in the same order and so does the content of the conformation GTTG⁻TT of the tetrad mmr which occurs necessarily whenever an isotactic sequence breaks off, relative to the other likely conformation, the GTGTTT [14]. The content of GTTG⁻TT conformation may be determined by means of substitution reaction at as low temperature as −15 °C. The values so obtained are denoted by *h* in Table 9.1. The implications of the substitution reaction to the tacticity arrangement are commented below.

With reference to the PMMA tacticity it appears clear that samples U and V are totally isotactic particularly sample U, while sample W is non Bernoullian syndiotactic (Table 9.1). The overall isotactic content decreases from sample U to sample, V. Therefore what has been said in the precedent paragraphs applies to PMMA. Sample W will exhibit more syndiotactic sequence content. On the other hand, the content of rrm tetrad, which occurs whenever one syndiotactic sequence breaks off, will be greater. More details concerning PMMA tacticity are given below.

To the above considerations it should be added the chain conformations, which each structure may adopt.

Figure 9.1 reproduces the picture of the two indicated repeating stereosequences as built up with CPK models. By simply manipulating them, it can be seen that the isotactic sequence cannot adopt a conformation other than …GTGTGT…. The occurrence of a conformation other than GTGT implies the planar coexistence of bulky atomic groups, which is highly unlikely and is to be discarded [15].

As widely conveyed for PVC [13, 16] the last triad (i.e., the mm in mmr tetrad) is the only likely triad to adopt, incidentally, the GTTG–conformation. In fact there is an equilibrium GTGT ↔ GTTG– which in both polymer PVC and PMMA, strongly lies over GTGT conformation. Nevertheless some important differences between both polymers are evident: PMMA shows large free volume, more hindered rotation and much lesser possibility of interchain hydrogen bonds. As a result the occurrence of GTTG– conformation at the end of isotactic sequences will be more restricted and of lower content than in PVC. In order to compare both conformations, which may occur in the mmr tetrad terminating any isotactic sequence, they are reproduced in Fig. 9.2.

As can be seen from Figs. 9.1 and 9.2, it holds that: (a) the mmr repeating sequence is shorter and of greater volume than mmm or rrr sequences. Moreover, these differences are considerably enhanced for GTTG–TT conformation relative to GTGTTT conformation. Considering that mmr is a local helix-coiled regularity disruption, the GTTG–TT involves higher local change in both free volume and rotational motion; (b) the rotational motions are favored and are less dependent on the adjacent isotactic

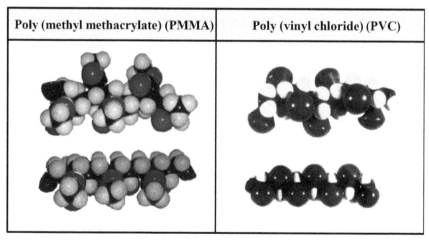

Poly (methyl methacrylate) (PMMA)	Poly (vinyl chloride) (PVC)

FIGURE 9.1   Some repeating sequences of equal length of PMMA (left side) and PVC (right side): all isotactic ~GTGT~ conformation (upper) and all ~TT~ conformation (bottom). (CPK molecular models).

Poly (methyl methacrylate) (PMMA)	Poly (vinyl chloride) (PVC)
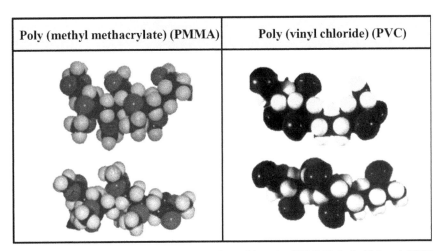	

FIGURE 9.2   Some repeating sequences of equal length of PMMA (left side) and (PVC (right side): ͜GTTG⁻TT conformation (upper) and ͜GTGTTT conformation (bottom). (CPK molecular models).

sequences in GTTG⁻TT conformation, therefore GTTG⁻TT motion exhibits less coupled nature than GTGTTT conformation; and (c) turning to the syndiotactic TTTT sequence, it appears evident that it is longer and exhibits lower free volume and more reduced ability to rotate as compared to isotactic GTGT sequence.

9.3.2   THE ROLE OF SOME STEREOCHEMICAL MICROSTRUCTURES ON THE PVC AND PMMA DYNAMO MECHANICAL PROPERTIES

The results obtained for the PVC samples prior to and after substitution (Table 9.1) are displayed in Figs. 9.3A–C. By simple inspection it is evident that the β relaxation shifts to lower temperature and its height tends to vanish progressively as the substitution progresses. A more detailed observation of both effects, in Figs. 9.3D and 3E, allows one to asses that they decrease both more markedly for the lower substitution extents (0–4%). Then, the trend is towards leveling off.

In this connection the decrease in peak height observed in Fig. 9.3D should be assumed as the result of the progressive disappearance of mmmr pentads associated with long isotactic sequences.

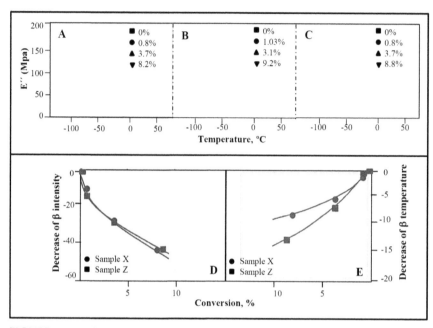

FIGURE 9.3 Effect of nucleophilic substitution on β relaxation of PVC modified in cyclohexanone solution: (A) sample X, (B) sample Y, (C) sample, Z. Evolution of β peak intensity (D) and β peak temperature (E) versus degree of substitution of PVC in cyclohexanone solution [(f= 10 Hz)].

With respect to the shifting to lower temperatures of β relaxation the more likely explanation lies in the fact that the length of the isotactic sequences bearing the reactive mmmr pentad is reduced as substitution advances. This agrees with some suggestions in the literature that β peak shifts to lower temperature as the length of the segment chain in motion decreases [17].

The higher decrease in both β temperature and peak height for the weakly substituted samples (1–1.5%) is also of high interest. It agrees with the specific disappearance of the mmmr pentads under GTGTTG‑TT conformation [12,13] thereby indicating that the local motion of this conformation is highly favored compared to GTGTGTTT conformation. This fundamental feature is consistent with the polymer response to a rather great number of strengths other than that involved in β relaxation [1–3, 5, 18–25]. In all cases there appears to exist a straight relation between any

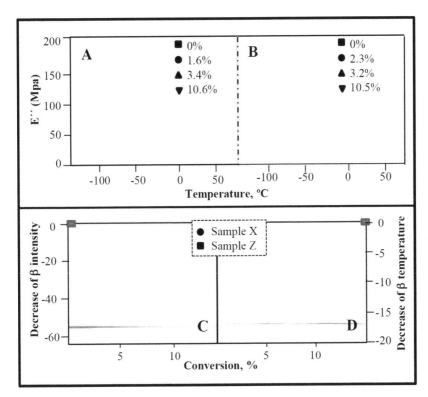

FIGURE 9.4   Effect of nucleophilic substitution on β relaxation of PVC modified in melt state: (A) sample X, (B) sample, Z. Evolution of β peak intensity (C) and β peak temperature (D) versus degree of substitution of PVC in melt state [(f= 10 Hz)].

physical behavior and both free volume and the rotational motion facilities of the segment chain involved.

The same PVC polymers after substitution in the melt with DOP proved to behave similarly from a qualitative point a view in that the b-transition shifts to lower temperatures and the height of the β peak becomes lower (Figs. 9.4A–B and 9.4C–D). Nevertheless these effects are somewhat less discriminated and significantly enhanced in the melt at low conversion rates of modification and at higher conversions respectively. This is consistent with our prior work on substitution in the melt [15]. It was actually shown that the mmmr pentads under GTGTTG⁻TT conformation is able to react exclusively even in the absence of plasticizer [15]. In the same way, when using moieties of plasticizer, the remaining mmmr pentads, which are under GTGTGTTT conformation, happen to react until complete consumption without the other reactive structures starting to react. Owing to the severe conditions used, the reaction through the latter conformation might overlap, to some extent, that through the former conformation.

From the above results it follows that β relaxation of PVC strongly responds to the tacticity dependent microstructure, in particular, the favored mobility of local configurations (mmmr pentads) termini of long isotactic sequences when they are taking the GTGTTG⁻TT conformation.

Figure 9.5A plots the DMTA diagrams for samples U, V and W of poly(methyl methacrylate). By inspection it is clear that the β position and the height for sample U are both higher than those of samples V and, W. A better comparison of both magnitudes can be obtained by plotting them against the overall isotactic content in Fig. 9.5B. The absence of curves symmetry suggests that the right-hand is influenced by a relaxation occurring at higher temperature, certainly Tg.

Sample U exhibits the higher intensity of the maximum (Fig. 9.5B) relative to samples V and, W. In addition the intensity of sample W is somewhat higher than that of sample, V. The results in Fig. 9.5B seem to be at variance with a simple correlation with the overall isotactic content (Table 9.1). A likely explanation can be given in the light of P_{mm} and P_{mr} values reported in Table 9.1. In sample W, the increase in average length of the isotactic sequences is accompanied by a decrease in frequency of the same sequences, so that the number of mmr structures does not increase

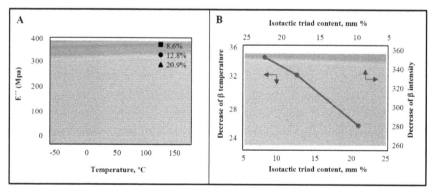

FIGURE 9.5   Effect of isotactic triad content (mm) on β relaxation of PMMA (A). Evolution of β peak intensity and β peak temperature versus overall isotactic triad content, mm of PMMA samples (B). [(f= 10 Hz)].

or even decrease somewhat with respect to sample, V. On these grounds it may be concluded, at least in a tentative way, that the number of mmmr microstructures should vary in the sense: $U > V \approx W$.

Assuming the above argument, the results in Fig. 9.5B would indicate the occurrence of a clear correlation between the intensity of β relaxation and the content of mmmr microstructures. This would account for the β transition to obey the capability of motion of mmmr, Fig. 9.5B depicts the evolution of β temperature with the overall isotactic content. Even through the shift of β temperature seems like the evolution of the intensity (Fig. 9.5B), it cannot be attributed to the amount of mmmr microstructures because β temperature value relates chiefly to the nature of the structure that is able to motion. Since, this structure is mmmr, the shift of β temperature must obey the differences in surrounding of mmmr in the samples. In this connection it may be assumed from P_{mm} and P_{mr} values that the length of the isotactic sequence, concomitant with any mmmr microstructure, increases from sample U to sample V and from this to W sample, the latter change being much more marked than the former. On the other hand the isotactic sequence is more rigid as its length increases. This might cause the capability to motion of the corresponding mmmr microstructure to increase in the same sense.

Actually, the degree of coupling of mmmr motion with that of the adjacent units should be higher as the rigidity of the latter units decreases. As a consequence, the mmmr with the little rigid adjacent units, as a whole, would merge into a new microstructure of lower capability of motion relative to the uncoupled mmmr microstructure.

The results on β transition of PMMA, despite preliminary nature, give evidence of the straight relation between β transition and the mmmr-based microstructures in PMMA. Further results are expected to shed light on the very nature of that relation as issued from the results obtained so far.

Comparing these behaviors and those previous reported for PVC prior to and after substitution it is evident that the initial drop of the β relaxation intensity and temperature can be attributed to the disappearance of mmmr under GTGTTG⁻TT conformation of very low degree of coupling with the vicinal isotactic sequence, and then, of enhanced capability of motion. As above indicated, the occurrence of that conformation in PMMA is very unlikely. Thus, its influence on β transition should be discarded.

On the contrary, it was well shown in PVC that β temperature decreases with the decreasing length of the isotactic sequences, which gives support to the above explanation of the results in Fig. 9.5B showing the β temperature shift with the overall isotactic content of PMMA.

9.4 CONCLUSION

From the above results it may be concluded that some normal stereochemical microstructures namely the mmrnr repeating isotactic sequence are a major driving force for the physical processes involved in the dynamic mechanical properties of PMMA and PVC. Actually, the β relaxation is the result of the progressive disappearance of mmmr, where the GTGTTG‾TT conformation exhibits a highly favored local motion and free volume compared to GTGTGTTT conformation [22, 23].

In conclusion, these results take the knowledge of the β relaxation process an important step further and shed further light on the property determining role of some local repeating stereochemical sequences as widely proposed in earlier work.

KEYWORDS

- chain conformation
- local motion
- molecular microstructure
- nucleophilic substitution
- PMMA
- PVC
- β-relaxation

REFERENCES

1. Guarrotxena, N., Elícegui, A., del Val, J., Millán, J. J Polym Sci Polym Phys 2004, 42, 2337.
2. Guarrotxena, N., Martínez, Millán, J. Polymer 2000, 41, 3331.

3. Guarrotxena, N., Vella, N., Toureille, Millán, J. Macromol Chem Phys 1997, 198, 457.
4. Guarrotxena, N., Contreras, J., Toureille, Millán, J., Polymer 1999, 40, 2639.
5. Guarrotxena, N., Millán, J., Vella, Toureille, T., Polymer 1997, 38, 4253.
6. Guarrotxena, N., Toureille, Millán, J., Mcromol Chem Phys 1998, 199, 81.
7. Guarrotxena, N., Millán, J., Sessler, Hess, G., Macromol Rapid Commun 2000, 21, 691.
8. Ferry, J., *Viscoelastic Properties of Polymers*, John Wiley & Sons, New York, 1980.
9. Ward, I., *Mechanical Properties of Solid Polymers*, John Wiley & Sons, Chichester,1983.
10. Ward, Hadley, D., *An Introduction to the Mechanical Properties of Solid Polymers*, John Wiley & Sons, Chichester, 1993.
11. Jones, D., Int. J. Pharm. 1999, 179, 167.
12. Guarrotxena, N., Martínez, Millán, J., J Polym Sci Polym Chem 1996, 34, 2387.
13. Guarrotxena, N., Martínez, Millán, J., Eur Polym J 1997, 33, 1473.
14. Tonelli, A. E., *NMR spectroscopy and polymer microstructure: The conformational connection*. A, T Bell Laboratories, VCH Publishers Inc., New York, 1990.
15. Moritani, Fujiwara, Y., J Chem Phys 1973, 59, 1175.
16. Guarrotxena, N., Schue, F., Collet, Millan, J., Polym Int 2003, 52, 429.
17. Wada, Y., Tsuge, K., Arisawa, K., Ohsawa, Y., Shida, K., Hotta, Y., et al., J Polym Sci Part C 1996,15, 101.
18. Guarrotxena, N., Martínez, Millán, J. Polymer 1997, 38, 1857.
19. Guarrotxena, N., del Val, J., Millán, J. Polym Bull 2001, 47, 105.
20. Guarrotxena, N., Tiemblo, P., Martínez, G., Gómez-Elvira, J., Millán, J., Eur Polym J 1998, 34, 833.
21. Guarrotxena, N., de Frutos, Retes, J. Macromol Rapid Commun 2004, 25, 1968.
22. Guarrotxena, N., Vella, N., Toureille, Millán, J. Polymer 1998, 39, 3273.
23. Guarrotxena, N., Contreras, J., Toureille, Martínez, G. Polymer 1999, 40, 2639.
24. Guarrotxena, N., Contreras, J., Martínez, Millán, J. Polym Bull 1998, 41, 355.
25. Guarrotxena, N., Vella, N., Toureille, Millán, J.Macromol Chem Phys 1996, 197, 1301.

CHAPTER 10

SURFACE-MODIFIED MAGNETIC NANOPARTICLES FOR CELL LABELING

BEATA A. ZASONSKA,[1] VITALII PATSULA,[1] ROSTYSLAV STOIKA,[2] and DANIEL HORÁK[1]

[1]Institute of Macromolecular Chemistry, Academy of Sciences of the Czech Republic, Heyrovskeho Sq. 2, 162 06 Prague 6, Czech Republic

[2]Institute of Cell Biology, National Academy of Science of Ukraine, Drahomanov St. 14/16, 79005 Lviv, Ukraine

CONTENTS

10.1 INTRODUCTION

A great effort has been recently devoted to the design and synthesis of new magnetic nanoparticles driven by the rapid development of the nanomedicine and nanobiotechnology [1]. Among them, iron oxide nanoparticles, in particular magnetite Fe_3O_4 and maghemite $\gamma\text{-}Fe_2O_3$, play a prominent role since iron is indispensable component of living organisms and has reduced toxicity [2]. Surface-modified iron oxide nanoparticles have been found very attractive for cell separation [3] and labeling [4], cancer therapy [5], drug delivery [6] and as contrast agents for magnetic resonance imaging (MRI).

There are many methods to obtain various types of iron oxide nanoparticles differing in shape, morphology, size and availability of the reactive groups on the surface. The oldest preparation involves the size reduction [7], for example, grinding of bulk magnetite in the presence of large amounts of surfactant in a ball mill for 500–1,000 h. Other synthetic approaches for development of magnetic nanomaterial include hydrothermal process [8], sol-gel method [9] or spray pyrolysis [10]. However, the most popular techniques for preparation of such particles include coprecipitation of Fe(III) and Fe(II) salts in the presence of an aqueous base (e.g., NH_4OH or NaOH)or thermal decomposition of organo-metalic complexes in high-boiling solvents [11]. For the latter, precursors, such as, Fe(III) acetyl acetonate [12], Fe N-nitrosophenylhydroxylamine [13] or $Fe(CO)_5$ were suggested [14].

Iron oxide nanoparticles possess a lot of unique properties, such as, small size (100 nm) allowing them to function at the cellular level, superparamagnetism, high magnetization and large specific surface area. However, neat (uncoated) particles show high nonspecific adsorption of biomolecules, undesirable in vitro and in vivo interactions, relative toxicity and tendency to aggregate [15]. This can be avoided by their surface modification with biocompatible polymers, which also determines ability of the nanoparticles to interact with living cells in a well-defined and controlled manner, as well as ensures immunotolerance and biocompatibility. Typical polymer shells are made from organic, like poly(ethylene glycol) (PEG) [16], poly(vinyl alcohol) [17], poly(N, N-dimethylacrylamide) (PDMAAm) [18], or inorganic materials, for example, silica [19]. This additional layer can render the particles with colloidal stability, avoids interactions with the surrounding environment and introduces specific functional groups on the surface.

In this chapter, synthesis, properties and some applications of new poly(*N*, *N*-dimethylacrylamide)-coated maghemite (γ-Fe$_2$O$_3$@ PDMAAm), silica-coated maghemite(γ-Fe$_2$O$_3$@SiO$_2$)and methyl-poly(ethylene glycol)-coated magnetite(Fe$_3$O$_4$@mPEG)nanoparticles are described. Both PDMAAm and silica are hydrophilic, chemically inert and biocompatibile materials, hence, they are attractive for drug delivery systems and applications in medical diagnostics. Moreover, the polymers can behave like transfection agents enabling efficient engulfment of the particles by the cells, for example, stem or neural cells and macrophages. Macrophages, that are formed in response to an infection and accumulate damaged or dead cells, are important in the immune system [20]. These large, specialized cells can recognize, engulf and destroy foreign objects. Through their ability to clear pathogens and instruct other immune cells, they play a pivotal role in protecting the host. They also contribute to the pathogenesis of inflammatory and degenerative diseases [21]. Labeling of macrophages with magnetic particles enables thus their tracing in the organism using MRI.

10.2 PREPARATION OF MAGNETIC NANOPARTICLES

Chemical and physical properties of magnetic nanoparticles, such as, size and size distribution, morphology and surface chemistry, strongly depend on selection of the synthetic method, starting components and their con-centration [11, 22]. Nanoparticles ranging in size from 1 to 100 nm exhibit superparamagneticbehavior [22]. In this report, two methods of iron oxide synthesis are presented.

10.2.1 COPRECIPITATION METHOD

Typical synthesis of magnetic nanoparticles is exemplified by formation of maghemite (γ-Fe$_2$O$_3$) during coprecipitation of Fe(II) and Fe(III) salts followed by oxidation of Fe$_3$O$_4$ with sodium hypochlorite [23]. Briefly,

0.2 Maqueousiron(III) chloride (100 mL) and 0.5 M iron(II) chloride (50 mL) were sonicated for a few minutes and mixed with 0.5 M aqueous ammonium hydroxide (100 mL). The mixture was then continuously stirred (200 rpm) at room temperature for 1h. Formed Fe_3O_4 nanoparticles were magnetically separated and seven times washed with distilled water. Subsequently, the colloid was sonicated with 5 wt.% sodium hypochlorite solution (16 mL) and again five times washed with water to obtain the final γ-Fe_2O_3 nanoparticles.

10.2.2   THERMAL DECOMPOSITION

Another possibility to produce super paramagnetic nanoparticles consists in thermal decomposition of iron organic compounds, for example, iron(III) oleate [24]. The method allows preparation of monodisperse Fe_3O_4 nanoparticles with controlled size. As an example, we describe preparation of iron(III) oleate by reaction of $FeCl_3 \times 6H_2O$ (10.8 g) and sodium oleate (36.5 g) in a water/ethanol/hexane mixture(60/80/140 mL) at 70 °C for 4 h under vigorous stirring. The upper organic layer was then separated, three times washed with water (30 mL each) and the volume reduced on a rotary evaporator. Obtained brown waxy product was vacuum-dried under phosphorus pentoxide for 6 h. The resulting Fe(III) oleate (7.2 g) and oleic acid (4.5 g) were then dissolved in octadec-1-ene (50 mL) and heated at 320°C for 30 min under stirring (200 rpm). The reaction mixture was cooled to room temperature, the particles precipitated by addition of ethanol (100 mL) and collected by a magnet. Obtained nanoparticles were then five times washed with ethanol (50 mL) and redispersed in toluene and stored.

10.3   MODIFICATION OF THE NANOPARTICLE SURFACE

Disadvantage of neat iron oxide colloids is that they induce undesirable interactions, for example, adhesion to the cells. To prevent this, it is recommended to coat their on oxide surface with a biocompatible polymer shell. Surface of theγ-Fe_2O_3nanoparticles was therefore firstly modified with an initiator and N, N-dimethylacrylamide was then polymerized from the surface. 2,2'-Azobis(2-methylpropionamidine) dihydrochloride (AMPA) served a suitable polymerization initiator. In contrast, if the particles were hydrophobic, for example, obtained from the thermal decomposition, they were dispersible only in organic solvents. To make them water-dispersible and suitable for biomedical applications, their surface was modified with mPEG derivatives via a ligand exchange.

10.3.1 COATING WITH POLY(N, N)-DIMETHYLACRYLAMIDE (PDMAAM)

Coating of the γ-Fe$_2$O$_3$ nanoparticles with PDMAAm via grafting from approach is schematically shown in Fig. 10.1. In the following, example of this synthetic approach is described in a more detail. The polymerization was run in a 30-mL glass reactor equipped with an anchor-type stirrer. First, the AMPA initiator (4.8 mg) was added to 10 mL of the colloid (47 mg γ-Fe$_2$O$_3$/mL) during 5 min, DMAAm (0.3 g) was dissolved and the mixture purged with nitrogen for 10 min. The polymerization was started by heating at 70 °C for 16 h under stirring (400 rpm). After completion of the polymerization, the resulting γ-Fe$_2$O$_3$@PDMAAm particles were magnetically separated and washed ten times with distilled water until all reaction byproducts were removed. Advantage of the γ-Fe$_2$O$_3$@PDMAAm particles consists in possibility to introduce additional functional comonomer into the shell to attach a highly specific bioligand, such as, antibody, peptide or drug.

FIGURE 10.1   Scheme of preparation γ-Fe$_2$O$_3$@PDMAAm nanoparticles via grafting from approach using 2,2'-azobis(2-methylpropionamidine) dihydrochloride (AMPA) initiator.

10.3.2   COATING   WITH   TETRAMETHOXYORTOSILICATE (TMOS)   AND   (3-AMINOPROPYL)TRIETHOXYSILANE (APTES)

Another frequently used coating of iron oxide particles is based on silica. Silica is generally synthesized by hydrolysis and condensation of tetraethylorthosilicate (TEOS) or tetramethylorthosilicate (TMOS) (Fig. 10.2). Neat silica particles are obtained by Stöber method in ethanol under the presence of ammonia catalyst [25] or in surfactant-stabilized reverse microemulsion containing two phases [26. Theγ-Fe$_2$O$_3$nanoparticles were coated by a silica shell using TMOS according to earlier published method [27. Shortly, solution containing 2-propanol (24 mL), water (6 mL) and 25 wt.% aqueous ammonia (1.5 mL) was mixed with γ-Fe$_2$O$_3$ colloid (1 mL; 50 mg γ-Fe$_2$O$_3$) for 5 min. TMOS (0.2 mL) was added and the mixture stirred (400 rpm) at 50°C for 16 h. Resulting γ-Fe$_2$O$_3$@SiO$_2$ colloid (Fig. 10.3 c, g) was then five times washed with ethanol using magnetic separation. In the next step, amino groups were introduced on the particle surface using (3-aminopropyl)triethoxysilane (APTES). In a typical experiment, γ-Fe$_2$O$_3$@SiO$_2$ nanoparticles were dispersed in ethanol (50 mL) under sonication for 15 min and APTES (0.15 mL), ethanol (20 mL) and water (1 mL) were added. After completion of the reaction, the resulting γ-Fe$_2$O$_3$@SiO$_2$-NH$_2$ particles (Fig. 10.3 d, h) were washed with water.

10.3.3   COATING   WITH   METHYL-POLY(ETHYLENE   GLYCOL) (MPEG)

In order to make the hydrophobic iron oxide particles dispersible in water, their surface was modified by a ligand exchange method [28]. As a hydrophilic ligand, mPEG was selected due to its nontoxicity, hydrophilicity and low opsonization in biological media. mPEG was terminated with

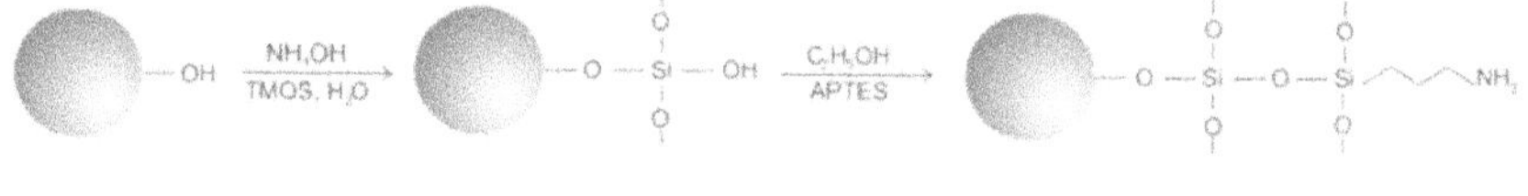

FIGURE 10.2 Scheme of salinization of γ-Fe$_2$O$_3$ with tetramethylorthosilicate (TMOS) and modification of γ-Fe$_2$O$_3$@SiO$_2$ nanoparticles with(3-aminopropyl) triethoxysilane(APTES).

groups, such as, phosphonic $(PO(OH)_2)$ [29] and hydroxamic (NHOH) acid [30], exhibiting strong interactions with the iron ions. Fe_3O_4 particles prepared by thermal decomposition were coated by mPEG terminated with phosphonic(PA-mPEG) or hydroxamic acid (HA-mPEG).In the following, the surface modification is described in a more detail. HA- or PA-mPEG (70 mg) and hydrophobic Fe_3O_4 nanoparticles (10 mg) were added to 4 mL of tetrachloromethane/toluene mixture (1:1 v/v) and sonicated for 5 min. The mixture was then heated at 70°C for 48 h under vigorous stirring. The Fe_3O_4@PEG nanoparticles were purified by repeated precipitation with petroleum ether (3 × 30 mL) at 0°C and diethyl ether (3 × 30 mL) and redispersed in water.

10.4 PROPERTIES OF THE SURFACE-MODIFIED IRON OXIDE NANOPARTICLES

The synthesized surface-modified iron oxide particles were thoroughly characterized by a range of methods including transmission (TEM) and scanning electron microscopy (SEM), atomic absorption spectroscopy (AAS), attenuated total reflectance Fourier transform infrared spectroscopy (ATR FTIR), dynamic light scattering (DLS) and magnetic measurements. Shape of the iron oxide particles prepared by the coprecipitation and thermal decomposition methods was spherical and cubic, respectively. The number-average diameter $(D_n)$of the γ-Fe_2O_3particles prepared by precipitation was 10 nm (TEM) and polydispersity index PDI (D_w/D_n) = 1.24 (D_w is the weight-average diameter) suggesting a moderately broad particle size distribution (Fig. 10.3a).Since it was rather difficult to control size and particle size distribution by the precipitation method, thermal decomposition approach was investigated. Size of the Fe_3O_4 particles was controlled in the 8–25 nm range and monodispersity was achieved (Fig. 10.4). For example, if the reaction temperature increased from 320 to 340°C, the D_n increased from 8 nm (Fig. 10.4a) to 17 nm (Fig. 10.4b) due to an increase in the growth rate of the nanoparticles. If the concentration of oleic acid stabilizer increased from 0.008 to 0.08 mmol/mL, the particle size decreased from 12 to 8 nm (Fig. 10.5) because more stabilizer stabilizes more particles. However, the particles prepared by this method

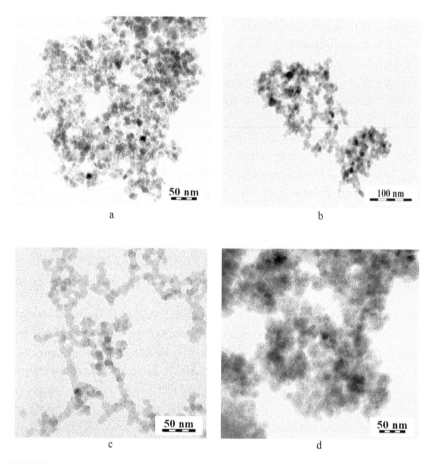

FIGURE 10.3 (a–d) TEM and (e–h) SEM micrographs of (a, e) neatsuperparamagnetic γ-Fe$_2$O$_3$ nanoparticles synthesized by coprecipitation method, (b, f) γ-Fe$_2$O$_3$@PDMAAm (via grafting from approach), (c, g) γ-Fe$_2$O$_3$@SiO$_2$ and (d, h) γ-Fe$_2$O$_3$@SiO$_2$-NH$_2$ nanoparticles.

were hydrophobic; the organic shell formed ~ 80 wt.% of the total mass according to AAS. Such particles formed very stable colloids in organic solvents, such as, toluene or hexane, but not in water. Magnetic properties of the nanoparticles were described earlier [28].

Compared with the neat nanoparticles, D_n of the dried γ-Fe$_2$O$_3$@ PDMAAm nanoparticles was larger (12 nm) due to presence of the shell, but the polydispersity substantially did not change (PDI 1.18; Fig. 10.3b).

FIGURE 10.3 (Continued)

The hydrodynamic diameter D_h of γ-Fe$_2$O$_3$@PDMAAm, PA- and HA-mPEG-coated Fe$_3$O$_4$ was substantially larger, for example, 206, 35 and 68 nm, respectively, than D_n. The reason consists in that the DLS provided information about D_h of the particle dimers and clusters in water, where hydrophilic PDMAAm chains swell. Zeta potential (ZP) of the γ-Fe$_2$O$_3$@PDMAAm, PA- and HA-mPEG-Fe$_3$O$_4$ was −53, 26.3 and 12.4 mV, respectively. Since ZP of the γ-Fe$_2$O$_3$@PDMAAm was highly negative, the nanoparticle dispersions were very stable (up to a few months) due to the electrostatic repulsion. Regardless of the low positive ZP of the Fe$_3$O$_4$@mPEG particles, their colloid solutions were also very stable due to steric repulsion provided by mPEG. PA-mPEG-Fe$_3$O$_4$ colloid (D_h~40 nm) was stable also at various NaCl concentrations ranging from 1 to 1000 mmol/L.

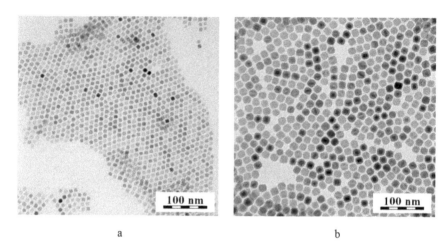

a   b

FIGURE 10.4 TEM micrographs of (a) 8 and (b)17 nm super paramagnetic Fe_3O_4 nanoparticles prepared by thermal decomposition method at (a) 320 and (b) 340°C.

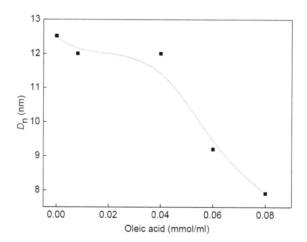

FIGURE 10.5 Dependence of number-average diameter D_n of Fe_3O_4 nanoparticles on oleic acid concentration. Particles were prepared in octadec-1-ene at 320°C for 30 min.

In contrast, HA-mPEG-coated $Fe_3O_4(D_h{\sim}65$ nm) demonstrated stability only at 1 and 10 mmol of NaCl/L. ATR FTIR and Fe analysis confirmed successful coating of the iron oxide nanoparticles with both PDMAAm by the grafting-from method and mPEG by ligand exchange method [23, 28].

Optionally, the γ-Fe$_2$O$_3$nanoparticles were covered with a silica shell at various γ-Fe$_2$O$_3$/TMOS ratios (0.1–0.8 w/w) to control the morphology and size of the nanoparticles observed by TEM (Fig. 10.3c) and SEM (Fig. 10.3 g). Size of the γ-Fe$_2$O$_3$@SiO$_2$ particles ranged from 12 to 192 nm depending on the γ-Fe$_2$O$_3$/SiO$_2$ ratio (Fig. 10.6). With increasing amounts of silica relative to the iron oxide and with introduction of amino groups by reaction with APTES, D_n of theγ-Fe$_2$O$_3$@SiO$_2$ and γ-Fe$_2$O$_3$@SiO$_2$-NH$_2$nanoparticles increased (Fig. 10.3d) due to their aggregation. This was accompanied with broadening of the particle size distribution. According to AAS, content of iron decreased from 66.1 in γ-Fe$_2$O$_{3to}$ 27.7 and 19.8 wt.% in γ-Fe$_2$O$_3$@SiO$_2$ and γ-Fe$_2$O$_3$@SiO$_2$-NH$_2$ nanoparticles, respectively. This was in agreement with increasing thickness of the silica shell surrounding the γ-Fe$_2$O$_3$ particles. Nevertheless, this amount of iron was sufficient to confer the particles with good magnetic properties. Coating of the γ-Fe$_2$O$_3$ particles with a thin silica shell hindered particles from aggregation and made them hydrophilic; as a result, the particles were well dispersible in water. Secondary coating obtained by reaction of γ-Fe$_2$O$_3$@SiO$_2$ particles with APTES made possible prospective attachment of a target biomolecule, for example, protein, antibody, enzyme or drug. However, γ-Fe$_2$O$_3$@SiO$_2$-NH$_2$ nanoparticles often formed aggregates at neutral pH suggesting that the initial γ-Fe$_2$O$_3$@SiO$_2$ particles agglomerated during the reaction with APTES.

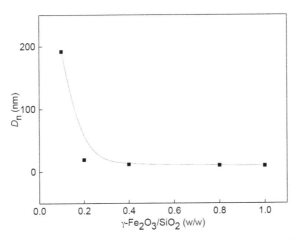

FIGURE 10.6   Dependence of number-average diameter D_n of γ-Fe$_2$O$_3$@SiO$_2$particles on γ-Fe$_2$O$_3$/SiO$_2$ ratio.

10.5 ENGULFMENT OF THE NANOPARTICLES BY STEM CELLS AND MACROPHAGES

Labeling of the cells with surface-functionalized iron oxide nanoparticles is increasingly important for diagnostic and separation of DNA [31], viruses [32], proteins [33] and other biomolecules [34]. A great deal of attention is recently devoted to stem cells and their ability to differentiate in any specialized cell type. Earlier, we have developed poly(L-lysine)-coated γ-Fe_2O_3 nanoparticles(γ-Fe_2O_3@PLL) and γ-Fe_2O_3@PDMAAm particles obtained by the solution radical polymerization in the presence of γ-Fe_2O_3 [18, 35]. Such particles were found to be highly efficient for *in vitro* labeling of human (hMSCs) and rat bone marrow mesenchymal stem cells (rMSCs). In this report, bothγ-Fe_2O_3@PDMAAm obtained by grafting from approach and γ-Fe_2O_3@SiO$_2$nanoparticleswere investigated in terms of their engulfment by macrophages (Fig. 10.7). This is an important task from the point of view of controlling introduction, movement and overall fate of the labeled cells after their implantation in the organism.

In a typical stem cell labeling experiment, the hMSCs or rMSCs were cultured in Dulbecco's Modified Eagle's Medium (DMEM) in a humidified 5% CO_2 incubator; the medium was replaced every 3 days until the cells grew to convergence. Uncoated, γ-Fe_2O_3@PLL, γ-Fe_2O_3@PDMAAm particles (via the solution polymerization) and the commercial contrast agent Endorem□ (dextran-coated iron oxide) were then used for labeling of the stem cells. After 72 h of labeling, the contrast agent was stained to produce Fe(III) ferrocyanide (Prussian Blue). The quantification of labeled and unlabeled cells was performed using TEM and inverted light microscope. Compared with Endorem□and unmodified nanoparticles, the PDMAAm- and PLL-modified particles demonstrated high efficiency of intracellular uptake into the human cells. Optionally, the labeled rMSCs cells were intracerebrally injected into the rat brain and magnetic resonance (MR) images were obtained. MR images of theγ-Fe_2O_3@PDMAAm (via the solution polymerization)- and γ-Fe_2O_3@PLL-labeled rMSCs implanted in a rat brain confirmed their better resolution compared with Endorem□-labeled cells [18, 35].

In our experiments, both γ-Fe_2O_3@PDMAAm (via grafting from approach), γ-Fe_2O_3@SiO$_2$ and γ-Fe_2O_3@SiO$_2$-NH$_2$ nanoparticles (4.4 mg/mL)

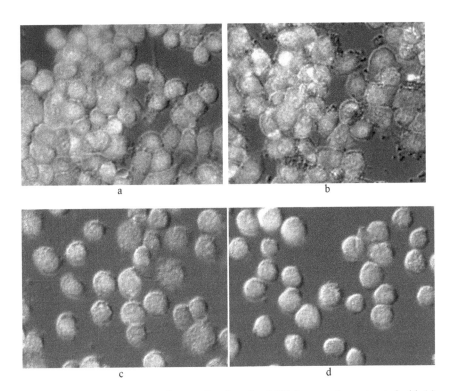

FIGURE 10.7   Fluorescence micrographs of murine J 774.2 macrophages treated with (a) γ-Fe$_2$O$_3$, (b) γ-Fe$_2$O$_3$@PDMAAm (via grafting from approach), (c) γ-Fe$_2$O$_3$@SiO$_2$ and (d) γ-Fe$_2$O$_3$@SiO$_2$-NH$_2$ nanoparticles.

were opsonized with FBS proteins at 37°C for 24 h. They were then incubated with murine J774.2 macrophages and stained with Acridine Orange and Hoechst 33342. Uptake of the particles by the cells and their morphological changes were analyzed using fluorescence microscopy. Cytotoxicity of the γ-Fe$_2$O$_3$@PDMAAm and neat γ-Fe$_2$O$_3$ nanoparticles was estimated using a hemocytometric chamber for counting number of the cells treated in the presence of nanoparticles (0.025, 0.5 and 1 wt.%) in the culture medium for 24 h.

The efficiency of the engulfment of the γ-Fe$_2$O$_3$@ PDMAAm and neat γ-Fe$_2$O$_3$ nanoparticles by the murine J 774.2 macrophages was determined after 30 min, 1, 2, 3 and 24 h cell cultivation in the presence of the particles. Figure 10.7 shows Acridine Orange and Hoechst 33342-stainedmacrophages treated with the nanoparticles for 3 h. After 30-min treatment of

J774.2 macrophages with γ-Fe$_2$O$_3$@PDMAAm nanoparticles, their majority remained unengulfed in the culture medium. Visible engulfment of the nanoparticles appeared after 1-h treatment. After 2-h treatment, granulation of the cytoplasm was observed due to accumulation of the γ-Fe$_2$O$_3$@PDMAAm nanoparticles in the peripheral region of the cytoplasm. After 3-h treatment, majority of the γ-Fe$_2$O$_3$@PDMAAm nanoparticles was engulfed by the macrophages and some cells demonstrated signs of lysosomal activation characterized by red Acridine Orange fluorescence. Only a minimal amount of the γ-Fe$_2$O$_3$@PDMAAm nanoparticles remained unengulfed indicating that the engulfment was very efficient. PDMAAm showed the affinity to cell membrane components facilitating thus the endocytosis.

As a control experiment, the engulfment of the neat γ-Fe$_2$O$_3$ nanoparticles in the macrophages was investigated. Within 1–3 h treatment, the number of vacuoles, their size, as well as the number of lysosomal clusters associated with large vacuoles, increased with time. Numerous unengulfed γ-Fe$_2$O$_3$ nanoparticles were accumulated on the surface of treated macrophages, while free γ-Fe$_2$O$_3$ nanoparticles were almost absent. The size of the cells treated with γ-Fe$_2$O$_3$ nanoparticles was also increased.

All described super paramagnetic nanoparticles were relatively nontoxic for the cultured cells. Apparently, for the efficient particle engulfment by the macrophages, the presence of positively charged amidine groups in γ-Fe$_2$O$_3$@PDMAAm nanoparticles is beneficial. The efficiency of engulfment of the γ-Fe$_2$O$_3$@PDMAAm nanoparticles was quite high since after 2-h treatment most cells engulfed the nanoparticles and only few nanoparticles remained in the culture medium. Fluorescence microscopy confirmed only weak activation of lysosomes, which manifested itself by a change in the color of Acridine Orange from green to red. Acridine Orange, a weakly basic amino dye, is known to be a lysosomotropic agent. In its stacked form, for example, in lysosomes, it emits red fluorescence, while in the cell nuclei at neutral pH it emits yellow-green fluorescence. Activation of macrophages during the engulfment of foreign extracellular materials was accompanied by an increase in the activity of digestive vacuoles and, thus, it caused a red fluorescence shift due to accumulation of the dye in lysosomes. Activation of lysosomal compartments accompanied intracellular processing of the engulfed particles (microorganisms, viruses, damaged cells, and foreign macromolecules) [23]. Chemical structure of uncoated γ-Fe$_2$O$_3$ nanoparticles thus provided potential toxicity for the treated cells, which manifested itself by time-dependent evolution of vacuoles in the cell cytosol.

10.6   CONCLUSIONS

In summary, two different types of iron oxide nanoparticles were synthesized, maghemite(γ-Fe$_2$O$_3$)and magnetite (Fe$_3$O$_4$). The first ones were prepared by coprecipitation of Fe(II) and Fe(III) salts with aqueous ammonia. Obtained magnetite was then oxidized with sodium hypochlorite to chemically stable maghemite. However, the particle size distribution of these particles was rather broad as determined by a range of physicochemical characterization methods including SEM, TEM and DLS measurements. In contrast, monodispersesuperparamagneticFe$_3$O$_4$nanoparticles with size controlled from 8 to 25 nm were produced by the thermal decomposition of Fe(III) oleate at different temperatures and oleic acid concentrations. The particles were successfully transferred in water by the ligand exchange method. As a hydrophilic ligand, derivatives of mPEG with specific functional groups were used that strongly chemically bonded with iron. Optionally, γ-Fe$_2$O$_3$ particles were surface-modified with PLL, PDMAAm (both by the solution radical polymerization and grafting from method) or SiO$_2$. The successful coating of the iron oxide nanoparticle surface was confirmed by both ATR FTIR spectroscopy and Fe analysis. The colloidal particles were stable in aqueous media for several months.

The biotargeting characteristics of the nanoparticles are mainly defined by the biomolecules conjugated to the particle surface. It is desirable that the particle shell contains either membranotropic molecules like phospholipids, poly(ethylene glycol) or macromolecules (proteins) present in biological fluids. In this work, surface of the formed nanoparticles was opsonized with proteins available in the fetal bovine blood serum. The γ-Fe$_2$O$_3$@PDMAAm and γ-Fe$_2$O$_3$@SiO$_2$nanoparticles, in contrast to the neat nanoparticles were shown to be noncytotoxic and intensively phagocytozed by the mammalian macrophages. Additionally, there was no cell irritation during the phagocytosis of the γ-Fe$_2$O$_3$@PDMAAm nanoparticles. In contrast, time-dependent vacuolization of neat γ-Fe$_2$O$_3$ nanoparticles in cytoplasm of the macrophages was observed suggesting cytotoxicity of the material.

Silica used as an inorganic inert coating of the γ-Fe$_2$O$_3$ nanoparticles proved to be also suitable modification agent preventing aggregation of the particles and enhancing their chemical stability. This inorganic material is also easily susceptible to chemical modifications, which make synthesis of particles for combined diagnosis and therapy possible. Biological experiments demonstrated that both γ-Fe$_2$O$_3$@PDMAAm and γ-Fe$_2$O$_3$@SiO$_2$ and γ-Fe$_2$O$_3$@SiO$_2$-NH$_2$ core–shell nanoparticles were recognized and engulfed by the macrophages. The uptake of the surface-coated iron oxide nanoparticles by phagocytic monocytes and macrophages could provide a valuable in vivo tool by which magnetic resonance imaging can monitor introduction, trace movement and observe short- and long-term fate of the cells in the organism.

In conclusion, high potential of the polymer-coated magnetic nanoparticles can be envisioned for many biological applications. The particles can be easily magnetically separated and redispersed in water solutions upon removing of the external magnetic field. Magnetically labeled cells can be steered and concentrated inside the body by a magnet. The iron oxide particles, modified with organic, as well as inorganic polymer coatings, seem to be very promising not only for cell imaging and tracking, but also for drug and gene delivery systems and capture of various cells and biomolecules required for diagnostics of cancer, infectious diseases and neurodegenerative disorders.

ACKNOWLEDGEMENT

The financial support of the Ministry of Education, Youth and Sports (project LH14318) is gratefully acknowledged.

KEYWORDS

- atomic absorption spectroscopy
- attenuated total reflectance Fourier transform infrared spectroscopy
- dynamic light scattering
- magnetic resonance

- **magnetic resonance imaging**
- **scanning electron microscopy**
- **tetraethylorthosilicate**
- **tetramethylorthosilicate**

REFERENCES

1. Akbarzadeh, A., Samiei, M., Davaran, S., Magnetic nanoparticles: Preparation, physical properties, and applications in biomedicine, Nanoscale Res. Lett. 7, 1–13 (2012).

2. Jeng, H. A., Swanson, J., Toxicity of metal oxide nanoparticles in mammalian cells, J. Environ. Sci. Health A Tox. Hazard Subst. Environ. Eng. 4, 12699–12711 (2006).

3. Gupta, A. K., Gupta, M., Synthesis and surface engineering of iron oxide nanoparticles for biomedical applications, Biomaterials 26, 3995–4021 (2005).

4. Arbab, A. S., Bashaw, L. A., Miller, B. R., Jordan, E. K., Lewis, B. K., Kalish, H., Frank, J. A.,Characterization of biophysical and metabolic properties of cells labeled with superparamagnetic iron oxide nanoparticles and transfection agent for cellular MR imaging, Radiology 229, 838–846 (2003).

5. Schleich, N., Sibret, P., Danhier, P., Ucakar, B., Laurent, S., Muller, R. N., Jérôme, C., Gallez, B., Préat, V., Danhier, F., Dual anticancer drug/superparamagnetic iron oxide-loaded PLGA-based nanoparticles for cancer therapy and magnetic resonance imaging, Int. J. Pharm. 15, 94–101 (2013).

6. AlexiouC., SchmidR. J., JurgonsR., KremerM., WannerG., BergemannC., HuengeseE., NawrothT., ArnoldW., ParakF. G., Targeting cancer cells: magnetic nanoparticles as drug carriers, Eur. Biophys. J. 35, 446–450 (2006).

7. Papell, S. S., Low viscosity magnetic fluid obtained by the colloidal suspension of magnetic particles, US Pat. 3, 215, 572 (1965).

8. Viswanathiah, M., Tareen, K., Krishnamurthy, V., Low temperature hydrothermal synthesis of magnetite, J. Cryst. Growth 49, 189–192 (1980).

9. Sugimoto, T., Sakata, K., Preparation of monodispersepseudocubic α-Fe$_2$O$_3$ particles from condensed ferric hydroxide gel, J. Colloid. Interface Sci. 152, 587–590 (1992).

10. Strobel, R., Pratsinis, S., Direct synthesis of maghemite, magnetite and wustite nanoparticles by flame spray pyrolysis, Adv. Powder Technol. 20, 190–194 (2009).

11. Cornell, R. M., Schwertmann, U., The Iron Oxides: Structure, Properties, Reactions, Occurrences and Uses, 2nd Ed., Wiley, Darmstadt 2000.

12. Willis, A., Chen, Z., He, J., Zhu, Y., Turro, N., O'Brien, S., Metal acetylacetonates as general precursors forthe synthesis of early transitionmetal oxide nanomaterials, J. Nanomater. 2007, 1–7 (2007).

13. Rockenberger, J.,Scher, E.,Alivisatos, P., A new nonhydrolytic single-precursor approach to surfactant-capped nanocrystals of transition metal oxides, *J. Am. Chem. Soc.*121, 11595–11596 (1999).

14. Woo, K.,Hong, J., Choi, S., Lee, H., Ahn, J., Kim, C., Lee, S., Easy synthesis and magnetic properties of iron oxide nanoparticles, *Chem. Mater.*16, 2814–2818 (2004).

15. Baalousha, M., Manciulea, A., Cumberland, S., Kendall, K., Lead, J. R., Aggregation and surface properties of iron oxide nanoparticles: Influence of pH and natural organic matter, Environ. Toxicol. Chem. 27, 1875–1882 (2008).

16. Barrera, C., Herrera, A. P., Rinaldi, C., Colloidal dispersions of monodisperse magnetite nanoparticles modified with poly(ethylene glycol), J. Colloid. Interface Sci. 329, 107–13 (2009).

17. Chastellain, M., Petri, A., Hofmann, H., Particle size investigations of a multistep synthesis of PVA coated superparamagnetic nanoparticles, J. Colloid Interface Sci. 278, 353–60 (2004).

18. Babič, M., Horák, D., Jendelová, P., Glogarová, K., Herynek, V., Trchová, M., Likavčanová, K., Hájek, M., Syková, E., Poly(N,N-dimethylacrylamide)-coated maghemite nanoparticles for stem cell labeling, Bioconjugate Chem. 20, 283–294 (2009).

19. Lu, Y., Yin, Y., MayersB. T., Xia, Y., Modifying the surface properties of superparamagnetic iron oxide nanoparticles through a sol−gel approach, NanoLett.2, 183–186 (2002).

20. Mosser, D. M. The many faces of macrophage activation, J. Leukocyte Biol. 73, 209-212 (2003).

21. Chawla, A., Nguyen, K. D., Goh, Y. P. S., Macrophage-mediated inflammation in metabolic disease, Nat. Rev. Immunol. 11, 738–749 (2011).

22. Lu, A.-H., Salabas, E. L., Schüth, F., Magnetic nanoparticles: Synthesis, protection, functionalization, and application, Angew. Chem. Int. Ed. 46, 1222–1244 (2007).

23. Zasonska, B. A., Boiko, N., Horák, D., Klyuchivska, O., Macková, H., Beneš, M., Babič, M., Trchová, M., Hromádková, J., Stoika, R., The use of hydrophilic poly(N,N-dimethylacrylamide) grafted from magnetic -Fe$_2$O$_3$ nanoparticles to promote engulfment by mammalian cells, J. Biomed. Nanotechnol. 9, 479-491 (2013).

24. Park, J., An, K. J., Hwang, Y. S., Park, J. G., Noh, H. J., Kim, J. Y., Park, J. H., Hwang, N. M., Hyeon, T.,Ultra-large-scale syntheses of monodispersenanocrystals, Nat. Mater. 3, 891–895 (2004).

25. Stöber, W., Fink, A., Controlled growth of monodisperse silica spheres in the micron size range, J. Colloid Interface Sci. 26, 62–69 (1968).

26. Finnie, K. S., Bartlett, J. R., Barbé, C. J. A., Kong, L., Formation of silica nanoparticles in microemulsions, Langmuir 23, 3017–3024 (2007).

27. Sakka, S., Sol-Gel Science and Technology, Springer, San Diego 2005.

28. Patsula, V., Petrovský, E., Kovářová, J., Konefal, R., Horák, D., Monodispersesuperparamagnetic nanoparticles by thermolysis of Fe(III) oleate and mandelate complexes, Colloid Polym. Sci., DOI:10.1007/s00396-014-3236-6.

29. Mohapatra, S., Pramanik, P., Synthesis and stability of functionalized iron oxide nanoparticles using organophosphorus coupling agents, Colloids Surf. A 339, 35–42 (2009).

30. Ramis, G., Larrubia, M., An FT-IR study of the adsorption and oxidation of N-containing compounds over Fe$_2$O$_3$/Al$_2$O$_3$ SCR catalysts, J. Mol. Catal. A Chem. 215, 161–167 (2004).

31. Saiyed, Z., Ramchand, C., Telang, S., Isolation of genomic DNA using magnetic nanoparticles as a solid-phase support, J. Phys. Condens. Matter 20, 204153 (2008).

32. Imbeault, M., Lodge, R., Ouellet, M., Tremblay, M., Efficient magnetic bead-based separation of HIV-1-infected cells using an improved reporter virus system reveals that p53 up-regulation occurs exclusively in the virus-expressing cell population, Virology 393, 160–167 (2009).

33. Lee, S., Ahn, C., Lee, J., Lee, J. H., Chang, J., Rapid and selective separation for mixed proteins with thiol functionalized magnetic nanoparticles, Nanoscale Res. Lett. 7, 279 (2012).

34. Haun, J. B., Yoon, T.-J., Lee, H., Weissleder, R., *Magnetic nanoparticle biosensors*, Wiley Interdiscip. Rev. Nanomed. Nanobiotechnol. 2, 291–304 (2010).

35. Babič, M., Horák, D., Trchová, M., Jendelová, P., Glogarová, K., Lesný, P., Herynek, V., Hájek, M., Syková, E., Poly(L-lysine)-modifiediron oxide nanoparticlesfor stem cell labeling, BioconjugateChem. 19, 740–750 (2008).

CHAPTER 11

IMPACT OF TRANSITION METAL COMPOUNDS ON FIRE AND HEAT RESISTANCE OF RUBBER BLENDS

V. F. KABLOV, O. M. NOVOPOLTSEVA, V. G. KOCHETKOV,
N. V. KOSTENKO, and K. A. KALINOVA

Volzhsky Polytechnical Institute (branch) Volgograd State Technical University, 42a Engelsa Street, Volzhsky, 404121, Russia, www.volpi.ru; E-mail: nov@volpi.ru

CONTENTS

ABSTRACT

Development of modern industry requires an increase in the temperature limit of exploitation of elastomer materials that is achieved by using new components that provide the flowing of physical and chemical transformations enhancing their operational stability. In this paper, a possibility of using metal oxides with variable valence to create elastomer compositions has been shown. Impact of the oxides on flame and heat resistance of rubbers based on general-purpose raw rubbers has been considered.

11.1 INTRODUCTION

Development of modern industry requires an increase in the temperature limit of exploitation of elastomer materials including extreme conditions in the field of thermal decomposition, *which* is achieved by using new components that provide the flowing of physical and chemical transformations enhancing their operational stability.

Additives that can change their structure under external influences (e.g., layered, intumescent additives, etc.) play an important role in improvement of rubber operational stability [1, 2].

In extreme operating conditions, at temperatures near and above the temperature working capacity of a material, functionally active fillers can play a stabilizing role in thermal destruction of the material [1, 2].

One of the perspective directions for solving the problem is to use intumescent and highly dispersed metal-containing fillers in elastomeric compositions, as also aluminum silicates, fillers with a catalytic activity, highly dispersed silicon carbide [3–5] including the compounds of the transition metals.

Some of the metals related to d-elements have properties which enable to apply them as protective and wear resistant coatings, a fireproof material for aircraft and rocket engines, a component of multilayer coatings for laser mirrors and beam splitters [6], for production of refractories for increasing campaign in furnaces for melting glass and aluminum. Such refractories are used in the metallurgical industry for gutters, glasses in the continuous casting of steel and crucibles for melting rare earth elements. They are also used in some ceramic-metal coatings, which have high hardness and resistance to many chemicals and keep short-term heating up to 2750°C.

It is also known the use of metal organic compounds containing these elements as a crosslinking agent for polymers [7].

11.2 EXPERIMENTAL RESULTS AND DISCUSSION

The research was devoted to a study of influence of transition metal compounds on the properties of rubber blends and their vulcanizates based on general-purpose raw rubbers.

Rubber blends were prepared according to the standard recipes based on styrene-butadiene raw rubber using the sulfur-vulcanizing group with partial replacement of carbon black on the analyzing compounds (Table 11.1).

Kinetic parameters of rubber blends were determined by using Monsanto 100S rheometer. Research has shown that introduction of transition metal compounds increases the induction period, but it practically does not change the vulcanization rate (Fig. 11.1). In addition, there is a reduction of elastic-strength properties, however, resistance to thermal-oxidative aging

TABLE 11.1   Filler Content in the Studied Compositions

Filler	Rubber blend number					
	Z-0	Z-1	Z-2	Z-3	Z-4	Z-5
Carbon black П-324	40	35	30	25	20	15
ZrO$_2$	-	5	10	15	20	25

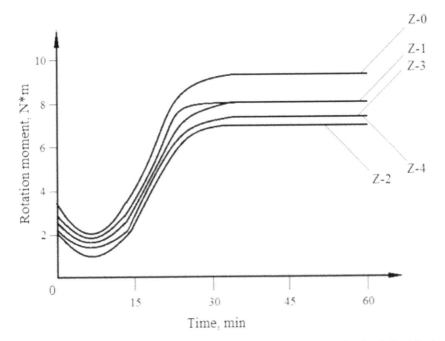

FIGURE 11.1   Kinetic curves of vulcanization: 1 – Control rubber blend Z-0; 2 – blend Z-1; 3 – blend Z-2; 4 – blend Z-3; blend Z-4.

and resistance to flame action rises (Table 11.2). To estimate fire resistance, the dependence of temperature on the unheated sample surface from the exposure time of the plasma torch flame was determined (Fig. 11.2). Temperature on the sample surface was generated about 2000°C. For the experimental samples the heating time increases and degradation occurs at higher temperatures compared with the control sample.

When the experimental samples are exposed to flame, a dense and flame retardant coke, which protects the sample from burning, is formed on their surface (Fig. 11.3).

11.3 CONCLUSION

So, the research has shown that the investigated transition metal compounds can be used to effectively enhance the fire resistance of elastomeric materials and reduce their cost.

TABLE 11.2 Physical and Mechanical Properties of Vulcanizates*

Parameter	Rubber blend number				
	Z-0	Z-1	Z-2	Z-3	Z-4
Tensile strength (f_p), MPa	18,0	14,0	15,0	12,2	13,3
Elongation at break, %	420	410	590	560	490
Relative residual elongation after fracture, %	12	9	11	9	9
Change of parameters after aging					
(100°C × 72 h), %:					
Δf_p	−45	−36	−40	−34	−31
$\Delta \varepsilon$	−67	−61	−63	−64	−59
Linear combustion velocity, mm/min	24,56	23,96	22,72	15,96	15,18
Warm-up time of the sample surface to 100°C, sec	60	60	60	80	90
Time of burning-out of the sample, sec	100	110	110	120	130

*Vulcanization mode: 145°C, 30 min

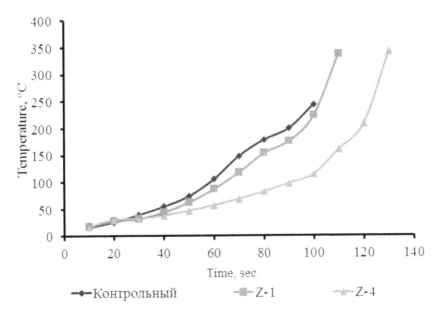

FIGURE 11.2 Dependence of temperature on the unheated sample surface on the heating time.

FIGURE 11.3 View of carbonized surface of an experimental vulcanizate sample (coke layer) under the flame exposure.

KEYWORDS

- **d-elements**
- **elastomers**
- **fillers**
- **fire resistance**
- **modifying additives**
- **rubbers**

REFERENCES

1. Kablov, V. F. System technology of elastomeric materials – integration processes in development of materials and products. Proceedings of XV International scientific and practical conference "Rubber industry. Raw materials, materials and technologies" (Moscow, 2009), P. 4.
2. Kablov, V. F., Novopoltseva, O. M., Kochetkov, V. G. [et al]. Impact of perlite filler on heat resistance of rubbers based on ethylene propylenediene raw rubber. Modern Problems of Science and Education. 2013. № 3, URL: www.science-education.ru/109–9370
3. Kablov, V. F., Novopoltseva, O. M., Kochetkov, V. G. [et al]. Heat protective coatings containing perlite. International Journal of Applied and Fundamental Investigations. 2012. № 1. P. 174.
4. Lifanov, V. S., Kablov, V. F., Novopoltseva, O. M. [et al]. Study of elastomeric materials with microdispersed silica carbide wastes. Modern Problems of Science and Education. 2013. № 4. URL: www.science-education.ru/110–9971
5. Lifanov, V. S., Kablov, V. F., Lapin, S. V., Kochetkov, V. G., Novopoltseva, O. M. Elastomeric materials with microdispersed silica carbide wastes. Kauchuk i rezina, 2013, №6, pp. 8–10.
6. Rakov, E. G., Golubkov, V. V., Hyu Van Nguen. Development of carbon nanomaterials. Herald of National Research Nuclear University "MEPhI", 2012. Vol. 1(2), pp. 167–171.
7. Monte, S. J. Kenrich Petrochemicals. Titanates and zirconates in thermoplastic and elastomer compounds. Rubber World, 2012, pp. 40–45.

CHAPTER 12

INVESTIGATION OF THERMAL POWER CHARACTERISTICS OF WOOD PULP

V. YALECHKO,[1] V. KOCHUBEY,[2] Y. HNATYSHYN,[1]
B. DZYADEVYCH,[1] and G. E. ZAIKOV[3]

[1]National Forestry University of Ukraine, General Chuprynky St.
103, 79057 Lviv, Ukraine, Tel. (380–322) 237-80-94,
E-mail: hnat_ya@inbox.ru, V0l0dyMyR@ukr.net

[2]Lviv Polytechnic National University, 12, Lviv, Lviv Oblast, 79000,
Ukraine

[3]M. Emanuel Institute of Biochemical Physics Russian Academy
of Sciences, Kosygin St., 4, Moscow, 119334, Russia,
E-mail: chembio@sky.chph.ras.ru

CONTENTS

ABSTRACT

For study the heating value of wood used integrated thermal analysis. We
have studied the aspen wood aged 10 years. For sample 3 (bark), which

is characterized by the largest coke residue, the heterogeneous oxidation process occurs most rapidly. From the thermal analysis shows that the pattern of the cortex has the highest heating value.

12.1   INTRODUCTION

The use of wood pulp as an energy feedstock is rather vital. Effective and ecologically combustion of biomass in fuel devices determined by the characteristics of combustion mode.

Important role today given the issues of using wood biomass, including fast growing, as energy resources.

In order to determine the combustion characteristics of wood biomass and setting the optimal parameters necessary the complex research.

Investigation methods of burning process are the physical modeling on the laboratory and semiindustrial plants, with further full-scale tests of the developed flow sheets and analytically through the mathematical models. Analytical researches requiring information about the kinetic and thermal power characteristics of wood biomass.

Should be noted that here possible significant reduction of substantial funds and resources expended in obtaining the necessary information from the relevant experimental setups [1–3].

12.2   RESEARCH METHODOLOGY

To study the calorific value of wood used integrated thermal analysis, including thermogravimetry (TG), differential thermogravimetry (DTG) and differential thermal analysis (DTA).

The objects of research were samples: stem wood of aspen (sample 1), stem wood of aspen, mixed in equal proportions from the bark of aspen (sample 2), aspen bark (sample 3). The age of aspen wood was 10 years.

Thermal analysis of samples of aspen wood was performed on deryvatohraf Q – 1500D system "F. Paulik – J. Paulik – L. Erdey " with the registration of the analytical signal of mass loss and thermal effects using a computer. Samples of wood were analyzed in dynamic mode with a heating rate of 10°C/min in air. The mass of each sample was 100 mg. The reference substance was aluminum oxide [4, 5].

12.3 RESULTS AND DISCUSSION

Thermograms of samples presented in Figs. 12.1–12.3, and the results of their treatment in Table 12.1. In Figs.12.4 and 12.5 shows the comparison of TG and DTA curves of samples of aspen wood.

At the 1st stage (20–187°C) occur endothermic processes due to evaporation of chemically bound water and constitutional water. Intensive mass loss observed on samples of TG curves at temperatures higher 200°C.

At the second stage of thermolysis, which according to the differential thermogravimetric analysis takes place in the temperature range 186–277°C, along with the endothermic dehydration and pyrolysis processes (cleavage of volatile degradation products), which are accompanied by a

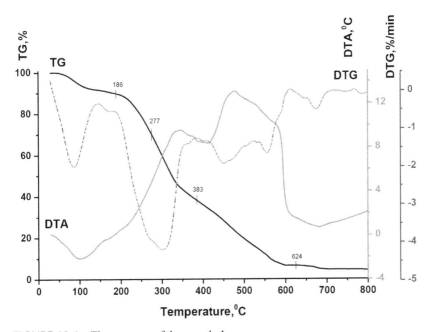

FIGURE 12.1 Thermogram of the sample 1.

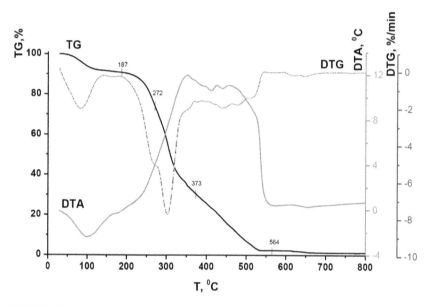

FIGURE 12.2   Thermogram of the sample 2.

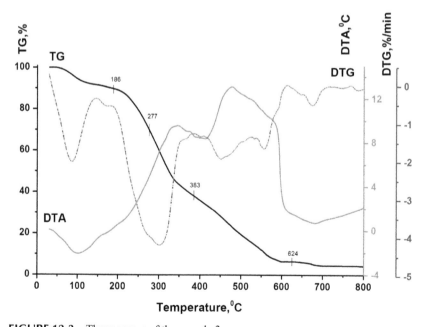

FIGURE 12.3   Thermogram of the sample 3.

TABLE 12.1 Results of Comprehensive Thermogravimetric and Differential Thermogravimetric Analyzes of Samples 1–3

Sample	Stage	Temperature range, °C	Loss of mass, %
Sample 1	1	20–187	7.6
	2	187–277	19.5
	3	277–372	43.3
	4	372–569	28.5
Sample 2	1	20–187	9.2
	2	187–272	17.5
	3	272–373	42.9
	4	373–564	28.1
Sample 3	1	20–186	10
	2	186–277	20.4
	3	277–383	31.2
	4	383–624	32

sharp decrease the degree of polymerization of cellulose, developing exothermic thermooxidative destructive processes, as DTA curve shows the course aspen samples (Fig. 12.1–12.3).

Unlike other samples of aspen, like sample 3 in the second phase of thermolysis most heavily lose mass (Fig. 12.4). This indicates that the sample of bark most heavily take place the processes of cleavage of volatile decomposition products. (Lipskis, A.L., Kvikli, A.V., Lipskene, A.M., Maciulis, A.N. 1976; Yegunov, V.P. 1992; Tsapko, Y.V. 2011)

The third stage of thermolysis (277–383°C), accompanied by the largest mass loss of the aspen samples (Table 12.1) and the appearance of bright exothermic effect on the curves DTA, there are active thermooxidative destructive processes, accompanied by flame combustion of volatile decomposition products. For sample 3 thermooxidative processes in the air phase flow less intense. This shows a small weight loss of the sample and the appearance of the smallest compared with other samples, exothermic effect at the DTA curve.

At the fourth stage of thermolysis (372–624°C) occurs burning of the carbonated residue of aspen samples. For sample 3, which characterized

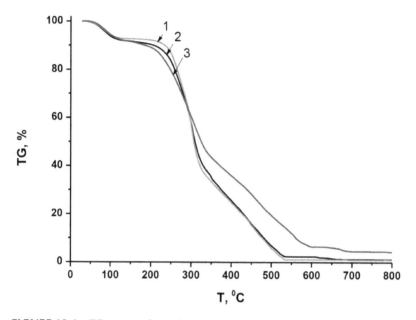

FIGURE 12.4   TG curves of samples: 1 – sample 1, 2 – sample 2, 3 – sample 3.

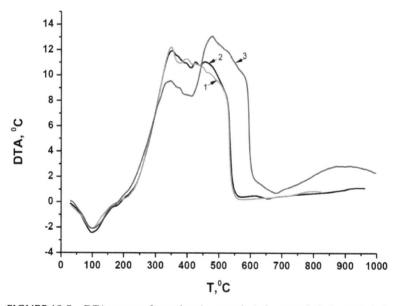

FIGURE 12.5   DTA curves of samples: 1 – sample 1, 2 – sample 2, 3 – sample 3.

by the largest coke residue heterogeneous oxidation process occurs most rapidly. This is evidenced by the appearance of the most striking in comparison with other samples exothermic effect on the DTA curve, which is shifted into the region of higher temperatures [5–7].

Chemical analysis of elemental composition of dry weight mixture of aspen wood: carbon – 52.65%, hydrogen – 4.38%, oxygen – 37.4%, nitrogen – 0.42%, ash – 1.5%, heating value is 18.4 MJ/kg.

12.4 CONCLUSIONS

As seen from conducted thermal analysis the sample of bark has the highest heating value. The processes of thermooxidative degradation and burn-out of carbonated residue is accompanied by largest exothermic effect.

The results of mathematical modeling make it possible use them to develop effective constructions of appropriate fuel devices for efficient utilization of wood waste and wood biomass. During the combustion of biomass in power plants or boilers emitted only CO_2 that will be absorbed by the plant during its growth.

KEYWORDS

- heating value
- mathematical modeling
- thermal analysis
- utilization
- wood biomass

REFERENCES

1. Madoyan, A. A. et al. (1991). More efficiently burn low-grade coal in power boilers. Energoatomizdat, Moscow.
2. Khzmalyan, D. M. (1990). The theory of combustion processes. Energoatomizdat, Moscow.
3. Zeldovich, Y. B., Bernblatt, G. I., Librovich, V. B., Makhviladze, G. M. (1980). The mathematical theory of combustion and explosion.

4. Shestak, Y. (1987). The theory of thermal analysis. Mir, Moscow, USSR.
5. Lipskis, A. L., Kvikli, A. V., Lipskene, A. M., Maciulis, A. N. (1976). The calculation kinetic parameters of thermal degradation of polymers. Macromolecular compounds. Vol (A) XVIII, 426–431.
6. Yegunov, V. P. (1992). Introduction to thermal analysis. Samara.
7. Tsapko, Y. V. (2011). The study of kinetic parameters during pyrolysis of fire protected wood by impregnating agents. Fire safety № 19, 163–169.

CHAPTER 13

RELAXATION PROCESSES IN THE SHRINKABLE POLYVINYLCHLORIDE FILMS WITH TACTILE MARKING OF SHRINKABLE LABELS

ALEXANDER KONDRATOV and ALEXEY BENDA

Moscow State University of Printing Arts, ul. Leninskiye Gory, 1, Moscow, 127550, Russia; E-mail: apk@newmail.ru

CONTENTS

ABSTRACT

Tactile marking of flexible packaging, labels, bottle labels and tags made of shrinkable films is a new direction of counterfeit and falsification protection of food and goods intended for the people with weak eyesight. The tactile marks and pictograms in accordance with international standards for relief height and width have been receiving by local thermo-mechanical modification of polymer film structure before printing of informational texts or images. It's impossible to detect the tactile marks on blanks before heat setting on the packing, because they arise from the thermoshrinkage. The example of a tactile line-shaped mark is shown in the Fig. 13.1. The localized film bowing arises in the zone of thermomechanical polymer structure modification after the label shrinkage on the bottle.

13.1   INTRODUCTION

The falsification of industrial consumer goods and the emergence of counterfeit productions are important dangerous economic problems of brand manufacturers in many developed countries. The methods of falsification of counterfeit goods and production of various kinds of counterfeit are based on the substitution of true goods for not valuable analogs under retaining the similarity of packages or the labels of true goods. The

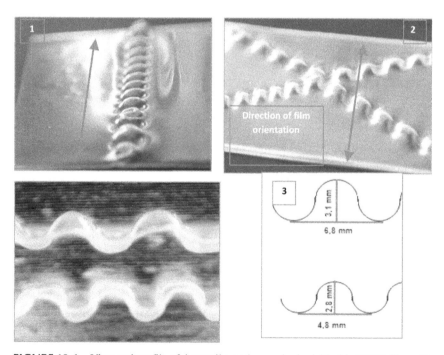

FIGURE 13.1   View and profile of the tactile mark on polyvinylchloride (RVC) film with thermo modified linear intervals located at an angle of 90 degrees (1.3) and of 45 degrees (2.4) to the shrinkage direction

counterfeiters try to make the packaging materials and printing design identical to the true one, deceiving the consumers with that. The manufacturers of the brands take measurements for counterfeit protection of the goods and use unique technologies and materials, complex network methods to identify the authenticity of articles. That significantly raises the prices of products and reduces their competitiveness. The falsifiers master the technologies of counterfeit protection of packaging over and over again and supply counterfeit goods to the markets. As a result, there is a constant need for low cost way of protection for products against counterfeit by using of the difficult-to-counterfeit packaging [1].

The creation of special counterfeit protected packaging providing the information about product authenticity and manufacturer, which is very important for the people with weak eyesight, is of greatest interest. The use of invisible thread for marking of goods by applying to the product or packaging is well known [2]. It may be invisible white threads and invisible black threads; they could be twisted together and combined with conventional threads of cotton, flax, silk or synthetic fibers. But for all that, the marking numbers, letters or symbols could be fastened to tags, labels, products or packaging by saddle stitching, sewing into the surface, sticking on the label surface, pasting or embedding in materials by other means. Such marking is easy-to-manufacture and allows its tactile authentication.

It's possible to apply protective packaging information by Braille printing [3] which is a pixel imaging or text embossed in label material or applied above the surface. At the creating of protective signs such as a combination of embossed points on cardboard by relief stamping the blank of future packaging is being passed through an embossing press [4]. At the thermolifting of a point mark or a text, the fine powder is being applied on the printing material. The powder particles expand at heating significantly, foaming polystyrene is being generated. As a result of heating the powdered polystyrene foams and the image becomes more relief. This method can be used for Braille printing on self-adhesive labels [4] or special protective tags for different packaging.

The tactile marking on polymer packaging or label can be hidden or visible depending on purpose and technology used for authentication of goods. The visible labeling is intended for identification of mass-produced goods by the customers with weak eyesight at the moment of choice or purchase of goods. The hidden marking is being realized by manufacturer or supplier of packaged product for counterfeit protection of products. It is intended for authentication of goods by consumers after its acquisition or revealing of trade falsification by supervisors.

The suggested tactile marking of flexible packaging, labels, bottle labels and tags made of shrinkable films is a new direction of protection against counterfeit and falsification of food and goods, which intended for the people with weak eyesight [2]. The tactile marks and pictograms in accordance with international standards for relief height and width had been received by local thermo-mechanical modification of polymer film structure before printing on it informational texts or images. It's impossible to detect the tactile marks on blanks before heat setting on the packing, because they arise only after the thermo shrinkage.

The examples of tactile line-shaped mark are shown in the Fig. 13.1. The localized film bowing arises in the zone of thermomechanical polymer structure modification after the label shrinkage on the bottle [7].

The form of a mark can comply with the configuration of letters, Roman numerals or Arabic figures, international tactile marks on the plates for the blind, as well as it can carry information about product and its authentication. The authentication procedure of goods by consumers after their acquisition involves the sinking of a film packaging or a protective label in hot water, or the placement it into the oven for products' heating and later visual and tactile identification of the emerged marks (Fig. 13.2).

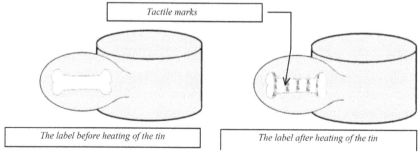

Tactile marks

The label before heating of the tin

The label after heating of the tin

FIGURE 13.2   Results of heating of packaging with complex separating polymer label – visual and tactile identification of the protective marks.

13.2   EXPERIMENTAL PART

Tactile marking technique (method of obtaining of a protective mark on a polymer packaging, a label, a tag).

The method of tactile marking [5] is based on obtaining of interval thermoshrinkable films [6] by local thermomechanical effect on thermoshrinkable material of a packaging, a label or a tag. The thermomechanical effects can be realized by using the heat of sealing or embossing printing equipment. The thermo mechanical effect is made by a short-term local heating under defined pressure on a film section in the form of a future tactile sign. After the hidden marking, the print is not visible or tangible prior to the heat treatment process of a packaging, a label or a tag. For the visible labeling, the print is heated immediately after mark creation on the film details of packaging or, for example, in fixation process of thermoshrinkable label on a bottle with vendible product. [7]

The calorimetric study of thermoshrinkable polyvinylchloride films was conducted to investigate the physical and chemical nature of the effect of tactile marking and to determinate the optimum quantitative parameters for the process of short-term local heating under pressure.

The research of protective packaging samples with controlled shrinkage For creating and researching of laboratory samples of interval materials were used the solid thermo shrinkable films. Samples of polyvinylchloride were manufactured in Europe and in Russia. The intervals with linear marking width of 2.2 mm were formed by local heating at 70°C – 112°C of the film samples under pressure of 70 psi. The constant pressure was created by short-term press of hot steel blades with thickness of 2.0 mm, where the temperature was maintained to 0.5°C. The heating temperatures are shown in the Table 13.1. The samples of interval PVC films processed by local thermomechanical modification at 90°C–112°C developed a "tactile wave" after general heating of films from 60°C to 100°C.

For identification of the shrinkage results the millimeter grid was printed before the heating on laboratory samples of interval films by screen-printing method. The gird was required for dilatometric measurements of the deformation processes and thermal shrinkage of films as well as for determination of the transition width of the zone between the intervals.

TABLE 13.1 Local Shrinkage of Modified Intervals of Different Quality Polyvinylchloride Films in Hot Water (95–100 °C), 15 s

No.	Temperature of local heating,°C	Minimal shrinkage of the interval film, % With local heating	Maximal shrinkage of the interval film, % Without local heating
Shrinkable polyvinylchloride "Eurofilm"			
1	70	16	42
2	75	10	41
3	80	6	40
4	85	11	41
5	90	5	42
6	92	0	42
7	94	0	41
8	96	0	40
Shrinkable polyvinylchloride "Don-Polymer", RF			
9	80	0	51
10	110	0	49
11	112	0	50

Also the dependence of the film shrinkage was investigated in the modified zone and adjacent areas of the film at the modification temperature range of 70°C ÷ 112°C and at the time of local influence of 5 ÷ 7 seconds. After the shrinkage the distorted gird in the transition zone between the intervals was measured by microscope. Depending from the results of measurements were defined the local shrinkage areas of unmodified intervals, the residual of the local shrinkage of modified areas, the width of the transition zone as well as there was calculated the shrinkage gradient in the transition zone.

The local shrinkage of modified intervals is showed at variable degrees depending on the temperature of marking. The obtained results are summarized in the Table 13.1.

At the temperatures above 80°C and 90°C, depending on the film quality, the linear dimensions reduction of zones of polyvinylchloride film, which were previously subjected to local heating, are zero, and the local shrinkage on adjacent areas of the film reaches its maximal values (40 and 50 percent depending on qualities of polyvinylchloride films).

The authors investigated the reasons and optimal conditions for emergence of the tactile-sensitive "wave" on the interval thermoshrinkable films by methods of differential calorimetry and thermomechanical analysis by using of physical modeling of thermoshrinkage process in the interval films made of hard polyvinylchloride.

As a model of behavior of a zone of the film subjected to the local thermal modification with the shrinkage stress of the film sample was investigated after heat treatment under isometric conditions (with fixed sizes) (Fig. 13.3).

Under these conditions were provided several interconnected relaxation processes depending on pressure and fixing of the sample sizes in the thermoshrinkable film. There was shrinkage (reduction of not fixed sizes of the film sample, decreasing of internal stress at the heating and increasing of internal stress in the heated zone after cooling).

The last process is internal stress increasing during film cooling after heating with the fixed length. This is anomalily. The drop of stress in the thermoshrinkable polyvinylchloride film sample during its cooling after

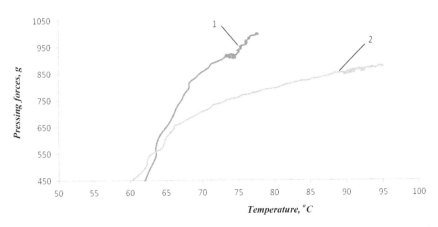

FIGURE 13.3 Temperature dependence of local pressing forces (local shrinkage) of polyvinylchloride film. 1 – film in original state 2 – film after heating under isometric conditions of marking.

heating to the temperature higher than the vitrification temperature does not correspond with the known concepts of relaxation mechanisms [8]. This unconventional "shape memory" effect is shown in the thermome-chanical diagram (Fig. 13.4), it's worthy of special additional researching.

The Fig. 13.4 shows a change of stress in a thermo shrinkable poly-vinylchloride film sample during heating process by hot air with subse-quent cooling in stationary tensometer clips [7].

The tensometric curve can be divided into some characteristic parts according to every heating-cooling cycle. There are two such parts in the first cycle. The first part is the curve of a rapid tension increasing in the film sample with a fixed length at heating to the temperature, conforming with the PVC vitrification temperature (80°C ± 8°C). The achieved level of shrinkage tension of 10.3 g/sm we took for one.

During the cooling of sample to 22 °C can be observed an anomalous increasing of internal strain in the film to the level of 1.4. This strain increasing by 40% compare to the "hot" state conflicts with the known theoretical views of relaxation mechanisms in the stressed deformed

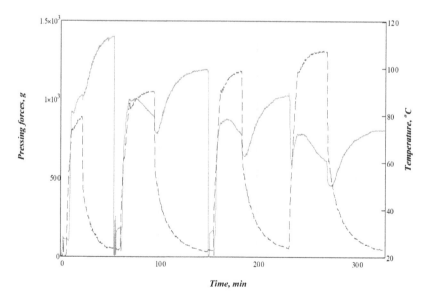

FIGURE 13.4 Stresses' relaxation in thermoshrinkable film during heating cycle of a ring-shaped sample of the film with the fixed annular length under conditions of a monotonous rise of maximum heating temperature.

polymer [8]. If the same sample is heated a few times, the relaxation processes and the subsequent stress would increase like was described above. The same processes would be repeated but with a lower speed and at a higher temperature.

Mismatch of the temperature ranges and speeds of shrinkage stress-relaxation zone of thermomechanical modification of the polymer structure (area of tactile marking) with an interval of temperature and stress relaxation rate in the whole label are the reasons for relief development, which is used for marking.

ACKNOWLEDGEMENT

The present work is supported by the Ministry of Education and Science of Russian Federation, Contract # 2014/87–1064 of the 30th of January 2014.

KEYWORDS

- packaging
- polyvinylchloride
- shape memory
- shrinkable polymer film
- tactile marking

REFERENCES

1. Kondratov, A. P., "Gradient and interval thermoshrinkable materials for counterfeit protection of printing products", "News of Institutions of higher education. Issues of printing and publishing", 4, 57–63 (2010).
2. Patent of RF, number 2498905, "Method of marking by invisible thread of, V.G. Savinovsky", V.G. Savinovsky is patentee, No. 2011142943/12, patent application on 24.10.2011, publication on 20.11.2013, Bulletin No. 32, 20.11.2013.
3. Braille printing, website of company "P -print" [electronic resource] URL: http://r-print.net/uslugi/trafaretnaya_pechat/nanesenie_shrifta_brajlya.
4. Tactile products for the people with weak eyesight, website of company "Blind artist" [electronic resource] URL: http://www.pavlova.ws/publication/25.php.

5. Kondratov, A. P., Bablyuk, E. B., Patent of RF, No. 2448366, G06K9/00, "Method of counterfeit protection of printing products, patentee is Federal state budgetary educational establishment of higher vocational training "Moscow State University of Printing Arts", No. 2010108169 /08, patent application on 04.03.2010, publication on 20.04.2012, Bulletin No.11, 20.04.2012.
6. Kondratov, A. P., Patent of RF, No. 2444543, C08J5/18, "Thermosensitive polymer material for printing industry", Patentee is Federal state educational establishment of higher vocational training "Moscow State University of Printing Arts", No. 2010123588/05, patent application on 09.06.2010, publication on 10.03.2012, Bulletin number 32 of 20.11.2013
7. Kondratov, A. P., Zachinyaev, G. M., Physical simulation of fixation process of thermoshrinkable labels on cylindrical packaging in automatic applicators. "News of Institutions of higher education. Issues of printing and publishing", 3, 31–39 (2013).
8. Zaikov, G. E., Kozlov, G. V., Shustov, G. B. In book: Physical Chemistry of Low and High Molecular Compounds. Ed. Zaikov, G., Dalinkevich, A. New York: Nova Science Publishers, Inc. 81–90 (2004).

CHAPTER 14

THE TRANSPORT PROPERTIES OF MODIFIED FILMS WITH CEFOTAXIME AND 4-METHYL-5-OXYMETHYLURACYL

T. SH. KHAKAMOV, D. V. FEOKTISTOV, L. A. BADYKOVA, and R. KH. MUDARISOVA

Institute of Organic Chemistry, Ufa Scientific Center, Russian Academy of Sciences, Ufa, Bashkortostan, Russia; Bashkir State Medical University, Ufa, Bashkortostan, Russia; E-mail: badykova@mail.ru

CONTENTS

ABSTRACT

Interaction of citrus pectin with was studied by spectrophotometric methods. New film coatings with these medicinal preparations for mesh implants were produced on the basis of the citrus pectin–polyvinyl alcohol matrix. It was demonstrated that supramolecular formations of the polymeric formulation affect the transport properties of the films. A high antibacterial activity of film coatings with cefotaxime was revealed.

14.1 INTRODUCTION

The problem of a prolonged release of medicinal preparations in medicine and, in particular, in surgery is rather topical and various methods have been used for its solution [1, 2]. One of these consists in development of new sheaths for mesh polypropylene implants with controlled release of medicinal preparations. The sheaths are polymeric films that can be used for dosed introduction of medicinal preparations.

Of indubitable scientific and practical interest in development of sheaths of this kind are pectin substances, natural polymeric derivatives of 1,4-α-D-galacturonic acid, containing partly esterified carboxyl groups [3].

The polymers used in medicine must satisfy a number of rather stringent requirements: nontoxicity, biocompatibility, and high purity and homogeneity. Pectin's satisfy these requirements and, in addition, exhibit a high physiological activity by themselves [4], including immunomodulatory and antimicrobial effects [5]. Just this circumstance served as a reason why pectin is used as a sheath for mesh implants.

However, the rapid solubility of pectin films in water is factor restricting their application. To improve the physical characteristics of these films, their modification with polyvinyl alcohol was performed.

Polyvinyl alcohol (PVA) is widely used at present as a drug carrier [6]. Addition of a medicinal molecule to a matrix based on citrus pectin and PVA will make it possible to obtain such effects as prolonged release, change in solubility, and lowered toxicity of an immobilized medicinal preparation. An also important circumstance is that this matrix is soluble in biological fluids of an organism and, the molecular mass of the polymers used being low, it will be well washed-out from the organism [7].

The goal of our study was to develop and examine polymeric films based on citrus pectin and polyvinyl alcohol, with controlled release of medicinal preparations, cefotaxime and 4-methyl-5-oxymethyluracyl.

14.2   EXPERIMENTAL PART

The analyzes (measurements, calculations) were made on the equipment of the Collective Usage Center "Chemistry," Institute of Organic Chemistry, Ural Scientific Center, Russian Academy of Sciences.

As objects of study served citrus pectin (CP) with molecular mass of 162000 and PVA with molecular mass of 7200 (content of vinyl acetate units 0.8–2.0%). Were chose for immobilization the cefotaxime (CFT) antibiotic and 4-methyl-5-oxymethyluracyl immune stimulating preparation (pentoxyl) [8].

The medicinal preparations were used without additional purification. UV spectra of aqueous solutions of CP with CFT and pentoxyl were measured with a Specord M-40 spectrophotometer at wavelengths in the range 220–350 nm in quartz cuvettes. IR spectra were recorded with a Shimadzu spectrophotometer at wave numbers in the range 700–3600 cm^{-1} in Vaseline oil. The size of supramolecular particles of the CP + CFT mixture was determined in aqueous solutions by the method of laser light scattering on a Shimadzu Sald 7101 instrument. The measurements were made in Sald-BC quartz cuvettes with vertical mechanical agitation. An inert atmosphere (dry purified argon) was created in the cuvette and measuring chamber.

Model films based on PVA and CP + PVA were formed by casting the solutions onto a glass substrate, followed by evaporation of the solvent in a vacuum over P_2O_5. Further, the films were dried in a vacuum at a temperature of 25 °C to a residual moisture content of 5–7%. The concentrations of CFT and pentoxyl in all the films were 0.2 mol^{-1} CP, and the film thickness was 0.5 mm. The CP:PVA ratios were 1:0.2, 1:0.4, and 1:0.6.

We determined the composition of the compounds being formed by the spectrophotometric methods of isomolar series and molar ratios [9]. The total concentration of the polysaccharide and medicinal substances in an isomolar series was constant (1×10^{-4} M). The [CP]:[CFT] and [CP]:[pentoxyl] molar ratios were varied from 50:1 to 1:20. In the solution series with constant CFT or pentoxyl concentrations of 1×10^{-4} M, the CP concentration was varied from 0.25×10^{-4} to 1×10^{-2}. The ionic strength was constant (0.1). This constancy was provided by introduction of a calculated amount of chemically pure NaCl into the solutions. The kinetics of CFT and pentoxyl release into the aqueous medium was studied spectrophotometrically on the basis of the optical density of the solutions at wavelengths of 245 and 312 nm for CFT and pentoxyl, respectively.

The polymeric formulations for determining the antibacterial activity were prepared as follows: a sterile vessel was charged with 100 mL of a 2% aqueous solution of CP and a prescribed amount of the PVA solution at a CP:PVA ratio of 1:0.2. The mixture was agitated for 1 h at a temperature of 25 °C, and then 0.5 g of CFT or pentoxyl was introduced. It was found experimentally that a larger amount of the medicinal preparation leads to loss of elasticity by the polymeric films. The mixture was again agitated with a magnetic rabble until the preparations were fully dissolved. A 0.1-g portion of glycerol was added to make the films elastic. The solution was agitated for 5 min. After that it was poured into a sterile mold in which a mesh polypropylene implant was preliminarily placed so that the solution fully covered the prosthesis. Further, the mold with the solution was placed in a drying box and was kept there at a temperature of 40 °C for three days. On being dried, the implant covered with a polymeric formulation was sterilized in an ozone chamber for 1 h. To study the antibacterial activity, we prepared our implant samples with new polymeric sheaths of the following composition: PVA + CFT, PVA + pentoxyl, CP + PVA + CFT, and CP + PVA + pentoxyl.

14.3   DISCUSSION OF RESULTS

Introduction of PVA into a CP solution makes it possible to obtain a sufficiently strong and elastic film. It is known that many properties of polymeric systems, including transport properties, are set in the stage of film formation and depend on the size of supramolecular formations in the polymeric mixture [10]. The possibility of controlling the structure and properties of films by varying, for example, the concentration of the solutions used will enable synthesis of films with controllable release of pharmacological preparations.

Figure 14.1 shows the differential size distribution of polymer molecules for aqueous solutions of CP and CP + PVA mixture with CP:PVA ratios of 1:0.2, 1:0.4, and 1:0.6.

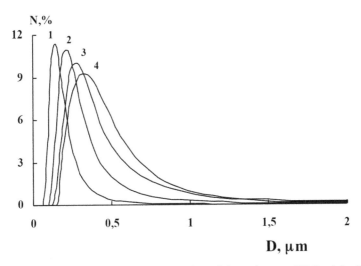

FIGURE 14.1   Macromolecule size distribution of the polymers. (*D*) Particle diameter and (N) number of particles. (*1*) Pure CP; CP:PVA ratio: (*2*) 1:0.2, (*3*) 1:0.4, and (*4*) 1:0.6. Solvent water.

The starting pectin is characterized by a rather narrow size distribution of its molecules (most have sizes within the range <0.15 μm). Adding PVA to the CP solution and further raising the CP:PVA ratio to 1:0.6 result in a wider particle size distribution, with formation 0.1 to 1.5 μm in size present in these systems. It is known that pectin macromolecules possess a high associative capacity [11]. Probably, processes involving changes of molecular associates of CP and PVA occur in polymer mixture solutions upon an increase in the PVA concentration, with the result that coarser structural formations are formed. The interaction of dissimilar molecules in mixtures of CP and PVA associates possibly yields stronger bonds than those in CP associates. The formation of these associates is due to the hydrogen bonding and hydrophobic interactions between macromolecules. According to [12], hydrogen bonds are formed in the CP + PVA mixture between hydroxyl groups of D-galacturonic acid and PVA or between carboxyl groups of D-galacturonic acid and hydroxyl groups of polyvinyl alcohol. Thus, varying the content of PVA in the system, we can change the size of the CP + PVA associates being formed and thereby affect the release of pharmacological preparations from the polymeric matrix.

In fabrication of prolonged-release medicinal preparations, it is necessary to take into account that, when a medicinal substance interacts with a polymeric matrix, their complexation is possible. The presence of hydroxyl and carboxyl groups in CP macromolecules favors the complexation of CP with CFT and pentoxyl. The interaction of CP with these medicinal substances was studied by UV and IR spectroscopies. We first examined the absorption spectra of CFT, pentoxyl, and mixtures of CFT and pentoxyl with pectin in aqueous solutions. Upon an interaction between the components, UV spectra show a shift of the absorption band by 5 nm in the bathochromic direction for CFT and by 10 nm in the hypsochromic direction for pentoxyl, with the absorption peaks becoming stronger. The increase in the intensity of the signals and the shift of the absorption peaks, observed in the electronic absorption spectra upon the interaction of the polysaccharide with CFT and pentoxyl, evidence that they react to give a complex compound.

We determined the composition of the complex compounds obtained by the methods of isomolar series and molar ratios. The results furnished by these methods suggest that complexes of 1:1 composition are formed in a dilute aqueous solution, for example, the complex being formed contains one molecule of a medicinal preparation per structural unit of the polysaccharide.

Based on the data obtained using the method of molar ratios, we calculated the stability constants β_k. CP forms with medicinal preparations medium-stability complexes ($\beta_k = 38.6 \times 10^{-4}$ M^{-1} for CP + CFT and 70.3×10^{-4} M^{-1} for pentoxyl).

The results obtained by analysis of the UV spectra are confirmed by IR spectroscopic data. We examined spectra of the starting substances and the resulting complexes, with assignment of the most important bands. The IR spectrum of the CP + pentoxyl complex show a shift of the absorption peak associated with stretching vibrations of OH groups (3600–3050 cm^{-1}) by 40–45 cm^{-1} and C–O groups of the glycoside (1135–1070 cm^{-1}) by 5–10 cm^{-1} to lower frequencies. The absorption band associated with stretching vibrations of the carbonyl group at 1732 cm^{-1}, characteristic of pectins, disappears in the spectrum of the complex. In addition, a shift to lower frequencies is observed absorption bands related to stretching vibrations of C=O (by 30 cm^{-1}), C–N (by

11 cm^{-1}, and N–H groups (by 30 cm^{-1}) of pentoxyl. Probably, the complexation occurs between the carboxyl and hydroxyl groups of CP and N–H and C–N groups of pentoxyl.

The interaction of CP with CFT is characterized by changes in the IR spectra at 3600–3050 cm^{-1}. The intensity of the absorption band associated with stretching vibrations of OH groups substantially decreases and the band is flattened, which indicates that free hydroxyl groups become associated when hydrogen bonds are formed. The absorption band at 1732 cm^{-1}, related to stretching vibrations of the C=O bond in the carboxyl group or polysaccharide ester, is shifted in the complex by 31 cm^{-1} to longer wavelengths. The group of bands at 1000–1200 cm^{-1}, characteristic of vibrations of the pyranose rings of pectin, has a lower intensity and is flattened in the case of the complex [13].

These changes in the spectra indicate that the polymeric carrier interacts with the medicinal preparations via hydrogen bonds, with the glycoside and hydroxyl groups of CP and CFT and the N–H and C–N groups of pentoxyl involved. Probably, the complexes are formed via coordination of a molecule of a medicinal substance and a carbohydrate unit of the polysaccharide, with intermolecular hydrogen bonds formed between these.

It is known from published data that the rate and extent of release of pharmacological preparations from polymeric systems can be directly controlled by varying the conditions in which the polymeric matrix is formed [14–16]. Because the sizes of the associates being formed vary with the CP PVA ratio, it can be assumed that the transport properties of films obtained at different ratios between the polymers will also be not the same.

The results of the investigations show that the antibiotic is released at the highest rate from the PVA-based film. Upon an increase in the PVA content of the composite to 0.6, the release of CFT from films based on CP + PVA becomes approximately two times slower. This effect can be attributed to the change in the supramolecular structure of the matrix upon modification of the films with polyvinyl alcohol. As shown above, introduction of PVA into a pectin film favors formation of hydrophobic associates and just this circumstance probably decelerates the release of CFT.

The maximum values of the release of CFT from CP + PVA films at various CP: PVA ratios are listed in the Table 14.1. Analysis of these results shows that the medicinal preparation is completely released in several hours from the PVA-based film and during a week from films based on CP + PVA.

Figure 14.2 shows dependences of the release of pentoxyl on the ratio between the components in the matrix.

It can be seen that the fundamental aspects of pentoxyl release from the given films are similar to those observed for the complex of CP + PVA with

TABLE 14.1 Maximum Release of Cefotaxime Relative to a Film Containing No CP

Release duration, days	Maximum yield of cefotaxime, %, at indicated CP: PVA ratio			
	1.0	0.2	0.4	0.6
1	97.2	80.1	53.5	48.1
2		64.3	57.7	51.3
3		79.2	68.8	57.2
5		89.6	73.6	64.2
7		97.7	79.2	69.2

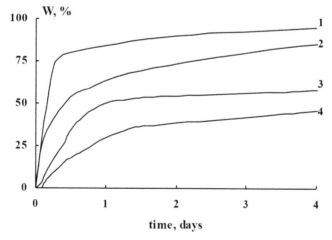

FIGURE 14.2 Dependence of the pentoxyl release from polymeric matrices on the PVA concentration in the polymeric matrix. Time and (*W*) release (*1*) Pure CP; CP: PVA ratio: (*2*) 1: 0.2, (*3*) 1: 0.4, and (*4*) 1: 0.6.

CFT: an increase in the content of PVA in the system makes the release of the antibiotic 2–2.5 times slower. Pentoxyl is released at the highest rate from the PVA-based film. Introduction of minor amounts of PVA into the formulation can make the prolongation duration of the medicinal preparation 1.5 times longer. At a CP:PVA ratio of 1:0.6, pentoxyl is released from the films during 10 days.

Thus, our experimental results demonstrate that modified films based on citrus pectin with cefotaxime and pentoxyl can prolong the action of immobilized medicinal preparations. The degree and rate of release of a medicinal preparation from the films directly depends on the size of the associates being formed.

A study of the antimicrobial properties, strength characteristics, and tissue response to implantation of these endoprosthetics demonstrated the antibacterial effect of the implant is retained during up to five days and the implant itself satisfies all the requirements imposed on modern plastic materials for hernioplastics. Results obtained by seeding of implant samples on test microflora cultures demonstrated that the largest growth retardation zone, compared with the standard disks with antiobiotics, is observed for samples based on CP + PVA + CFT. The samples based on CP + PVA + pentoxyl were less effective, but still exhibited a more pronounced effect than free antibiotics.

The results of the tests we performed show that the sheaths we developed as coatings for implants are promising because of providing reliable satisfactory results of curing.

14.4 CONCLUSIONS

1. It was found that the interaction of pectin with cefotaxime and 4-methyl-5-oxynethyluracyl yields complex compounds of 1:1 composition.

2. Modified polymeric films based on citrus pectin with cefotaxime and 4-methyl-5-oxynethyluracyl were obtained. It was found that the pharmacokinetics of release of the medicinal preparations from the films is determined by the content of polyvinyl alcohol in a film. It was unambiguously demonstrated for all the systems under

study that raising the content of polyvinyl alcohol in a formulation is accompanied by a regular decrease in the rate and degree of release of the immobilized medicinal preparations from the films.
3. Microbiological tests revealed a high antimicrobial activity of the modified film based on citrus pectin with cefotaxime.

KEYWORDS

- 4-Methyl-5-Oxymethyluracyl
- cefotaxime
- citrus pectin
- pentoxyl
- polymeric matrix

REFERENCES

1. Solovskii, M. V., Dubkova, V. I., Krut'ko, N. P., et al., *J. Applied Biochemistry and Microbiology (in Rus)*, 2009, vol. 45, no. 2, pp. 248–251.
2. Shmak, G., Duchk, V., and Pisanova, E., *Fibre Chemistry (in Rus)*, 2000, no. 1, pp. 39–45.
3. Ovodov, Yu. S., *Chem. nat. compaunds (in Rus)*, 1975, no. 3, pp. 300–315.
4. Kochetkov, N. K., *Methods of Carbohydrate Chemistry,* Moscow: World, (in Rus), 1967.
5. Komisarenko, S. N. and Spiridonov, V. N., *Vegetable resources (in Rus)*, 1998, vol. 34, no. 1, pp. 111–119.
6. *Polymers in Pharmacy*, Tentsova, A. A. and Alyushin, M. T., Eds., Moscow: Medicine, (in Rus), 1985.
7. Chazov, E. I., Smirnov, V. N., Mazaev, A. V., and Torchilin, V. P., *J. of the national Chemical Society of the name, D.I. Mendeleev (in Rus),* 1985, vol. 30, no. 4, pp. 365–372.
8. Mashkovskii, M. D., *Medicinal Preparations*, Moscow: New wave, (in Rus), 2008.
9. Bulatov, I. P. and Kalinkin, M. I., *Manual of Photometric Analysis Techniques*, Leningrad: Chemistry, (in Rus), 1986.
10. Mudarisova, R. Kh., Badykova, L. A., Koptyaeva, E. I., and Monakov, Yu. B., *Proceedings of the institutes of higher education. Series chem. and chem. technology (in Rus),* 2011, vol. 54, no. 5, pp. 78–81.
11. Shelukhina, N.P., *Scientific Foundations of the Technology of Pectin*, Frunze: Ilim, (in Rus), 1988.

12. Rashidova, S. Sh., Voropaeva, I. L., Mukhamedzhanova, M. Yu., et al., *Russ. J. Appl. Chem., (in Rus),* 2002, vol. 75, no. 7, pp. 1136–1140.

13. Ioffe, B. V., Kostikov, R. R., and Razin, V. V., *Physical Methods for Determining the Structure of Organic Molecules, Leningrad: Publ. of the Leningrad. Univ. (in Rus),* 1976.

14. Zakharova, E. L. and Oktyabr'skii, F. V., *Pharm. Chem. J. (in Rus),* 1993, vol. 27, no. 11, pp. 14–21.

15. Kulish, E. I., Kuzina, L. G., Chudin, A. G., et al., *Russ. J. Appl. Chem., (in Rus),* 2007, vol. 80, no. 5, pp. 810–812.

16. Suleimenov, I. E., Budtova, T. V., and Iskakov, R. M., *Polymeric Hyrogels in Pharmaceutics: Physicochemical Aspects,* Almata-SPb (in Rus), 2004.

CHAPTER 15

A STUDY ON THE PROCESS OF ENZYMATIC DEGRADATION OF CHITOSAN IN ACETIC ACID SOLUTION IN THE PRESENCE OF AMIKACIN SULFATE

E. I. KULISH, I. F. TUKTAROVA, V. V. CHERNOVA, and G. E. ZAIKOV

Bashkir State University, Russia, Republic of Bashkortostan, Ufa, 450074, St. Zaki Validi, 32, E-mail: tuktarova_irina@rambler.ru

CONTENTS

ABSTRACT

The process of enzymatic degradation of chitosan in acetic acid solution was analyzed within the method of capillary viscosimetry to evaluate the influence of including amikacin sulfate on the rate of the process. It was found that the process of enzymatic degradation of chitosan in the presence of the drug is well described within the scheme of the Michaelis–Menten.

15.1   INTRODUCTION

Natural polymer chitosan (CHT) is a promising material for producing bioactive protective biodegradable polymer materials for medical applications. One of the promising areas of using CHT is obtaining from CHT solutions bioactive film materials for medical applications, including for the treatment of burns, which are capable to biodegradation [1, 2]. Adding into the CHT film antibiotic amikacin sulfate can reduce the possibility of suppuration and contributes to the suppression of infection. In the case when CHT used as compositions with drugs must be considered the fact that the drugs can often have a major influence on the rate of enzymatic reaction that dictates the necessity of individual kinetic studies. The necessity for this work due also to the fact that, in contrast to the relatively well-developed learning tasks of enzymatic degradation of CHT under the action of specific enzymes – chitinase and chitosanase [2], the process of the enzymatic degradation of CHT under the influence of nonspecific enzymes (such as hyaluronidase which is present at the wound surface), not quantitatively studied.

15.2   EXPERIMENTAL PART

The objects of investigation were the CHT specimen produced by the company "Bioprogress" (Russia) with a molecular weight with M_{sd}=113,000 and antibiotic amikacin sulfate (AMS) produced by the company "Sintez" (Russia). As the enzyme preparation was used hyaluronidase enzyme preparation production "Microgen" (Moscow, Russia). The concentration of the enzyme preparation was 0.1, 0.2, and 0.3 g/L. Acetic acid of 1% concentration was used as a solvent. CHT concentration in solution for the enzymatic degradation process C_{ed} ranged from 0.1 to 5.0 g/dL. Antibiotic AMS, dissolved in a small amount of water was added to the CHT solution in a molar ratio CHT:AMS equal to 1:0.01 and 1:0.1.

The degree of enzymatic degradation was evaluated by the degree of intrinsic [η] viscosity decrease according to the viscosimetry data. Intrinsic viscosity in solution of acetic acid was determined at 25 °C using the method of Irzhak and Baranov [3].

To determine the initial values of the intrinsic viscosity $[η]_0$ of CHT a solution of concentration C=0.15 g/dL was used. To determine the values [η] of intrinsic viscosity during the enzymatic degradation, CHT solution in acetic acid to which was added a solution of an enzyme preparation maintained for a certain time. Thereafter, the process of enzymatic degradation was quenched by boiling the starting solution for 30 min in a water bath. Further, the initial concentration of the solution prepared by diluting a solution of C_{ed} to determine the intrinsic viscosity, the concentration C = 0.15 g/dL. The process of enzymatic degradation was carried out at a temperature of 36 °C. The initial speed of the enzymatic degradation of CHT V_0 was evaluated in the linear region for the fall of its intrinsic viscosity [η] and calculated by the formula [4]:

$$V_0 = \frac{C_{ed} K^{\frac{1}{\alpha}} ([η]_t^{-\frac{1}{\alpha}} - [η]_0^{-\frac{1}{\alpha}})}{t} \tag{1}$$

where is C_{ed} – concentration of chitosan in solution subjected enzymatic degradation, g/dL; t – degradation time, min., K and α – constants in the equation of Mark-Houwink-Kuhn $[η] = KM^{\alpha}$, M – molecular weight of CHT.

To determine the constants in the equation of Mark-Houwink-Kuhn it is necessary to calculate the values of the initial velocity of the enzymatic degradation according to Eq. (1), the sample was fractionated on 10 fractions ranging in molecular weight from 20,000 to 150,000 Daltons. The absolute value of molecular weight of CHT fractions was determined by a combination methods sedimentation velocity and viscosimetry. The molecular weight of the fractions was determined by the formula:

$$M_{sη} = \left(\frac{S_0 η_0 [η]^{1/3} N_A}{A_{hi} (1 - \bar{v} \rho_0)} \right)^{3/2} \tag{2}$$

TABLE 15.1   The Values of the Constants in the Equation of Mark-Houwink-Kuhn for CHT Solution in 1% acetic acid (Temperature 25 °C)

The molar ratio of components CHT:AMS	α	$K \times 10^4$
1:0	1.02	0.56
1:0.01	0.89	1.73
1:0.10	0.83	3.19

where is S_0 – sedimentation constant; η_0 – dynamic viscosity of the solvent, which is equal to 12.269×10^{-2} PP; $[\eta]$ – intrinsic viscosity, dL/g; N_A – Avogadro's number, equal to 6.023×10^{23} mol^{-1}; $(1 - \nu \rho_0)$ – Archimedes factor or buoyancy factor, ν – the partial specific volume, cm^3/g, ρ_0 – density of the solvent g/cm^3; A_{hi} – hydrodynamic invariant equal to 2.71×10^6. The constants determined for CHT in 1% acetic acid solution are shown in Table 15.1.

15.3   RESULTS AND DISCUSSION

Figure 15.1 (curves *1* and *2*) shows the dependence of the intrinsic viscosity of the CHT solution on the time of standing with enzyme.

It is evident that with increasing exposure time to the enzyme solution CHT, the viscosity decreases regularly, indicating a decrease in the molecular weight of CHT. The most significant drop in viscosity occurs in the initial period. Increasing the concentration of the enzyme preparation leads to a natural increase in the rate of incidence of the intrinsic viscosity.

As it was shown by the study, the observed dependence of the initial rate of enzymatic degradation of the substrate concentration can be described within the scheme of the Michaelis–Menten (Fig. 15.2). Submission to double-check the coordinates (graphical method of Line weaver-Burk) can accurately determine the values of the Michaelis constant K_m and maximum speed of enzymatic degradation V_{max} (Table 15.2).

Should pay attention to a very large value of the Michaelis constant $K_m = 3.4$ g/dL, which is significantly higher than value of $K_m = 0.03$ g/dL, as it was defined in [4] for carboxymethyl cellulose at the presence of cellulose system of Geotrilium candidum. This fact is obviously due to the fact

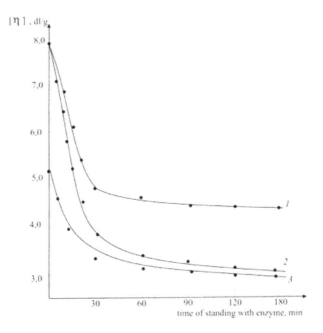

FIGURE 15.1 Dependence of the intrinsic viscosity of the CHT solution on the time of standing with enzyme preparation concentration 0.2 *(1)* and 0.3 *(2)* g/L, and CHT-AMS with enzyme preparation concentration 0.3 *(3)* g/L.

that the enzyme is used is the nonspecific with respect to the CHT and the fact that the conditions of degradation did not correspond to the temperature and the pH optimum of the enzyme hyaluronidase.

Addition of AMS to the CHT solution does not affect the appearance of the decline curve of the intrinsic viscosity CHT on the exposure time with the enzyme (Fig. 15.1, curve *3*). Also does not change the curve of the dependence initial rate of enzymatic degradation of CHT concentration in solution. However, comparison of the values V_{max} and K_m obtained for the system CHT-AMS, with values V_{max} and K_m for individual CHT (Table 15.2) shows that the addition of the AMS affects both the value V_{max} and on the value of K_m.

Moreover, the values of V_{max}/K_m, which actually have the meaning of pseudofirst order rate constants for the reaction of the HTZ + enzyme → product + enzyme also change regularly with increasing content of AMS in the mixture.

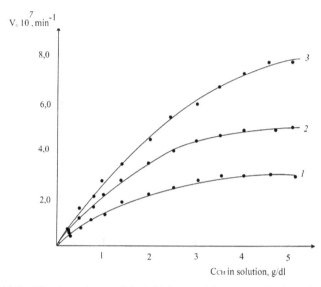

FIGURE 15.2   The dependence of the initial rate of the enzymatic degradation of CHT *(1)* with an enzyme preparation at a concentration 0.1 *(1)*, 0.2 *(2)* and 0.3 *(3)* g/L of CHT concentration in solution.

TABLE 15.2   The values of the Constants of the Enzymatic Degradation in the Equation of Michaelis–Menten For CHT Solutions in 1% Acetic Acid

The molar ratio of components CHT:AMS	C_e, g/L	K_m, g/dL	V_{max}, 10^6, min^{-1}	V_{max}/K_m, (g/dl×min)
	0.1	3.37	0.50	0.15
1:0	0.2	3.47	0.90	0.26
	0.3	3.42	1.50	0.44
	0.1	4.09	0.43	0.10
1:0.01	0.2	4.01	0.80	0.19
	0.3	4.03	1.20	0.30
	0.1	4.46	0.31	0.07
1:0.1	0.2	4.48	0.58	0.13
	0.3	4.50	0.88	0.19

Thus, for the first time identified by the kinetic characteristics of the activity of the hyaluronidase enzyme for the enzymatic destruction reaction CHT solution of 1% acetic acid in the presence and absence of antibiotic amikacin sulfate. The fact of reducing V_{max} and increased K_m and regular decreasing in V_{max}/K_m, which takes place when amikacin sulfate added to the CHT solution, indicates the possibility of delayed catalytic conversion of the substrate in a ternary complex enzyme-CHT-AMS.

ACKNOWLEDGEMENTS

This work was financially supported by the Ministry of Education of the Russian Federation on the theme "Development of physicochemical bases of creation of new polymeric materials for biomedical applications with controlled sorption, rheological and structural and physical characteristics on the basis of natural and synthetic polymers."

KEYWORDS

- amikacin sulfate
- chitosan
- enzymatic degradation
- hyaluronidase

REFERENCES

1. Tapan Kumar Giri, Amrita Thakur, Amit Alexander, Ajazuddin, Hemant Badwaik, Dulal Krishna Tripathi (2012). Modified chitosan hydrogels as drug delivery and tissue engineering systems: present status and applications. Acta Pharmaceutica Sinica B, 2(5), 439.
2. Hiroshi Ueno, Takashi Mori, Toru Fujinaga (2011). Topical formulations and wound healing applications of chitosan. Advanced Drug Delivery Reviews, 52(2) 105.
3. Baranov, V. G., Brestkin Yu. V., Agranova, S. A., Pinkevich, V. N. (1986). Behavior of macromolecules in polystyrene "thickened" good solvent. Polymer Science, 28B(10), 841.
4. Rabinovich, M. L., Klesov, A. A., Berezin, I. V. (1977). Kinetics of action of cellulitics enzyme action Geotrilium candidum. Viscometric analysis of the kinetics of hydrolysis of carboxymethyl cellulose.. Bioorganic chemistry, 3(3), 405.

CHAPTER 16

A STUDY ON CARBON BLACK LOCALIZING AT THE INTERFACE BETWEEN TWO POLYMERIC PHASES

A. E. ZAIKIN, V. V. MOLOKIN, and S. A. BOGDANOVA

Department of Plastics Technology, Kazan State Technological University, Kazan, Russia

CONTENTS

ABSTRACT

The causes and conditions necessary for carbon black accumulation at the interface in heterogeneous binary polymer blends have been studied, and some thermodynamic and kinetic facets of the phenomenon have been elucidated. The migration of carbon black particles from a phase to the interface has been to be governed by the thermodynamics of wetting the particles by polymer phases and is affected by the thermodynamic and kinetic peculiarities of macromolecule adsorption on a solid surface.

16.1 INTRODUCTION

The phenomenon of carbon black (CB) particles gathering at the interface in binary heterogeneous polymer blends is not only of fundamental interest, but has a practical aspect as a method for improving the electric conductivity of composite polymer materials [1–12] and the mechanical properties of polymer blends with a low interfacial adhesion [13–15]. The works examining the causes of this phenomenon are few in number. Most of authors only note that some part of a filler tends to accumulate at the interface. Contradictory assumptions of the conditions necessary for such localization are made.

Some workers [1–2] argue that a necessary condition for such localization is the essential difference between the energies of interactions of polymer components with a surface of powder particles. Others [5–8], in contrast, suppose that this phenomenon takes place only when the polymeric components of a blend have low and approximately equal energies of the interaction with the filler surface. Some authors [3] hold that a filler is driven out to the interface due to the crystallization of polymer components of a mixture.

Sumita et al. [9] reason, that the causes of filler localizing at the interface reduce to the well-known thermodynamic conditions for solid particles to reside at the interface between two nonmiscible liquids.

16.2 EXPERIMENTAL PART

The polymers used are characterized in Table 16.1. High-pressure polyethylenes (PE) with different values of the melt-flow index (MFI) also were used. The viscosity values of PE are given in Fig. 16.4. Poly(methyl methacrylate) (PMMA) of various viscosity was prepared by fractionating the initial polymer by molecular weight. The fractionation was carried out by stepwise polymer sedimentation from a chloroform solution using hexane. The viscosity values of the obtained PMMA samples are given in Fig. 16.4.

Carbon black under study (No.254) had specific surface area 250 m^2/g, average size and density of primary aggregates 28 nm and 1.8 g/cm^3,

TABLE 16.1   The Characteristic of Used Polymers

Polymer	Density g/cm³	Viscosity Pa·s (413 K, 15 s⁻¹)	Molecular weight, ×10⁻³	Comments
High Pressure Polyethylene (PE)	0.922	2900	37 (η)	MFI =2.1 g/10 min (at 463 K; 2.16 kg)
Polypropylene (PP)	0.91	2500 (at 463 K)	-	MFI =3.2 /10 min (at 503 K; 2.16 kg)
Polyurethane (PU)*	1.21	3500	-	OH: NCO =1.01
Polyisobutylene (PIB)	0.91	11,300	118 (η)	
cis-1,4-Polybutadiene (PBD)	0.92	-	104 (η)	Mooney – 48 (373 K)
Polystyrene (PS)	1.05	9800 (433 K)	190 (w)	
Polymethylmethacrylate (PMMA)	1.19	11200 (at 433 K)	150 (η)	
Copolymer of Ethylene and Vinylacetate (EVA):	0.950	980	15.5 (w)	Containing 28.9 wt.% Vinyl-acetate
Copolymer of Butadiene and Acrylonitrile (BNR)	0.986	Moony – 54 (at 373 K)	220 (w)	Containing 40 wt.% Acrylonitrile
Polyvinylacetate (PVA)	1.19	-	140 (η)	
Polydimethylsiloxane (PDMS)	0.98	-	560 (η)	
Polychloroprene (PCP)	1.22	-	170 (η)	Mooney – 62 (373 K)

MFI is the melt-flow index, η is the viscosity-average molecular mass, w is the weight-average molecular mass.

*PU was prepared on the basis of polyoxytetraethylene glycol, 4,4 – diphenylmethane diisozyanat and 1,4 – buthane diol.

correspondingly, and specific volume electric resistance 1.5×10^{-3} Ohm×cm (with a density of 0.5 g/cm³).

Low-molecular weight liquids were purified and distilled. The values of heat of carbon black wetting by liquids {ΔH} were measured by means of a DAK1–1A calorimeter at 298 K.

Polymers were mixed with carbon black in melt at 433±5K (for PP at 453 K) using laboratory rolls. The mixing was carried out in two stages. At the first stage the whole portion of CB was mixed with one of polymers, and then the second polymer was added. The mixing time at each stage was five minutes.

Polymer samples for measurements of the specific volume electric resistance (ρ) were prepared by molding using a hydraulic press at 443 $\pm$ 3 K (for PP at 463 K) under a pressure of 30 MPa for 300$\pm$5 seconds. If ρ was less than 1×10^6 Ohm$\times$cm, it was measured potentiometrically at 293 K (ISO 1853–75) using stripes 100 mm long, 10 mm wide and 1.2 mm thick. The potential difference was recorded for a 20 mm site so that contact resistance was eliminated. The results scattering between parallel experiments was $\pm$18%. For the electric resistance exceeding 1×10^6 Ohm$\times$cm ρ was measured according to the ISO 2878–78 procedure on plates 1.2 mm thick with an area of 16 cm^2. In this case stainless steel electrodes were pressed over the whole surface to both sides of a plate. The measurements were made using an E6–13 tera-ohmmeter with a 10 V potential difference between the electrodes. The scattering in ρ values between parallel experiments was $\pm$31%.

The carbon black distribution in polymer blends was examined by means of optical microscopy in transmitted light on thin (1–5 μm thick) sample slices according to the procedure described in Ref. [11].

The bond strength between a filler and polymer was assessed by the exfoliation force (F) of the polymer under study from filler particles fixed in a matrix of another polymer, polypropylene [14]. In this order carbon black was stirred into molten PP, and a plate was molded from this composition. The surface of the plate was treated with an abrasive to remove a polymer layer and expose the carbon black surface. Then the plate was coated from solution by a layer of the polymer under study. After removing a solvent the force of the coating polymer exfoliation from the plate surface was measured.

16.3   RESULTS AND DISCUSSION

It is known that carbon-black filled polymers conduct electric current only with the concentration of carbon black exceeding the threshold of percolation (φ_p). In heterogeneous polymer blends carbon black is distributed nonuniformly between polymer phases. If the concentrations of carbon black in both phases of a blend are lower than φ_p, the blend can conduct electric current only subject to the condition that the part of carbon black is localized at the interface and its concentration here reaches the percolation threshold [12]. So if the concentration of carbon black in both phases of a blend is only slightly lower than φ_p, even a minor accumulation of carbon black at the interface confer conductivity on the polymer blend. This enables the extent of carbon black aggregation at the interface to be judged by the conductivity value.

A variety of blends (Table 16.2) was analyzed under the condition that the carbon black is localized in a single polymer phase and its concentration in this phase is slightly below φ_p. To confine carbon black within one of the phases of a blend, the sequence of carbon black mixing with polymers was altered. Carbon black was first introduced into one of polymer components of a blend, and only then another component was added. A filler is known to retain almost entirely in that phase of a heterogeneous blend where it was introduced initially [1, 2, 10–12].

The localization of carbon black at the interface takes place not in all of the blends (Table 16.2). For most of blends the presence or absence of this phenomenon depends on the sequence of components mixing.

All the blends in Table 16.2 may be divided into three groups based on the presence of the super additive electrical conductivity and the effect of blending sequence on electric conductivity.

The first group consists of blends where the effect of super additive electric conductivity is observed only with a certain sequence of components blending.

The second group involves blends where a lowered ρ value is observed only with preliminary carbon black introduction into either of two polymer components, but the sequence of components mixing strongly affects the degree of ρ lowering.

The blends, which do not conduct electric current with any sequence of components blending, constitute the third group.

The study of the compositions by optical microscopy shows that increased electric conductivity takes place only for those blends and mixing procedures for which the carbon black localization at the interface is observed (Fig. 16.1).

TABLE 16.2 Electrical Conductivity of the (P1+CB) +P2 Systems

Phase P1 (50 vol. %)	Phase P2 (50 vol. %)	φ_p CB for a phase P1, vol. %	The contents CB in a phase P1, vol. %	ρ of a phase P1, Ohm·cm	ρ of a phase P2, Ohm·cm	ρ of a blend, Ohm·cm
Blends in which CB localization at the interface takes place only from one of two phases.						
PE+CB	PU	6.8	5.3	$>1\times10^{12}$	3×10^{10}	1.3×10^4
PU+CB	PE	10	8.5	7×10^7	$>1\times10^{12}$	9×10^8
PE+CB	BNR	6.8	5.3	$>1\times10^{12}$	3×10^9	5.9×10^4
BNR+CB	PE	16.5	15.5	4×10^7	$>1\times10^{12}$	2×10^9
PE+CB	PCP	6.8	5.3	$>1\times10^{12}$	3×10^{10}	2×10^4
PCP+CB	PE	18	16	3×10^8	$>1\times10^{12}$	1×10^{11}
PE+CB	PMMA	6.8	5.3	$>1\times10^{12}$	$>1\times10^{12}$	2×10^4
PMMA+CB	PE	15.5	14	$>7\times10^{11}$	$>1\times10^{12}$	$>1\times10^{12}$
PS+CB	PMMA	11.5	9.3	$>1\times10^{12}$	$>1\times10^{12}$	3.2×10^3
PMMA+CB	PS	15.5	14	$>1\times10^{12}$	$>1\times10^{12}$	$>1\times10^{12}$
EVA+CB	PS	5	2.7	1.5×10^{11}	$>1\times10^{12}$	$>1\times10^{12}$
PS+CB	EVA	11.5	9.3	$>1\times10^{12}$	7×10^{11}	7.2×10^2
PE+CB	PDMS	6.8	5.3	$>1\times10^{12}$	$>1\times10^{12}$	3×10^4
PDMS+CB	PE	4	3	$>1\times10^{12}$	$>1\times10^{12}$	$>1\times10^{12}$
PS+CB	PDMS	11.5	9.3	$>1\times10^{12}$	$>1\times10^{12}$	2.2×10^3
PDMS+CB	PS	4	3	$>1\times10^{12}$	$>1\times10^{12}$	$>1\times10^{12}$
PIB+CB	PDMS	8.6	6.5	$>1\times10^{12}$	$>1\times10^{12}$	2×10^7
PDMS+CB	PIB	4	3	$>1\times10^{12}$	$>1\times10^{12}$	$>1\times10^{12}$

Phase P1 (50 vol. %)	Phase P2 (50 vol. %)	φ_p CB for a phase P1, vol. %	The contents CB in a phase P1, vol. %	ρ of a phase P1, Ohm·cm	ρ of a phase P2, Ohm·cm	ρ of a blend, Ohm·cm
Blends in which CB localization at the interface takes place from both phases.						
PE+CB	PS	6.8	5.3	$>1\times10^{12}$	$>1\times10^{12}$	6×10^{9}
PS+CB	PE	11.5	9.3	$>1\times10^{12}$	$>1\times10^{12}$	4×10^{4}
PP+CB	PS	5.3	4.2	$>1\times10^{12}$	$>1\times10^{12}$	6.3×10^{8}
PS+CB	PP	4.2	3.6	$>1\times10^{12}$	$>1\times10^{12}$	6.9×10^{3}
PE+CB	EVA	6.8	5.3	$>1\times10^{12}$	8×10^{11}	1×10^{7}
EVA+CB	PE	5	2.7	1.5×10^{11}	$>1\times10^{12}$	5×10^{8}
Blends in which CB localization at the interface is very small.						
PE+CB	PIB	6.8	5.3	$>1\times10^{12}$	$>1\times10^{12}$	$>1\times10^{12}$
PIB+CB	PE	8.6	6.5	$>1\times10^{12}$	$>1\times10^{12}$	$>1\times10^{12}$
PE+CB	PP	5.9	4.8	$>1\times10^{12}$	$>1\times10^{12}$	$>1\times10^{12}$
PP+CB	PE	5.3	4.2	$>1\times10^{12}$	$>1\times10^{12}$	$>1\times10^{12}$
PE+CB	PBD	6.8	5.3	$>1\times10^{12}$	$>1\times10^{12}$	$>1\times10^{12}$
PBD+CB	PE	9	8	$>1\times10^{12}$	$>1\times10^{12}$	$>1\times10^{12}$

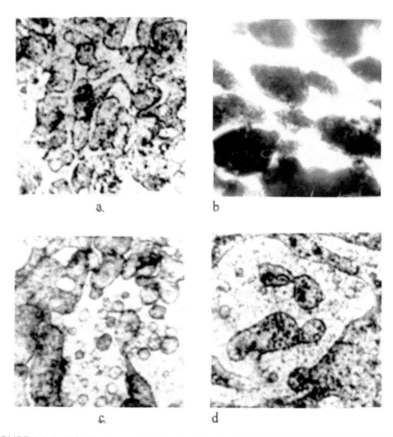

FIGURE 16.1   Distribution of CB in the blends: (PE+CB)+PU (a); (PU+CB)+PE (b), (PS+CB)+PE (c), (PE+CB)+PDMS (d) filled with 2.5 vol. % CB (Optical micrograph). The ratio of polymers is 1:1. Taking the components in brackets means their premixing. Magnification: 1100.

It is reasonable to propose that the process of filler localization at the interface between polymeric phases as well as between low-molecular weight phases is fully controlled by the thermodynamics of the competitive wetting of a solid particle by these phases, which is consistent with Sumita [9].

The phenomenon of solid particles aggregation at the interface between low-molecular weight liquid phases has been much studied, for example, for emulsion stabilized by high-dispersity powders or for powder flotation

[16]. The thermodynamic condition of localization of solid high-dispersity particles between low-molecular weight liquid phases stems from the Young law [16]:

$$-1 < (\sigma_{13} - \sigma_{23})/\sigma_{12} < 1 \qquad (1)$$

where σ_{13} is the interfacial tension between the first liquid and the particle surface, σ_{23} is the interfacial tension between the second liquid and the particle surface, and σ_{12} is the interfacial tension between the liquids.

Such a localization is thermodynamically efficient and will take place with any angle of solid surface wetting by liquid phases except for the angle equal to zero. With the unsatisfied Eq. (1), the particles of a filler would be fully wetted by one of the liquid phases (the phenomenon of spreading) and could not gather at the interface.

The validity of the last proposal is supported by the fact that carbon black localization at the interface is most pronounced for polymer blends with a high surface tension between polymers, that is, in those polymer pairs where thermodynamic gain of particle transfer to the interface is essential [17].

The applicability of the Eq. (1) for polymer blends is difficult to test because of lacking data on interfacial tension between a filler and polymer.

Nevertheless, there is an experimental observation contradictory to the condition above. This is dependence of carbon black localization at the interface on the sequence of components mixing. Indeed, the satisfied Eq. (1) inevitably results in the satisfied thermodynamic conditions of the transfer of particles to the interface both from the first phase ($\sigma_{13} > \sigma_{23} - \sigma_{12}$), and from the second phase ($\sigma_{23} > \sigma_{13} - \sigma_{12}$) [16].

Therefore, from the thermodynamic viewpoint, with the unsatisfied Eq. (1) the aggregation of solid particles at the interface must be observed for any sequence of components mixing. So, for example, the localization of carbon black at the interface of two low-molecular weight liquids does not depend on the sequence of components mixing. Altering the sequence of components mixing can affect only the thermodynamic efficiency of process of filler redistribution from the bulk to the interface, but not the parameters in Eq. (1). It should be noted that in most of publications cited above carbon black was introduced into the blend of polymers, and the

problem of the influence of components blending sequence on the localization of carbon black at the interface was not considered [2, 4–10].

To reveal the causes and conditions necessary for carbon black to localize at the interface there was a need in evaluating the energy of adsorption
interaction of polymers with the filler surface.

However, now there is no simple and reliable method to assess the
efficiency of interaction of polymers with a surface of high-dispersity
powders.

At the same time it is well understood that the energy of adsorption
interaction of low-molecular weight analogs of polymers with a solid surface, which depends on the surface energy of a solid body and the chemical
nature of an adsorbent, is comparable with such interaction for polymers
[18, 19]. This makes it possible to judge qualitatively the relative energy
efficiency of adsorption interaction of corresponding polymers with the
carbon black surface by the interaction energies of low-molecular weight
analogs. The efficiency of the interaction of liquid low-molecular weight
analogs of polymers with the carbon black surface was estimated by the
value of wetting heat (ΔH) (Fig. 16.2).

The studies show (Fig. 16.2) that ΔH of CB is in a certain manner
dependent on the polarity ξ. The polarity of polymers and liquids was calculated as follows:

$$\xi = (\delta_p^2 + \delta_h^2)/\delta^2 \qquad (2)$$

where: δ_p and δ_h are polar and hydrogenous components of the solubility
parameter of a liquid, δ is the solubility parameter of a liquid [20, 21].

With increasing ξ the wetting heat of CB No.254 passed a maximum at
$\xi = 0.33$. This value is consistent with the magnitudes of surface polarity
for two other types of CB calculated on the basis of Hansen's three-dimensional solubility parameter. So, for two different types of CB ξ was equal
to 0.43 and 0.38 [20, 21]. The ΔH increase up to a certain ξ value is due to
the formation of not only dispersion bonds but hydrogen and polar bonds
with oxygen- containing groups on the CB surface as well. The further
fall in ΔH is obviously connected with the decreasing energy of dispersion
interaction between liquids. The fact that the dependence of interaction of
organic liquids with the surface of some metal oxides on the polarity of a

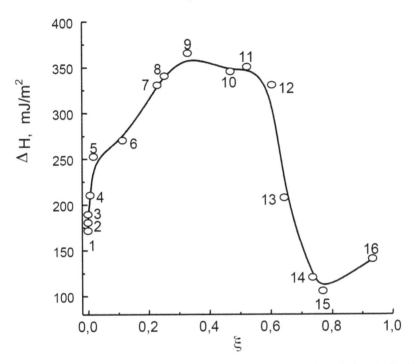

FIGURE 16.2 The heat of CB No. 254 wetting by liquids versus the polarity of a liquid. Carbon tetrachloride (1), n-hexane (2), 2,2,4 – trimethylpentane (3), ethylbenzene (4), toluene (5), 1 – chlorbutane (6), MMA (7), tetrahydrofuran (8), ethylacetate (9), hexanol – 1 (10), dimethylformamide (11), acetonitrile (12), ethanol (13), ethylene glycol (14), glycerin (15), water (16).

liquid passes an extremum was noticed by the researchers [22] (the interaction was assessed by the impregnation of powders by liquids).

The data obtained indicate that the increase in the polarity of polymers up to $\xi = 0.33–0.35$ must result in rising energy of their adsorption interaction with CB.

Besides, the relative efficiency of the interaction between polymer and a filler was judged indirectly by the specific force of exfoliation (**F**) of polymer from a filler [14]. The permissibility and reliability of such estimation for elastic polymers is evidenced by the fact that **F** of these polymers rises linearly with the increasing CB volume fraction in PP [14].

The F values show that with increasing polarity of polymer (Table 16.3) its adhesion to a filler rises. This correlates well on a qualitative level with the estimates of their interaction with CB from the wetting heat data (Table 16.3).

According to the results of these measurements, PE, PIB, PBD, PP, PS, and PDMS have the lowest and closest magnitudes of the energy of adsorption interaction with the CB and aerosil surface. The rest of polymers, judging by wetting heat and exfoliation force (Table 16.3), arrange in the following series according to the increasing efficiency of their interaction with CB and aerosil: EVA < BNR < PCP < PU < PVA.

The analysis shows that when an essential difference in adhesion of polymer components of a blend to CB takes place (PE+PU, PE+BNR, PE+PMMA, PE+PCP, PS+PMMA), CB moves to the interface only from the phase characterized by lower adhesion to CB.

However, this rule is invalid for explaining the effect of the sequence of components blending on CB localization at the interface in the blends of

TABLE 16.3   The Characteristics of the Interaction of Polymers and ç Weight Analogs with CB №254

Polymer			Low-molecular weight analog		
Nota-tion	ξ [17, 20, 21, 23]	ΔF, × kN/m	Name	ξ [20, 21, 23]	ΔH, mJ/m^2
PE	0	0.025	n – hexane	0	170
PIB	0	0.02	2,2,4-trimethylpentane	0	175
PS	0.012	0.015	Ethylbenzene	0.0076	205
PBD	0.003	0.15	Hexene-1	-	180
PDMS	0.04**	0.03	Cyclic tetramer of dimethylsiloxane	0.04	200
EVA	-	0.3	-	-	-
BNR	-	0.46	-	-	-
PCP	0.11**	0.65	1-chlorobutane	0.11	270
PU	-	1.17	-	-	-
PVA	0.33	2	Ethylacetate	0.33	365

* ΔF is F gain for the CB content increase in the substratum (PP) from 0 to 42 vol. %

** calculated from the ratio of polar and dispersion components of surface tension [17].

TABLE 16.4   The Characteristics of Cohesion Energy of Polymers at 433 K [17, 23]

Polymer	σ, mN/m	Polymer	σ, mN/m
PDMS	13.0	EVA	26.2
PP	20*	PVA	26.3
PU	23**	PS	29.9
PIB	24.8	PMMA	30.5
PE	25.1	PCP	31.5
PE	24.0*		

* at 463 K

** given for the low-energy block of copolymer.

polymers having close energies of adhesion to a filler, such as, PE+PDMS, PS+PDMS, PIB+PDMS, PE+PS, PS+PP.

The results obtained show that in these blends the removal of a filler from the bulk to the interface occurs from the phase of polymer characterized by higher value of cohesion energy (Table 16.4) (from PS to the interface with PO and PDMS, from PO to the interface with PDMS). As for the phase with low cohesion energy, such redistribution here either is absent (from PDMS to the interface with PE, PS, PIB) or is insignificant if the cohesion energy difference for polymer components is small (from PE or PP to the interface with PS).

It is known [16] that the wetting force of liquids increases with increase in their adhesion to the surface as well as with decrease in their cohesion energy. Taking into account Dupre's equation, inequality (1) may be written in the following form:

$$-1 < (\sigma_1 - \sigma_2 + w_{a1} - w_{a2})/\sigma_{12} < 1 \qquad (3)$$

where: σ_1 and σ_2 are the surface tension values of the first and the second liquids at the interface with air, correspondingly; w_{a1} and w_{a2} are the adhesion energies of the first and the second liquids to the surface of a particle.

The liquid with greater adhesion to a surface and lower surface tension has a preference in wetting this surface. When the energies of polymer-surface interaction are close, the polymer with lower surface tension (smaller cohesion) exhibits better wetting force. [16].

This suggests that the migration of carbon black towards the interface in most of blends proceeds effectively from the polymer phase with lower capability of wetting the carbon black surface. The redistribution of carbon black from the phase of polymer with higher wetting force to the interface takes place only for a small difference in wetting force of polymer components and is less pronounced.

This conclusion determines the conditions necessary for a filler to localize at the interface, but cannot explain the causes of such strong influence of the sequence of components mixing on a process, and hence, does not reveal the essence of a phenomenon.

It is impossible to understand the effect of blending sequence on filler localization at the interface from the single viewpoint of thermodynamics of particle wetting by two liquids. It is due to the fact that in polymer mixtures, for well-known reasons, the most energetically efficient distribution of a filler described by Eq. (1) is not achieved.

The study of a filler redistribution towards the interface shows that this process for all blends is completed after 2–3 min of mixing and does not depend on the sequence of components adding (Fig. 16.3). The increasing time of mixing to 30 min does not lead to the filler localization at the interface in those cases, where it has not taken place within 3 min after the start of mixing. To the contrary, where the mixing was prolonged to 10 min, ρ for many blends slightly rises.

The essential difference in electrical conductivity values for blends of identical formula but different sequence of preparation, and the stability of conductivity during the mixing process suggest that the concentration of a filler at the interface is governed by the equilibrium between the number of particles arriving at the interface and those removed back to the phase. Indeed, taking into account that, although the redistribution of a filler from a phase to the interface gives no energy gain and is associated with overcoming a high activation barrier of macromolecule desorption, it does occur. So, according to the statistics, the concentration of filler particles at the interface must rise in time, but it remains constant (Fig. 16.3).

A great difference in electrical conductivity of blends prepared by two different mixing procedures indicates that in a blend with low conductivity there exists a potential for greater number of filler particles to be localized

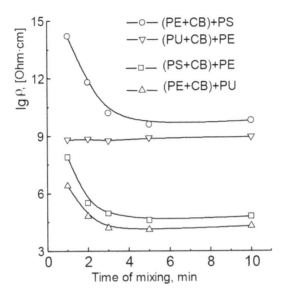

FIGURE 16.3 ρ of polymer blends versus time of mixing. The CB content: 5 vol. % in the PE phase, 8.5 vol. % in the PU phase, 9.3 vol. % in the PS phase. The ratio of polymers is 1:1.

at the interface than it is observed. However, this possibility is not realized. This can be explained only by the equilibrium existing between the number of particles arriving at the interface and those leaving the interface for a phase.

Let us analyze this equilibrium in more detail. Firstly, specify that the localization of a dispersed particle at the interface means that the macromolecules of both polymers are adsorbed on this particle. The velocity of particle transfer from a phase to the interface is determined by the number of their successful collisions dependent on the filler concentration in a phase. The collision is said to be successful when a part of macromolecules of the preliminarily filled polymer previously adsorbed on a filler is substituted by the macromolecules of another polymer component. The replacement of a very small number of macromolecules may be treated as a transfer of a filler particle to the interface.

The replacement is associated with overcoming an energy barrier of the desorption of macromolecules from a solid surface, which can be rather high for some macromolecules [24]. However, at the surface of a filler there always exists some part of macromolecules having a relatively low adsorption energy (small number of contacts with a surface). These macromolecules can be with relative ease substituted after a collision of particles for macromolecules of another polymer. High shear stress acting during mixing promote a macromolecule to overcome the activation barrier of desorption.

Obviously, the average activation energy of desorption is higher for macromolecules of the polymer having the higher wetting force. Consequently, the rate of filler particles migration to the interface must be lower from a phase with higher wetting force than from a phase with lower wetting force.

However, with a satisfied Eq. (1), after a long period of mixing the interface would be filled with particles without regard to the phase, which they left. Nevertheless, the experimental data are contradictory to this concept (Fig. 16.3). Therefore, the essential role in establishing the equilibrium concentration at the interface is played by the rate of filler particles leaving the interface and arriving at a phase.

Consider the factors controlling the rate of particles leaving the interface for a phase.

The removal of a particle from the interface is a result of the shear stress exerted on the particle by the polymer environment and is described by the Stokes law [25]. The shear stress affecting a particle is in direct proportion to its size, the viscosity of a polymer medium, and the shear rate. In certain situations this shear can abstract a particle from the interface and return it to one of polymer phases. The adhesion of a particle (W_a) to the unfilled polymer phase counteracts its removal from the interface. Only those particles stay at the interface for which the force of binding to the opposite phase exceeds the force of their separation from this phase. The ratio of these forces will determine the equilibrium concentration of a filler at the interface.

Let us estimate these forces. In both phases of a heterogeneous polymer blend the shear stress is the same, so the same is the force of particle removal from the interface. Hence, the local equilibrium concentration of particles at the interface will be determined by their adhesion to a phase. In this case it is important to specify to what of two polymer phases the

adhesion is considered. It is known [1] that the transfer of filler particles from one phase of polymer blend to another is observed very rarely.
Therefore, in most cases the particles of a filler after abstraction from the
interface return back in that phase from which they came to the interface. Otherwise a fast transfer of a filler would be observed from a phase
of lower wetting force to opposite polymer phase. So the force holding
particles at the interface is a result of their adhesion to the unfilled polymer phase. Besides, this suggests that dispersed particles at the interface
occupy nonequilibrium positions and are confined predominantly in the
preliminarily filled phase.

The energy of particle adhesion to a phase is determined by the following expressions [16]:

$$W_{a31} = \sigma_{23} - \sigma_{13} + \sigma_{12} \tag{4}$$

$$W_{a32} = \sigma_{13} - \sigma_{23} + \sigma_{12} \tag{5}$$

where: W_{a31} is the work of particle adhesion to the phase 1; W_{a32} is the work
of particle adhesion to the phase 2.

The adhesion of filler particles to the phase of polymer with the greater
wetting force is higher than to the phase with smaller wetting force.

Analyzing the ratio between the rates of particles arriving at and leaving the interface, it may be concluded that the smaller is the wetting force
of a polymer phase where particles are located, the greater is their concentration at the interface. This is consistent with experimental observations
(Table 16.2, Fig. 16.3).

However, the proposed mechanism of establishing the equilibrium
concentration of carbon black at the interface between polymers does not
explain why carbon black does not migrate to the interface from one of
two phases whereas such transfer occurs from another phase. Indeed, if the
Eq. (1) is satisfied, the transfer of carbon black from either of two phases
to the interface is thermodynamically efficient. Even with a low adhesion
of a filler to the interface the local concentration of a filler at the interface
would slightly exceed its concentration in phase. However, such an excess
is not observed for a number of polymer pairs attributed to the first group
according to the data in Table 16.2.

A broad spectrum of the energies of adsorption of macromolecules on a solid surface suggests that under certain conditions the localization of a filler at the interface is feasible even if the Eq. (1) is not fulfilled. This may take place if the location of the whole of a filler within a phase of one of polymers and not at the interface is thermodynamically efficient ($\sigma_{13} > \sigma_{12} + \sigma_{23}$ or $\sigma_{23} > \sigma_{12} + \sigma_{13}$) and a filler was introduced initially into a phase of another polymer. In this case the transfer of CB from one phase to another is thermodynamically favorable. It is necessary for a filler particle to transfer from one phase to another that the macromolecules of polymer previously adsorbed on its surface should be fully replaced by the macromolecules of the second polymer. However, such a replacement is unlikely because of a very high adsorption energy of some part of macromolecules. The experimental data confirm that such replacement is very rare to occur [1, 10, 11]. The partial replacement of macromolecules is more probable since there is a portion of macromolecules with low adsorption energy [24]. Such partial replacement just implies the localization of a filler at the interface.

Since the interfacial tension between polymers is minor [17] and the adhesion of polymers to a filler differs essentially (Table 16.3), it may be supposed that the filler localizing between polymer phases by the latter mechanism is most probable. The filler localization at the interface in all blends assigned to the first group according to Table 16.2 is likely to follow the last mentioned scheme. In those blends the local concentration of a filler at the interface is also governed by the equilibrium between the number of particles arriving at the interface and leaving it for a phase.

From the above discussion it follows that with increasing difference in wetting forces of polymers the equilibrium concentration of CB at the interface must rise when a filler transfers to the interface from the phase of a lesser wetting force and must fall when it comes here from the phase of higher wetting force. Besides, the rise of interfacial tension between polymers must promote the increase in local concentration of CB at the interface when it is redistributed from any phase.

Taken together the experimental observations (Table 16.2) confirm the validity of the latter conclusion.

Thus, dependence of the interfacial filler concentration on the sequence of components mixing is due to the peculiarities of macromolecule adsorption on a solid surface. Because of a high activation energy of desorption from a solid surface for a major part of macromolecules [24], the redistribution of dispersed particles from one polymer phase to another practically does not occur and those cannot occupy equilibrium position at the interface. Under these circumstances when a difference in wetting forces of phases takes place, the sequence of blending has a determining effect on the possibility and extent of the localization of dispersed particles at the interface. Let us consider the blends assigned to the third group according to Table 16.2. In the PE+PIB, PE+PP, PE+PBD blends the polymer components have close values of cohesion energy and similar interaction intensity with the carbon black surface and hence, close wetting force. Besides, these pairs exhibit very low interfacial energy [17]. Consequently, even though the Eq. (1) is fulfilled, the equilibrium CB concentration at the interface in these polymer blends will only slightly exceed the concentration in phase. The experimental data are fully consistent with this assumption. So, if CB is confined only in a single polymer component and its concentration here is 1.5 volume % below φ_p, the concentration of CB at the interface does not reach the percolation threshold (Table 16.2). However, when the concentration of CB in phase is 0.3% below φ_p, the PE+PP and PE+PIB blends conduct electric current and have ρ equal to 8×10^3 and 2×10^4 Ohm×cm correspondingly. Because of close wetting forces of the polymers in these blends, the migration of a filler to the interface is possible from both phases. Owing to this fact, when CB is preliminarily introduced into both polymer components and CB migrates to the interface from both phases, the fall in electric resistance of these blends below the 'additive' values is observed even if the CB concentration in both phases is 1.5% below the percolation threshold.

The essential role played by shear stress in the process of filler localizing at the interface suggests that the process is strongly affected by the viscosity of polymer components. Particular attention was given to the systems (PE+CB) +PMMA and (PS+CB) +PE, a case of the most interest since a filler is here redistributed from a phase of polymer with lesser wetting force. In these systems the viscosity of PMMA and of PE correspondingly was varied (Fig. 16.4). The variation of viscosity within 300 to 1500–1700 Pa·s range for PMMA and from 250 up to 1500 Pa·s for PE only moderately increases ρ of the (PS+CB) +PE blend (Fig. 16.4, the

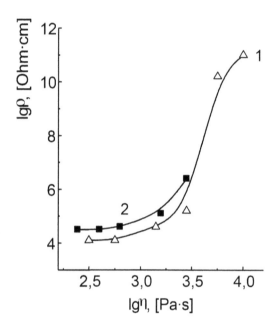

FIGURE 16.4 ρ of the (PE+CB) +PMMA (1) and (PS+CB) +PE (2) systems versus the effective viscosity of PMMA and PE correspondingly. Concentration of CB: 5.4 vol. % in the PE phase (1), 9.3 vol. % in the PS phase (2).

viscosity is given for the conditions close to those of mixing: shear rate is 100 s^{-1}). When PMMA and PE viscosity exceeds the above limits, ρ of blends rises essentially.

Such viscosity effect can be understood from the viewpoint of kinetics. The wetting of a filler by high-viscosity polymers occurs predominantly under the action of external forces straining polymer and indenting the filler particles into polymer bulk. When the polymers differ greatly in viscosity, the rate of strain will be far less for a high-viscosity polymer phase than for that of low-viscosity. So the high-viscosity component has no time to wet the filler particles surrounded by the low viscosity, easy to deform polymer. As a result, the CB localization at the interface is not observed. Besides, the rise of the viscosity of either of two polymer components increases viscosity of the system as a whole, and in the blending conditions this causes the shear stress and the force of particle break-off from the interface to rise. As a result, the equilibrium concentration of particles at the interface decreases.

The ratio of polymer viscosity values, under which a substantial drop in local interfacial CB concentration is observed, differs for various polymer pairs. The greater is the difference in wetting forces of polymers, the more times the viscosity of the second (having higher wetting force) polymer, can exceed the viscosity of the preliminarily filled polymer. So, this ratio is about 4 for the PE+PMMA blend, and about 2 for the PS+PE blend. The best conditions for particles to localize at the interface are provided when viscosity of the second polymer is slightly lower than that of a preliminarily filled polymer component.

16.4 CONCLUSIONS

1. The possibility of dispersed particles to be redistributed from a polymer phase to the interface between two polymers and the extent to which this event proceeds is substantially determined by the relative capacity of polymer components to wet the surface of filler particles and by the sequence of blending the components. When a difference in the wetting force of polymers increases, the concentration of dispersed particles at the interface rises in the event that they have preliminarily been introduced into the phase of polymer with a lesser wetting force, and reduces where they have preliminarily been introduced into the phase of polymer with a higher wetting force.

2. The localization of a filler at the interface in polymer blends is also possible in the case when the residence of particles at the interface is thermodynamically inefficient, but when their transfer from a filled polymer phase to unfilled one gives energy gain.

3. The local concentration of a filler at the interface is determined by the equilibrium established in the process of blending between the number of filler particles arriving to the interface from a polymer phase and those moving in the opposite direction under the action of external mechanical forces. This concentration rises with increasing energy of particle adhesion to the phase of unfilled polymer and with decreasing shear stress experienced by a particle.

KEYWORDS

- **carbon black**
- **melt-flow index**
- **poly(methyl methacrylate)**
- **polyethylenes**

REFERENCES

1. Marsh, P. A., Voet, A., Price, L. D., Mullens, T. J. *Rubber Chem. Technol.*, 1968, v.41, N2, 344–355.
2. Sircar, A. K. *Rubber Chem. Technol.*, 1981, v.54, N4, 820–834.
3. Lipatov Yu. S., Mamunya, E. P., Gladyreva, I. A., Lebedev, E. V. *Vysokomolekulyarnye Soedineniya*, 1983, v.A25, №7, 1483–1489.
4. Lipatov Yu.S., Mamunya, E. P., Lebedev, E. V., Gladyreva, I. A., *Kompozitsionnye polimernye materiaiy*, 1983, v.17, 9–14.
5. Gubbels, F., Jerome, R., Teyssie Ph., Vanlathen, E., Deltour, R., Calderone, A., Parente, V., Bredas, J. L. *Macromolecules*, 1994, v.27, N7, 1972–1974.
6. Gubbels, F., Blacher, S., Vanlathem, E., Jerome, R., Deltour, R., Brouers, F., Teyssie Ph. *Macromolecules*, 1995, v.28, N4, 1559–1566.
7. Soares, B. G., Gubbels, F., Jerome, R., Teyssie Ph., Vanlathen, E., Deltour, R. *Polymer Bulletin*, 1995, v.35, N1–2, 223–228.
8. Soares, B. G., Gubbels, F., Jerome, R., Vanlathen, E., Deltour, R. *Rubber Chem. Technol.*, 1997, v.70, N1, 60–70.
9. Sumita, M., Sakata, K., Asai, S., Miyasaka, K., Nakagawa, H. *Polymer Bulletin*. 1991, v.25, N2, 265–271.
10. Sumita, M., Sakata, K., Hayakawa, Y., Asai, S., Miyasaka, K., Tanemura, M. *Colloid Polymer. Sci.*, 1992, v.270, N2, 134–139.
11. Pavliy, V. G. Zaikin, A. E., Kuznetsov, E. V., Michailova, L. N. *Izvestiya vysshikh uchebnykh zavedeniy. Khimiya i khim. Telhnologiya*. 1986, v.29. vi5, 84.
12. Zaikin, A. E., Mindubaev, P.Yu., Arkhireev, V. P. *Polymer Sci.*, 1999, v.B41, №(1–2). 15–19.
13. Pavliy, V. G. Zaikin, A. E., Kuznetsov, E. V. *Vysokomolekulyarnye Soedineniya*, 1987, v.A29, №3, 447–450.
14. Zaikin, A. E., Galikhanov, M. F., Arkhireev, V. P. *Mekhanika kompozitsionnykh materialov i konstruktsiy*. 1998. v.4. №3. 55–61.
15. Zaikin, A. E., Galikhanov, M. F., Zverev, A. V., Arkhireev, V. P. *Polymer Sci.*, 1998, v.A40, №5, 847.
16. Adamson, A. The Physical Chemistry of Surface. New York: Wiley, 1982, 4th ed.
17. Wu, S. in "Polymer Blends," Edited by, D. R. Paul, S. Newmen, V.1, Chapter 6, Academic Press, New York, 1978.

18. Kraus, G. In "Reinforcement of Elastomers", Edited by Kraus, G., New York, Wiley Interscience, 1965.
19. Dannenberg, E. M. *Rubber Chem. Technol*, 1975, v.48, N2, 410.
20. Hansen, C. M. *J. Paint. Technol*. 1967, v.39, N.505, 104–113.
21. Hansen, C. M. *J. Paint. Technol*. 1967, v.39, N.511, 505–510.
22. Stepin, S. N., Bogachev, F. V., Svetlacov, N. V. *Zhurnal prikladnoy khimii*, 1991, №10, 2107–2110.
23. Kinlock, E. Adhesion and adhesives. Science and technology, N. Y.-L.: Univ. Press. 441.
24. Flir, G., Liklema, J. Adsorption from Solution at the Solid/Liquid, Parfitt, G. D. and Rochester, C. H. Editors, London: Academic, 1983.
25. McKelvey, J. M. Polymer Processing, Chapter 12, John Wiley, New York, 1963.

CHAPTER 17

SORPTION OF INDUSTRIAL DYES BY INORGANIC ROCKS FROM AQUEOUS SOLUTIONS

F. F. NIYAZI[1] and A. V. SAZONOVA[2]

[1]Head of the Department, Fundamental Chemistry and Chemical Technology, Southwest State University, st. 50 years October, 94, 305040, Kursk, Russia
E-mail: farukhniyazi@yandex.com

[2]Candidate of Chemical Sciences, Lecturer, Fundamental Chemistry and Chemical Technology, Southwest State University, st. 50 years October, 94, 305040, Kursk, Russia
E-mail: ginger313@mail.ru

CONTENTS

ABSTRACT

The authors of this chapter have defined chemical composition of inorganic polymeric rocks. They have also studied kinetics of industrial dyes sorption from aqueous solutions by the rocks under investigation. There have been stated kinetic curves of the sorption. The work also considers possibilities to use kinetic equation of pseudo-first and pseudo-second orders for the description of sorption process.

17.1 INTRODUCTION

Production and use of dyes is connected with the use of huge quantity of water a great part of which is released in polluted state.

On the average, about 225 tons of water is spent per 1 ton of dye. During production of each ton of dyes together with industrial sewage enterprises release dozens and hundreds tons of different mineral and organic compounds in the form of waste [1].

At present different types of dyes are widely used for production of paper, paints and pigments in textile, shoe-manufacturing and printing industries. During production processes from 10 to 40% of dyes, being used, get into sewage, which, coming not fully purified into natural pools, cause serious violation of biocenose in them [2].

Getting into water pools dyes have a negative effect on associations of water organisms and also violate their oxygen regime. Most of organic dyes are high-toxic and allergic effect. Besides, as a rule, cationic dyes are more toxic for water ecosystems that anionic ones [3]. Thus, it is actual to search for new approaches to the problem of sewage treatment containing dyes.

Existing methods of sewage treatment from industrial dyes can be divided into regenerating, destructive and biological. Among them regenerating methods is more perspective since they allow to extract substance, being constrained in sewage, for their further use. This problem may be solved by using such methods of concentration as extraction, ionic replacement, sorption and others.

Adsorption on mesoporous activated coals is one of effective methods of sewage treatment contaminated by dyes [4]. However, such coals are expensive adsorbents and are produced in small quantities. Perspective direction cheaper is utilization, on the one hand, cheap, and, on the other hand, available sorption materials as adsorbents [5].

Thick chalk beet stretches through European continent, including the north of France, south part of England, Poland, goes through the Ukraine, Russia and is displaced into Asia – Syria and Libyan Desert.

Kursk region possesses exceptional by volume and variety natural resources being able to provide needs of the region by carbonate rocks. That is why, inorganic polymeric rocks of Kursk region have been used as sorption material because they are high-dispersion systems.

Carbonate rocks, under investigation, were analyzed concerning content of: calcium and magnesium carbonates – by complex onometric methods; sulfate-ions and ferrous and aluminum oxides – by gravimetric method; chloride-ions – by methods of turbidimetry. There has been found the content of insoluble residue in hydrochloric acid, and electric conduction of carbonate rocks saturated aqueous solution has been also measured. Electric conduction was defined with the help of conductivity apparatus KSL-101. Received results of carbonate rocks chemical composition are given in Table 17.1.

The analysis of sorbent shows that calcite form a great share of systems, being studied. Deposits of carbonate rocks in Kursk region differ by low content of insoluble residue and high content of carbonates.

Structural and microscopic characteristics polymeric rocks were deferent by method of homogeneous field with the help of polarizing-interference microscope BIOLAR. The investigation has shown that original rocks consist of dolomite trigonal and rhombic crystals [6].

TABLE 17.1   Chemical Composition of Carbonate Rocks

Deposits of Kursk region	Area of sampling	
	Konyshovka	Medvenka
Granulometric composition, % (0.2–0.06)	73.37–68.49	88.13–79.84
Moisture content, %	9.6225	6.3125
Content of $CaCO_3$ и $MgCO_3$ in terms of $CaCO_3$, %	49.50	41.95
Insoluble residue in HCI	1.075	13.566
Sesquilateral content ferrous and aluminum oxides	0.273	1.530
Water-soluble substances, %	3.1823	2.15997
Electric conduction, мкСм/см	79.04	88.26
Salt content in terms of NaCl, mg/L	37.22	41.59
Sulphate-ions, %	0.00445	0.01403
Chloride-ions, мг/л	0.005	0.0015

The aim of this chapter is studying the process of industrial dyes of different nature sorption by carbonate rocks.

Dyes (cationic blue, cationic red, acidic brightly green and antraqui-none blue), widely used in industry were used as adsorptive.

Experiments in sewage treatment from industrial dyes were carried out using model aqueous solutions of industrial dyes and sewage from dye-finishing shop of knitted fabric incorporation "Seim" (Kursk). Dyes of "pure for analysis" qualification without additional cleaning were used for preparation of aqueous solutions. There were investigated aqueous solutions of dyes with concentration of 0.01 g/L at the temperature 298 °C with the solution volume V=50 mL.

Effect of sorbent mass on dyes sorption and kinetics of the process of sewage treatment from industrial dyes have been studied in order to find optimal parameters of sorption and also to state sorption properties of carbonate rocks. Method of one-step static sorption was been used in this work. Sorption was carried out by adding carbonate rocks samples, grinded to granules with the size of 0.06–2.0 mm, to the dyes solution being studied. Then all this was stirred by magnetic mixer and in definite intervals there were taken samples and was defined residual concentration of dyes by spectrophotometric method.

There have been taken readings of absorption spectra on coordinates: optical density (A) – wave length (λ) on the device СФ-26; there have been chosen wave lengths of maximum light absorption for cationic blue – 610 nm, cationic red – 490 nm, acid brightly green – 670 nm, antraquinone blue – 590 nm. There have been also found subordination bound of dyes solution to the mail law of light absorption – Beer-Lambert-Bonguer law.

Data on carbonate rocks mass affect on the sorption of industrial dyes are given in Table 17.2.

TABLE 17.2   Effect of Carbonate Rocks Mass on the Sorption of Industrial Dyes at Phase's Relation: t = 30 min, V = 50 mL, C = 0.01 g/L

Mass/Dye, g	0.05	0.1	0.25	0.5	1.0	2.0
Cationic blue	93.0	99.0	99.5	100	100	100
Cationic red	66.0	84.4	95.98	100	100	100
Antraquinone blue	76.0	88.0	100	100	100	100
Acid brightly green	0	20.0	38.6	40.2	46.6	49.6

Data of the given researches allowed finding sorption capacitance of carbonate rocks: for cationic dyes – 1 mg/g of sorbent, for antraquinone blue dye – 2 mg/g of sorbent, for acid brightly green dye – 0.1 mg/g of sorbent.

Cleaning efficiency (CO, %) shows the share of absolute substance quantity, which is caught by sorbent, and gives rather complete notion of the process and is character [7]. This factor is an important criterion while defining optimal condition of the sorption process and is calculated by the following formula

$$CO = \frac{(C_0 - C) \cdot 100\%}{C_0}$$

where C_0 – initial concentration, g/L; C – residual concentration, g/L.

These data allow choosing optimal mass of sorbent for definition of industrial dyes kinetics by carbonate rocks. Results of these studies are given by kinetic curves in Fig. 17.1.

Sorption lasts for the first 2–3 min after the beginning of contact phases regardless of the dye nature at the first section of the stepped kinetic curve.

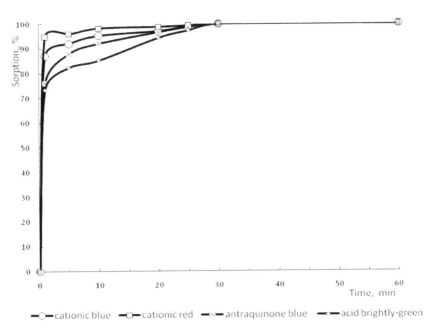

FIGURE 17.1   Kinetic curves of industrial dyes sorption by carbonate rocks.

But this sorption quickly increases at the second section during the following 3–30 min. Complete (100%) sorption of dyes takes place after 30 minuets of interaction and there is sorption equilibrium. Further increase of contact time is pointless.

Comparison of studies literature data received by us allows to come to a conclusion stepped character of sorption kinetic cures is the result of the fact that adsorption of large ions takes place on micro porous carbonate rocks. Mass transfer on the phase's boundary and interaction of dyes with carbonate rock surface play important role at the first stage of sorption; and at the second stage there is internal diffusion of dye into sorbent pores of accessible size, allowing the dye to be sorbet again on the external surface of sorbent [8].

It is stated that for all stepped kinetic curves the first section can be described by the equation of pseudo first order; but the second section is not described by integral equation of kinetic models neither of the pseudo first, nor pseudo second order [9].

Is renown that hydrogen index of environment (pH) has a great influence on the process of sorption and selection of sorbent for sewage treatment. It is connected with the fact that functional groups on the sorbent surface and functional groups of dye molecules may change depending on environment pH [10]. Change of pH after the process of sorption depending on the time of the contact of sorbents, being studied, and industrial dyes is given in Table 17.3.

Increase of pH value is observed at mixing carbonate rocks with water can be described by hydrolysis of calcium and magnesium carbonates. On the contrary, decrease of pH values is observed during the process of industrial dyes sorption and this may be explained by chemical interaction of carbonate rocks and dyes.

TABLE 17.3 Change of pH Depending on the Time of Contact of Carbonate Rocks and Industrial Dyes

Industrial dye	pH_0	Time of contact, min					
		1	5	10	20	25	30
Blank test	6.5	7.25	7.67	7.92	8.10	8.54	8.83
Cationic red	4.26	7.29	7.08	7.55	7.17	7.18	7.00
Cationic blue	3.41	7.57	7.60	7.64	7.67	7.44	7.38
Antraquinone blue	5.67	7.45	7.33	7.45	7.38	7.36	7.32
Acid brightly green	5.38	7.32	7.20	7.18	7.05	6.95	6.82

This can be very important when we chose sorbent for sewage neutralization during the process of acid sewage treatment. Received results of research showed high adsorption ability of carbonate rocks regarding industrial dyes of different classes.

Suggested method of sewage sorption treatment from dyes increases the variety of sorbents, being used during treatment, and allows using local carbonate rocks as sorbents.

It is necessary to note that kinetic curves differ from each other to a small extent while comparing sorption ability of carbonate rocks in relation to dyes. However, consumption of carbonate sorbent is lowering for cationic and antraquinone dyes at the same phase's relation and efficiency of aqueous solution treatment.

As to technological execution it is necessary to carry out the process of sorption in contact absorbers, equipped by mechanical mixers of batch action. Separation of solid and liquid phases is carried out by the way of decantation and filtration. It is economically pointless to recover waste sorbent that is why it is necessary to use it.

Once of the ways to use this sorbent is to use it as filling agent. This is connected with the fact that calcium carbonate is a widely used material in the world industry today. Development of the branches of rubber-engineering, electric-power, glass, paper, polymeric, paint and varnish and other industries requires to increase production of high-quality filling agents, and in the first place it is chalk. Characteristic feature of this natural material is connected with the fact that it is ease to quarry and process it at rather small expenditures. Its quarrying and processing do not cause serious ecological violations, and its reserves are practically unlimited in many European countries, countries of former UIS and in Russia.

KEYWORDS

- dye
- inorganic rocks
- kinetic curve
- pH
- sorption
- utilization

REFERENCES

1. Stepanov, B. I. Introduction in chemistry and technology of organic dyes. M, "Chemistry," 1977, 488.
2. Ramesh, D. D., Parande, A. K., Raghu, S., Prem Kumar, T. J. Cotton Sci. 2007, 11, 141–153.
3. Hao, O. J., Chiang, P. C. Crit. Rev. Environ. Sci Technol, 2000, 30, 141–153.
4. Soldatkina, L. M., Sagajdak, E. V., Menchuk, V. V. Adsorbtion of cationic dyes from water solutions on sunflower. Chemistry and technology of water, 2009, V. 31, №4, 417–426.
5. Litvina, T. M., Kushnir, I. G. Problem of reset, processing and waste recyclings. Odessa, 2000, 258.
6. Niyazi, F. F., Maltsevf, V. S., Burykina, O. V., Sazonov, A. V. Kinetics of sorption of ions of copper by cretaceous breeds. News of Kursk State Technical University, №4, Kursk, 2010, 28–33.
7. Gocharuk, V. V., Puzyrnaja, L. N., Pshinko, G. N., Bogolepov, A. A., Demchenko, V. J. Removal of heavy metals from water solutions montmorillonite, modified polyethyleneimine. Chemistry and technology of water, 2010, V. 32, №2, 125–134.
8. Soldatkina, L. M., Sagajdak, E. V. Kinetics of adsorption of water-soluble dyes on the active coals. Chemistry and technology of water, 2010, V. 32, №4, 388–398.
9. Janos, P., Buchtova, H., Ryznarova, M.. Water Res. 2003, 37, №20, 4938–4944.
10. Bagrovskaja, N. A., Nikiforova, T. E., Kozlov, V. A. Influence acidities of the environment on equilibrium sorption of ions Zn (II) and Cd (II) polymers on the basis of cellulose. Magazine of the general chemistry. 2002, 72, V. 3, 373–376.

CHAPTER 18

INVESTIGATION OF VARIOUS METHOD FOR STEEL SURFACE MODIFICATION

IGOR NOVÁK,[1] IVAN MICHALEC,[2] MARIAN VALENTIN,[1] MILAN MARÔNEK,[2] LADISLAV ŠOLTÉS,[3] JÁN MATYAŠOVSKÝ,[4] and PETER JURKOVIČ[4]

[1]Department of Welding and Foundry, Faculty of Materials Science and Technology in Trnava, 917 24 Trnava, Slovakia

[2]Slovak Academy of Sciences, Polymer Institute of the Slovak Academy of Sciences, 845 41 Bratislava, Slovakia

[3]Institute of Experimental Pharmacology of the Slovak Academy of Sciences, 845 41 Bratislava, Slovakia

[4]VIPO, Partizánske, Slovakia; E-mail: upolnovi@savba.sk

CONTENTS

18.1   INTRODUCTION

The surface treatment of steel surface is often used, especially in an auto-motive industry that creates the motive power for research, design and production. New methods of surface treatment are also developed having major influence on improvement the surface properties of steel sheets while keeping the price at reasonable level [1–3].

The nitrooxidation is one of the nonconventional surface treatment methods, which combine the advantages of nitridation and oxidation processes. The improvement of the mechanical properties (Tensile Strength, Yield Strength) together with the corrosion resistance (up to level 10) can be achieved [5–8]. The fatigue characteristics of the nitrooxidized material can be also raised [6].

Steel sheets with surface treatment are more often used, especially in an automotive industry, which creates the motive power for research, design and production. New methods of surface treatment are also developed having major influence on improvement the surface properties of steel sheets while keeping the price at reasonable level.

Previous outcomes [1, 3, 4, 9, 10] dealt with the welding of steel sheets treated by the process of nitrooxidation by various arc and beam welding methods. Due to high oxygen and nitrogen content in the surface layer, problems with high level of porosity had occurred in every each method. The best results were achieved by the solid-state laser beam welding, by which the defect-free joints were created. Due to high initial cost of the laser equipment, the further research was directed to the joining method that has not been tested. Therefore the adhesive bonding was chosen, because the joints are not thermally affected, they have uniform stress distribution and good corrosion resistance.

The goal of the paper is to review the adhesive bonding of steel sheets treated by nitrooxidation and to compare the acquired results to the non-treated steel.

18.2   EXPERIMENTAL PART

For the experiments, low carbon deep drawing steel DC 01 EN 10130/91 of 1 mm in thickness was used. The chemical composition of steel DC 01 is documented in Table 18.1.

TABLE 18.1   Chemical composition of steel DC 01 EN 10130/91

EN designation	C [%]	Mn [%]	P [%]	S [%]	Si [%]	Al [%]
DC 01 10130/91	0.10	0.45	0.03	0.03	0.01	-

18.2.1 CHEMICAL MODIFICATION

The base material was consequently treated by the process of nitrooxidation in fluidized bed. The nitridation fluid environment consisted of the Al_2O_3 with granularity of 120 μm. The fluid environment was wafted by the gaseous ammonia. After the process of nitridation, the oxidation process started immediately. The oxidation itself was performed in the vapors of distilled water. Processes parameters are referred in Table 18.2.

18.2.2 ADHESIVES

In the experiments, the four types of two-component epoxy adhesives made by Loctite Company (Hysol 9466, Hysol 9455, Hysol 9492 and Hysol 9497) were used. The properties of the adhesives are documented in Table 18.3.

TABLE 18.2   Process of Nitrooxidation Parameters

	Nitridation	Oxidation
Time [min]	45	5
Temperature [°C]	580	380

TABLE 18.3   The Characterization of the Adhesives

	Hysol 9466	Hysol 9455	Hysol 9492	Hysol 9497
Resin type	Epoxy	Epoxy	Epoxy	Epoxy
Hardener type	Amin	Methanethiol	Modified Amin	
Mixing ratio (Resin:Hardener)	2:1	1:1	2:1	2:1
Elongation [%]	3	80	0.8	2.9
Shore hardness	60	50	80	83

18.2.3   METHODS

The experiments were done at the Faculty of Materials Science and Technology, Department of Welding and Foundry in Trnava. The adhesive
bonding was applied on the grinded as well as nongrinded surfaces of the
material to determine the grinding effect on total adhesion of the material
so as on ultimate shear strength of the joints. The grinded material was
prepared by grinding with silicone carbide paper up to 240 grit.

Before the adhesive bonding, the bonding surfaces (both grinded as
well as nongrinded) were decreased with aerosol cleaner. The overlap area
was 30 mm. To ensure the maximum strength of the joints, the continuous
layer of the adhesive was coated on the overlap area of both bonded materials. The thickness of the adhesive layer was 0.1 mm and it was measured
by a caliper. The joints were cured under fixed stress for 48 h at the room
temperature. The dimensions of the joints are referred in Fig. 18.1.

The mechanical properties of the joints were examined by the static
shear tests. As a device, the LaborTech LabTest SP1 was used. The conditions of the tests were set in accordance with STN EN 10002–1. The static
shear tests were repeated on three separate samples and an average value
was calculated.

The fracture areas were observed in order to obtain the fracture character of the joints. The JEOL JSM-7600F scanning electron microscope was
used as a measuring device.

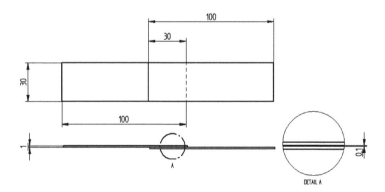

FIGURE 18.1   The dimension of the bonded joints.

The differential scanning calorimetry (DSC) was performed on Netzsch STA 409 C/CD equipment. As the shielding gas, Helium with purity of 99.999% was used. The heating process starts at the room temperature and continued up to 400°C with heating rate 10°C/min. The DSC analysis on Hysol 9455 was done on Diamond DSC Perkin Elmer, capable of doing analyzes from −70 °C.

18.3   RESULTS AND DISCUSSION

The material analysis represents the first step of evaluation. There are many factors having an influence on the joint quality. The properties of the nitrooxidized material depends on the treatment process parameters. For the adhesive bonding, the surface layer properties are important because of that, the high adhesion is needed to ensure the high strength of the joint.

The overall view on the microstructure of the nitrooxidized material surface layer is referred in Fig. 18.2a. On the top of the surface, the oxide layer (see Fig. 18.2b) was created. This layer had a thickness of approx. 700 μm. Beneath the oxide layer, the continuous layer of ε-phase, consisting of nitrides $Fe_{2-3}N$ and with the thickness of 8–10 μm was observed.

The surface energy measurements were performed due to obtain the properties of the material, which are important for adhesive bonding. For observing the grinding effect on the total surface energy, the measurements were done on the base as well as on the grinded material.

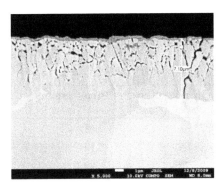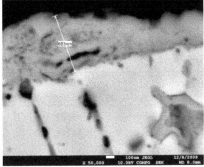

FIGURE 18.2   The microstructure of the surface layer. (a) overall view; (b) detail view on the oxide layer.

To determine the surface energy, the portable computer-based instrument SeeSystem was used. Four different liquids (distilled water, formamide, diiodomethan and ethylene glycol) were instilled on the material surface and contact angle was measured. The Owens-Wendt regression model was used for the surface energy calculation. The total amount of six droplets were analyzed of each liquid. The results (see Table 18.4) proved that the nitrooxidation treatment had a strong affect on material surface energy, where the decrease by 28% in comparison to nonnitrooxidized material had occurred. The surface energies of grinded and nongrinded material without nitrooxidation were very similar, while the increase of surface energy of grinded nitrooxidized material by 35% in comparison to nongrinded material was observed. In the case of barrier plasma modified steel is the surface energy higher compared unmodified material and namely its polar component is significantly higher than polar component of surface energy for unmodified sample as well as for sample modified by nitrooxidation.

The mechanical properties of the material were obtained by the static tensile test. Total amount of three measurements were done and the average values are documented in Table 18.5. By the results, it can be stated, that after the process of nitrooxidation, the increase of Yield Strength by 55% and Tensile Strength by 40 % were observed. The barrier plasma did not influence the mechanical properties after surface modification of steel that remained the same as for unmodified sample.

The tensile test of the adhesives were carried out on the specimens, which were created by curing of the adhesives in special designed polyethylene forms for 48 h. The results are shown in Table 18.6. In three of the

TABLE 18.4   The Surface Energy of Materials

Material type	Total surface energy [mJ/m²]	Dispersion component [mJ/m²]	Acid-base component [mJ/m²]
DC 01	38.20	35.62	2.58
DC 01 grinded	38.80	33.66	5.14
Nitrooxidized	27.60	25.79	1.80
Nitrooxidized grinded	37.31	32.54	4.77
Barrier plasma treated	39.99	33.69	6.30

adhesives (Hysol 9466, Hysol 9492 and Hysol 9497), very similar values were observed while in case of Hysol 9455, only tensile strength of 1 MPa was observed.

The differential scanning calorimetry was performed due to obtain the glass transition temperature as well as the melting points of the adhesives. The results are in Table 18.7. To measure the glass transition temperature of Hysol 9455, the measurements had to be started from the cryogenic temperatures. The results of such a low glass transition temperature explained the low tensile strength of the Hysol 9455, where at the room temperature, the mechanical behavior changed from rigid to rubbery state. The results of DSC analyzes are given in Figs. 18.3–18.6.

The adhesive joint evaluation consisted of observing the mechanical properties and fracture surface respectively.

In order to obtain mechanical properties of the joints, the static shear tests were carried out. Results (Table 18.8) showed that the highest shear strength was observed in grinded nitrooxidized joints. The Hysol 9466 provided joints with the highest shear strength

TABLE 18.5 Mechanical Properties of the Base Materials

	Yield Strength [MPa]	Tensile Strength [MPa]
DC 01	200	270
Nitrooxidized	310	380
Barrier plasma treated	200	270

TABLE 18.6 The Mechanical Properties of the Adhesives

	Hysol 9466	Hysol 9455	Hysol 9492	Hysol 9497
Tensile Strength [MPa]	60	1	58	65

TABLE 18.7 The Glass Transition Temperatures of Adhesives

	Hysol 9466	Hysol 9455	Hysol 9492	Hysol 9497
Glass transition temperature (°C)	52.6	5.0	61.2	62.4
Melting point (°C)	315.7	337.0	351.8	327.3

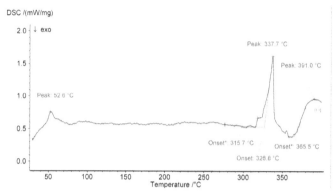

FIGURE 18.3   The DSC analysis of the Hysol 9466.

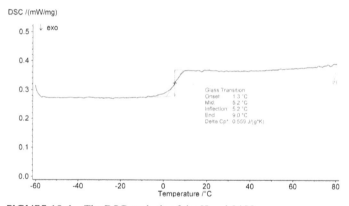

FIGURE 18.4   The DSC analysis of the Hysol 9455.

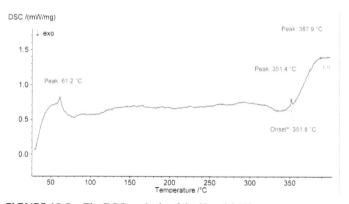

FIGURE 18.5   The DSC analysis of the Hysol 9492.

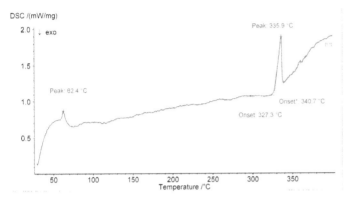

FIGURE 18.6   The DSC analysis of the Hysol 9497.

TABLE 18.8   The Results of Shear Test of the Joints

Material	Shear strength [MPa]			
	Hysol 9466	Hysol 9455	Hysol 9492	Hysol 9497
DC 01	9.0	2.8	7.1	6.0
DC 01 grinded	8.9	2.9	7.2	6.1
Nitrooxidized	12.9	3.1	7.8	5.4
Nitrooxidized grinded	12.9	5.9	12.7	7.0
Barier plasma treated	13.8	6.1	14.2	7.8

The mechanical properties of the joints made on nonnitrooxidized material did not depended on the surface grinding. The mechanical properties of the joints made on nitrooxidized material, in comparison to DC 01, were higher by 43% in case of Hysol 9466, 11% in case of Hysol 9455 and 10% in case of Hysol 9492. In the case of adhesive Hysol 9497, the decrease of the shear strength had occurred. The joints produced on grinded nitrooxidized material, in comparison to DC 01, had a higher shear strength by 43% in case of Hysol 9466, 110% in case of Hysol 9455, 79% in case of Hysol 9492 and 15% in case of Hysol 9497. The shear strength of adhesive joint was for barrier plasma modified steel for all kinds of adhesive Hysol higher compared unmodified and nitrooxidized steel.

Results of fracture morphology of the joints made of nonnitrooxidized material are shown in Fig. 18.7a. Only the adhesive type of fracture morphology (see Fig. 18.7b) was observed in every type of the adhesive. Cohesive and combined fracture type were not observed. It can be stated that the adhesion forces were not strong enough so that the joints were fractured between the material and adhesive.

The results of the fractographic analysis of the nongrinded nitro-oxidized steel are documented in Fig. 18.8. No oxide layer peeling was observed. The cleavage fracture pattern was observed only as well as adhesive fracture morphology. The close-up view on the cleavage fracture is shown in Fig. 18.8b.

Differential Scanning Calorimetry revealed that three of the adhesives had a very similar glass transition temperature, so the meshing of the adhesives will start in the same way.

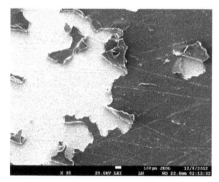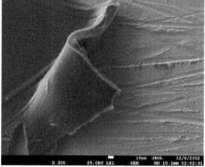

FIGURE 18.7 The fractographic analysis of the fractured adhesive joint of non-nitrooxidized material. (a) overall view; (b) close-up view.

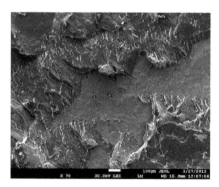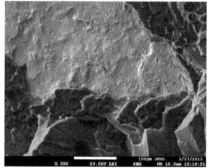

FIGURE 18.8 The fractographic analysis of the fractured adhesive joint of nitrooxidized material. (a) overall view; (b) close-up view.

The results of mechanical properties evaluation of the joints proved that the material after the nitrooxidation process had a better adhesion to the epoxy adhesives than plain material DC01. Due to this fact the higher shear strength was achieved. It can be explained by the surface oxide layer porosity, which helped the adhesive to leak in.

On the other hand, increase of mechanical properties of joints prepared from grinded nitrooxidized material can be explained by removing the surface oxide layer and thus resulting into rapid increase of surface energy.

The only adhesive type of fracture was observed and the fractographic analysis showed, that only cleavage type of fractures has been created. It can be stated, that the surface energy of the materials was not appropriate for the cohesive fracture pattern.

18.4 CONCLUSION

Joining of steel sheets treated by the process of nitrooxidation represents an interesting technical as well as technological problem. The fusion welding methods with high energy concentration, e.g. laser beam welding are one of the possible options, however even with high effort of minimizing the surface layer deterioration, it's not possible to completely avoid it.

Adhesive bonding of nitrooxidized steels presents thus the second alternative, when the surface layer is not damaged and the adhesive joint keeps its properties after it has been cured. Adhesive bonding of metallic substrates, often requires removing the surface oxide layer from the areas to be bonded. In case of materials treated by nitrooxidation, this is possible, however the damage of the formed surface layer will occur. The goal of this paper was to review the effect of surface layer, created by the process of nitrooxidation and or by barrier discharge plasma treatment, on final mechanical properties of the joints, evaluation of fracture morphology and results comparison for both, treated and untreated material.

The acquired results have been revealed, that the presence of nitrooxidation surface layer caused decrease of free surface energy by 28%. The surface energy and namely its polar component were for barrier plasma modified steel higher than for unmodified and nitrooxidized steel. On the other hand, this surface layer brings on the joints shear strength increase by 10–43% in dependence of the adhesive used. In case of Hysol 9497, the decrease of the joint shear strength by 10% was observed. In the case of barrier plasma modified steel were the strength of adhesive joint higher compared unmodified and nitrooxidized material. We can presume that the increase of the shear strength was mainly due to porous structure of the surface layer, which enabled the adhesive to leak in.

The adhesive type of fracture morphology was observed during fractographic analysis. Regarding the characteristics of used adhesives, the cleavage fracture morphology of the joints was occurred.

There can be concluded on the base of received results, that the epoxy adhesive bonding represents the suitable alternative of creating the high quality joints of steel sheets treated by nitrooxidation as well as treated by barrier discharge plasma.

ACKNOWLEDGEMENTS

This chapter was prepared within the support of Slovak Research and Development Agency, grant No. 0057–07 and Scientific Grant Agency VEGA, grant No. 1/0203/11 and 2/0199/14.

This publication was prepared as an output of the project 2013–14547/39694:1–11 "Research and Development of Hi-Tech Integrated Technological and Machinery Systems for Tyre Production – PROTYRE" cofunded by the Ministry of Education, Science, Research and Sport of the Slovak Republic pursuant to Stimuli for Research and Development Act No. 185/2009 Coll.

KEYWORDS

- **differential scanning calorimetry**
- **steel sheets**
- **tensile strength**
- **yield strength**

REFERENCES

1. Michalec, I. CMT Technology Exploitation for Welding of Steel Sheets Treated by Nitrooxidation. Diploma thesis, Trnava 2010.
2. Konjatić, Pejo; Kozak, Dražan; Gubeljak, Nenad. The Influence of the Weld Width on Fracture Behaviour of the Heterogeneous Welded Joint.Key Engineering Materials. 488–489 (2012); 367–370.
3. Bárta, J. Welding of special treated thin steel sheets: Dissertation thesis, Trnava, 2010.
4. Marônek, M. et al. Laser beam welding of steel sheets treated by nitrooxidation, 61st Annual Assembly and International Conference of the International Institute of Welding, Graz, Austria, 6–11 July 2008.
5. Lazar, R., Marônek, M., Dománková, M. Low carbon steel sheets treated by nitrooxidation process, Engineering extra, 2007, No. 4, p. 86.
6. Palček, P. et al. Change of fatigue characteristics of deep-drawing sheets by nitrooxidation.: Master Journal List, Scopus. In: Chemické listy. ISSN 0009-2770. Vol. 105, Iss. 16, Spec. Iss (2011), pp. 539–541.
7. Bárta, J. et al. Joining of thin steel sheets treated by nitrooxidation, Proceeding of lectures of 15th seminary of ESAB + MTF-STU in the scope of seminars about welding and weldability. Trnava, AlumniPress, 2011, pp. 57–67.
8. Marônek, M. et al. Welding of steel sheets treated by nitrooxidation, JOM-16: 16-th International Conference On the Joining of Materials and 7-th International Conference on Education in Welding ICEW-7, May 10–13th, Tisvildeleje, Denmark, ISBN 87-89582-19-5.
9. Viňáš, J. Quality evaluation of laser welded sheets for cars body. In: Mat/tech automobilového priemyslu: Zborník prác vt-seminára s medzinárodnou účasťou: Košice, 25.11.2005. Košice: TU, 2005. pp. 119–124. ISBN 80-8073-400-3.
10. Michalec, I. et al. Resistance welding of steel sheets treated by nitrooxidation. In: TEAM 2011: Proceedings of the 3rd International Scientific and Expert Conference with simultaneously organized 17th International Scientific Conference CO-MAT-TECH 2011, 19th −21st October 2011, Trnava Slovakia. Slavonski Brod: University of Applied Sciences of Slavonski Brod, 2011. ISBN 978-953-55970-4-9. pp. 47–50.

CHAPTER 19

IMPACT OF OZONATION AND PLASMA TREATMENT OF UNSATURATED RUBBERS ON THEIR ADHESION CHARACTERISTICS

V. F. KABLOV,[1] D. A. PROVOTOROVA,[1] A. N. OZERIN,[2]
A. B. GILMAN,[2] M. YU. YABLOKOV,[2] N. A. KEIBAL,[1]
and S. N. BONDARENKO[1]

[1] Volzhsky Polytechnical Institute, branch of Volgograd State Technical University, Volzhsky, Volgograd Region, Russia; E-mail: mystery_21_12@mail.ru; www.volpi.ru

[2] N.S. Enikolopov Institute of Synthetic Polymeric Materials of RAS

CONTENTS

ABSTRACT

This chapter considers the research related to modification of unsaturated rubbers by ozonation and plasma treatment for improving adhesion properties. Mechanism of macroradicals formation in the ozonation process as well as the influence of the contact time on adhesion properties of the rubbers has been studied. It has been shown that the best characteristics comparing with the initial ones can be obtained at 0.5–1 h ozonation. The gluing strength increases by 10–70% on average at that. Using a goniometric method, a change in the contact properties of the unsaturated rubber surface after treatment of the samples in dc discharge has also been studied. It has been shown that the plasma modification leads to a significant improvement in the contact properties of the rubber samples. The structure of film surface has been investigated by Fourier-IR-spectroscopy.

19.1   INTRODUCTION

Nowadays, modification is of a higher priority and the most rational method compared to the synthesis of new materials, and allows both to improve performance characteristics of rubbers and save the base complex of their properties.

Epoxydation, being one of the variants of chemical modification, represents a process of introduction of epoxy groups to a polymer structure that improves properties of this polymer. Thanks to these properties they find application as coatings for metals and plastics, adhesives, mastics and potting compounds in electrotechnics, microelectronics and other areas of engineering [1].

It is known that epoxy compounds are good film-formers in glue compositions and increase the overall viscosity of the last ones. Besides, high reactivity of the epoxy groups provides the best adhesion characteristics [2].

One of the variants of introducing epoxy groups to a rubber structure is ozonation as ozone is highly reactive towards double bonds, aromatic structures and C-H groups of a macrochain.

Low-temperature plasma is used for surface modification of polymer films, membranes, fibers, medical polymers, fabrics, and for obtaining fine layer coatings of different chemical nature as well [3, 4].

The important feature of polymer plasma-chemical treatment, which defines a special interest towards this method, is that only treated surface of a material is changed, and the thickness of a very fine layer of the material, according to different estimates, ranges from 100 A° to several microns. The bulk of a polymer does not change, while maintaining the mechanical, physical, chemical, and electrical properties of the modified material [5].

The modern plasma-chemical modification methods benefit greatly in terms of ecological compatibility compared to the chemical modification that typically uses aggressive chemicals (acids, hydroxides, alkaline earth metals and their compounds).

In the work we considered a possibility of application of chlorinated natural rubber and isoprene rubber ozonation and additional plasma treatment for improving their adhesion properties.

Chlorinated natural rubber (CNR) is applied as an additive in glue compositions based on chloroprene and nitrile rubbers [6] that, in its turn, are widely used both in industry for gluing of different vulcanized rubbers together with metals or each other, and in everyday life for gluing various materials. As an individual brand glues based on CNR are rarely produced.

Isoprene rubber is an analog of natural rubber, but due to its low cohesive strength applied in their formulations much more rarely.

19.2 EXPERIMENTAL PART

As was mentioned above, we considered a possibility of application of chlorinated natural rubber and isoprene rubber ozonation, and additional plasma treatment for improving their adhesion characteristics. Chlorinated natural rubber of CNR-10, CNR-20, S-20 brands, and SKI-3 isoprene rubber were chosen for ozonation.

Formation of epoxy groups, which promote an increase in adhesion characteristics of glue compositions, is supposed to take place in the ozonation process.

Ozonation time 0.5–2 h was varied during modification. Ozone concentration (5×10^{-7} vol %) and temperature ($23°C$) were kept constant.

Further, glue compositions based on the ozonized rubbers were prepared. Glue compositions based on the ozonized CNR were 20% solutions of the rubber in an organic solvent, which was ethyl acetate. The compositions based on the isoprene rubber were 5% solutions of the ozonized rubber in petroleum solvent.

The gluing process was conducted at 18–25°C with a double-step deposition of glue and storage of the glued samples under a load of 2 kg for 24 h. Glue bonding of vulcanizates was tested in 24 (±0.5) hours after constructing of joint by method called "Shear strength determination" (State Standard 14759–69), in quality of samples there were used poly-isoprene (SKI-3), ethylenpropylene (SKEPT-40), butadiene-nitrile (SKN-18), and chloroprene (Neoprene) vulcanized rubbers.

The modifying in dc glow discharge was conducted on a vacuum plasma-chemical installation according to Ref. [4]. The rubber samples were mounted on the anode; the operating gas was the filtered air with pressure ~13 Pa, dc 50 mA, and the plasma exposure time was 60 s.

The wettability estimation of a polymer before and after ozonation as well as after plasma treatment was carried out in the study. The samples were films of CNR of CR-20 brand with the thickness of about 100 μm.

The surface properties were characterized by values of contact angles of wettability (Θ) measured on two operating fluids, which were deionized water and glycerin (error ±1°). The values of the work of adhesion (W_a) and the total surface energy (γ) as well as its polar (γ^p) and dispersion (γ^d) components were calculated according to the technique described in Ref. [6] using the Θ values obtained in the experiment.

The surface hydrophilization is likely connected with a change in the rubber chemical structure. A surface structure was studied with Fourier-IR-spectroscopy. Absorption band assignment was made according to Ref. [7].

19.3 RESULTS AND DISCUSSION

In ozonation, a partial double bonds breakage in the rubber macromolecules that leads to the macroradicals formation occurs (Fig. 19.1). Ozone molecules are attached to the point of the rubber double bonds breakage with the formation of epoxy groups [8]:

The formed macroradicals are probably interacting with macromolecules of the rubber, which is a substrate material, thereby higher adhesion strength is provided.

At first, the CNR of three brands was investigated. The results obtained in ozonation of three rubber brands are shown on the Fig. 19.2.

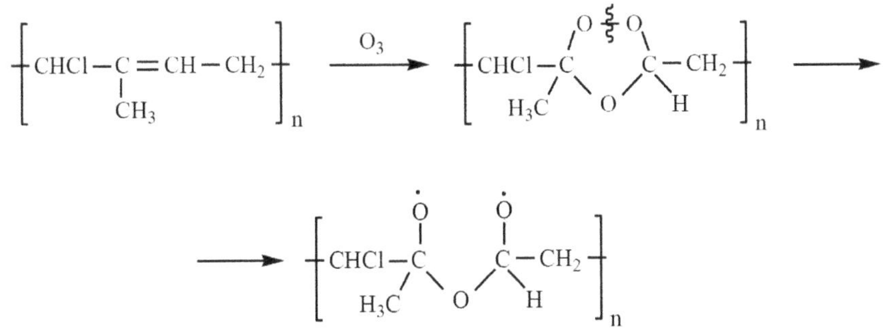

FIGURE 19.1 Reaction of the macroradicals formation (on the example of CNR).

A decrease in the adhesion strength at $\tau = 0,5$hr may be connected with preliminary destruction of macromolecules under action of reactive ozone.

From the Fig. 19.2 we can see that the maximal figures correspond to 1 h ozonation. The adhesion strength for rubbers based on different caoutchoucs increases by 10–70% at that.

The results of the shear strength change depending on the rubber brand and adherend type are shown on the Fig. 19.3.

It should be noticed that the extreme nature of the above-mentioned dependences can be explained by the diffusion nature of the interaction between adhesive and substrate. As it's shown from the figures, with the increasing content of functional groups the strength began to reduce, having reached of certain limit in the adhesive. In this case, only adhesive molecules have ability to diffusion [8].

Isoprene rubber was also treated with ozone at the same parameters maintained for CNR. The results are on the Fig. 19.4.

The data in Fig. 19.4 show that if the ozonation time is equal to 15 min (as in the case of CNR epoxydation), possible preliminary destruction of the rubber macromolecules takes place, which on the graphs is proved by almost simultaneous reduction in the shear strength values. Concurrently, formation and subsequent growth of macroradicals go on, as evidenced by the increase in adhesive characteristics at ozonation time 0.5 and 1 hr. Here the shear strength at gluing of different vulcanized rubbers increases by 10–60% on average, and then it starts to reduce again.

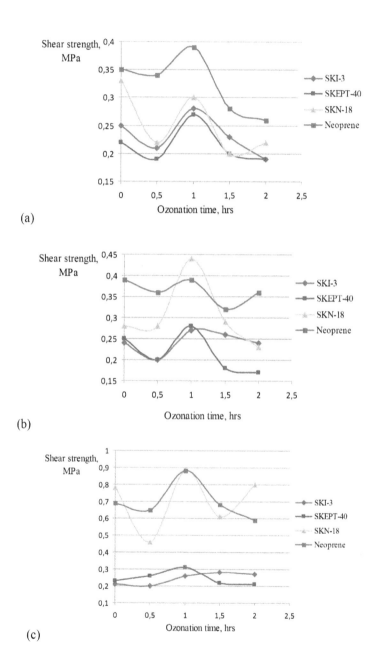

FIGURE 19.2   Influence of ozonation time on shear strength at gluing of vulcanizates with glue compositions based on CNR of the S-20 (a), CR-10 (b) and CR-20 (c) brands accordingly.

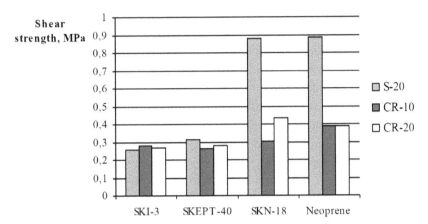

FIGURE 19.3 A change in shear strength for different CNR brands depending on adherend type (ozonation time τ = 1 h).

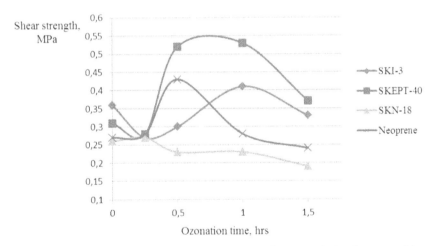

FIGURE 19.4 Influence of ozonation time on adhesive strength for the compositions based on isoprene rubber.

With further increase in ozonation time, the values of adhesion strength decline. That is apparently related to saturation of the polymer chain with epoxy groups and decrease in mobility of the macromolecules, and, consequently, the degree of interaction of the substrate with the adhesive composition as well as with destruction of the polymer chains.

The significant impact on the adhesion characteristics has a contact area between an adhesive and a substrate. An increase in the contact area leads to increased adhesion. The main role in this case is the not only the presence of microroughness in the substrate, but also an index of surface wettability of an adhesive.

The results obtained in defining the polymer wettability index before and after ozonation represent (Table 19.1) that the initial surfaces of a rubber are hydrophobic whereas ozonation allows obtaining θ_{water} values typical for the hydrophilic edge [9].

So, for enhancing the efficiency of ozonation, the additional treatment of the initial and ozonized rubber surfaces in dc charge was conducted.

Plasma impact on a polymer surface promotes mainly a change of its contact properties, such as, wettability, adhesion, permeability, biocompara-bility, etc.) [4, 5]. As a rule, an improvement of polymer adhesion properties by plasma treatment is connected not only with surface purification, but also explained by formation of hydrophilic groups of different chemical nature that provide higher adhesion properties for the modified surface. Composition, structure and properties of the polar groups depend both on the polymer nature and plasma parameters and plasma-supporting gas nature.

The research results revealed that plasma exposure on the CR-20 both before and after ozonation leads to the essential decrease in the Θ values on the water and glycerin, the significant increase in the work of adhesion, the total surface energy and predominant increase in its polar component as well (Table 19.2). The surface of rubber samples becomes hydrophilic [4]. For the CR-20 modified by plasma the decrease in values of contact angles on the water from 87° to 14° is examined; the total surface energy

TABLE 19.1   The Values of Contact Angles of Wettability on the Water and Glycerin (θ), the Work of Adhesion (W_a) and the Surface Energy (γ) For Samples of CNR CR-20 Initial and Ozonized

Sample	θ, deg.		W_a, mJ/m²		γ, mJ/m²		
	Water	Glyc.	Water	Glyc.	γ	γ^p	γ^d
CR-20 initial	77	77	76.6	77.6	24.2	7.0	17.2
CR-20 ozonized	55	65	114.6	90.2	52.3	50.1	2.0

Notes. (1) Measurement error for contact angles was ±1°. (2) γ^p – polar component of the surface energy, γ^d – dispersion component of the surface energy.

TABLE 19.2   The Values of Contact Angles of Wettability on the Water and Glycerin (θ), the Work of Adhesion (W_a) and the Surface Energy (γ) For Samples of CNR CR-20 Initial and Ozonized, Treated in Plasma

	θ, deg.		W_a, mJ/m²		γ, mJ/m²		
Sample	**Water**	**Glyc.**	**Water**	**Glyc.**	**Water**	**Glyc.**	**Water**
CR-20 initial	14	13	143.4	125.2	71.1	54.9	16.2
CR-20 ozonized	11	13	144.3	125.2	72.2	56.8	15.4

Notes. (1) Measurement error for contact angles was ±1°. (2) γ^p – polar component of the surface energy, γ^d – dispersion component of the surface energy.

increases by almost 3 times and its polar component rises by 7.8 times compared to the initial rubber.

The obtained results indicate the hydrophilic nature of the surface of the samples modified by plasma and a significant improvement in their contact properties.

The surface hydrophilization is likely connected with a change in the rubber chemical structure. In Fig. 19.5, the spectra are given for initial CR-20 (a) and modified by plasma (b). It is illustrated that current discharge exposure changes essentially the absorption band ratio in the 2800–3000 cm⁻¹ area, which corresponds to stretch overtones of CH₃и CH₂–groups; it can be explained by crosslinking on the modified surface. A broad absorption band in the 3500–3600 cm⁻¹ area indicates an improvement of the water adsorption.

Besides, it is known that a change of the contact properties of the polymers modified by plasma may be connected with the formation of an excessive surface charge on a surface and in a boundary layer [5].

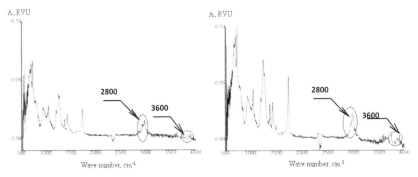

FIGURE 19.5   IR spectra of CR-20 samples before (a) and after treatment by plasma (b).

Thus, ozonation can be applied as an effective method of increasing adhesion properties of rubbers at the modification of film-forming polymers that are a main component in glues. Plasma treatment allows enhancing the effect obtained in ozonation at the expense of rising the wettability index for films based on the treated rubbers.

KEYWORDS

- **adhesion**
- **gluing**
- **modification**
- **ozonation**
- **plasma treatment**
- **unsaturated rubber**

REFERENCES

1. Semenova, E.V. Film-forming compositions based on butadiene containing elastomers modified in latex form: thesis of Ph.D. in Technical Sciences: 05.17.06. Semenova Elena Vladimirovna. Voronezh, 2002, 17 p.
2. Solovyev, M. M. Local dynamics of oligobutadienes of different microstructure and their modification products: thesis of Ph.D. in Chemistry Sciences: 02.00.06. Solovyev Mikhail Mikhailovich. Yaroslavl, 2009, 201 p.
3. Yasuda, X. Plasma Polymerization. X. Yasuda. – M.: Mir, 1988. 374 p.
4. Gilman, A. B. Exposure of dc discharge on properties and structure of polyimide films. A. I. Drachev, A. A. Kuznetsov, G. V. Lopukhova, V. K. Potapov, Eds. High-energy chemistry, 1997. Vol. 31. № 2, pp. 141–145.
5. Encyclopedia of low-temperature plasma. Themed volume XI-5. Edited by Yu.A. Lebedev, N.A. Platea, V.E. Fortova. Moscow, "Yanus-K" Publishing House, 2006, pp. 173–186.
6. Dontsov, A. A., Lozovick, G. Ya., Novitskaya, S. P. Chlorinated polymers. Moscow, "Chemistry" Publishing House, 1979, 232 p.
7. Kuptsov, A. X. Handbook of Fourier transform raman and infrared spectra of polymers. A.X. Kuptsov, Zhizhin, G.N. Moscow, Eds. "Fizmatlit" Publishing House, 2001, 581 p.
8. Berlin, A. A., Basin, V. E. Basics of Polymer Adhesion. Moscow, "Chemistry" Publishing House, 1969, 320 p.
9. Wade, W. L. Surface properties of commercial polymer films following various gas plasma treatments. Mannuone, R.J., Binder, M.; J. Appl. Polym. Sci., 1991. Vol. 43. № 9, pp. 1589–1591.

CHAPTER 20

INFLUENCE OF HYDROLYSIS ON DIFFUSION OF ELECTROLYTES

ANA C. F. RIBEIRO,[1] LUÍS M. P. VERÍSSIMO,[1]
ARTUR J. M. VALENTE,[1] ANA M. T. D. P. V. CABRAL,[2]
and VICTOR M. M. LOBO[1]

[1]Department of Chemistry, University of Coimbra, 3004–535
Coimbra, Portugal, Tel: +351-239-854460; Fax: +351-239-827703;
E-mail: anacfrib@ci.uc.pt; luisve@gmail.com; avalente@ci.uc.pt;
vlobo@ci.uc.pt

[2]Faculty of Pharmacy, University of Coimbra, 3000-295 Coimbra,
Portugal, E-mail: acabral@ff.uc.pt

CONTENTS

ABSTRACT

Experimental mutual diffusion coefficients of potassium chloride, magnesium nitrate, cobalt chloride and beryllium sulfate in aqueous solutions have been compared with those estimated by the Onsager-Fuoss. From the observed deviations, we are proposing a model that allows to estimate the percentages of H_3O^+ (aq) resulting of hydrolysis of some ions (i.e., beryllium, cobalt, potassium and magnesium ions), contributing in this way to a better knowledge of the structure of these systems. The diffusion of beryllium sulfate is clearly the most affected by the beryllium ion hydrolysis.

20.1 THEORETICAL ASPECTS

20.1.1 CONCEPTS OF DIFFUSION

Diffusion data of electrolytes in aqueous solutions are of great interest not only for fundamental purposes (providing a detailed comprehensive information – both kinetic and thermodynamic), but also for many technical fields such as corrosion studies [1–3]. The gradient of chemical potential in the solution is the force producing the irreversible process, which we call diffusion [1–3]. However, in most solutions, that force may be attributed to the gradient of the concentration at constant temperature. Thus, the diffusion coefficient, D, in a binary system, may be defined in terms of the concentration gradient by a phenomenological relationship, known as Fick's first law,

$$J = -D \text{ grad } c \tag{1}$$

or, considering only one dimension for practical reasons,

$$J = -D \frac{\partial c}{\partial x} \tag{2}$$

where J represents the flow of matter across a suitable chosen reference plane per area unit and per time unit, in a one-dimensional system, and c is the concentration of solute in moles per volume unit at the point considered; Eq. (1) may be used to measure, D. The diffusion coefficient may also measured considering Fick's second law, in one dimensional system,

$$\frac{\partial c}{\partial t} = \frac{\partial}{\partial x}\left(D \frac{\partial c}{\partial x}\right) \tag{3}$$

In general, the available methods are grouped into two groups: steady and unsteady-state methods, according to Eqs. (2) and (3). Diffusion is a three-dimensional phenomenon, but many of the experimental methods used to analyze diffusion restrict it to a one-dimensional process [1–4].

The resolution of Eq. (2) is much easier if we consider D as a constant. This approximation is applicable only when there are small differences of concentration, which is the case in our open-ended condutimetric technique [3, 5], and in the Taylor technique [3, 6, 7]. In these conditions, it is legitimate to consider that our measurements of differential diffusion coefficients obtained by the above techniques are parameters with a well defined thermodynamic meaning [3–7].

20.1.2 EFFECT OF THE HYDROLYSIS ON DIFFUSION OF ELECTROLYTES IN AQUEOUS SOLUTIONS

A theory of mutual diffusion of electrolytes in aqueous solutions capable of accurately predicting diffusion coefficients has not yet been successfully developed, due to the complex nature of these systems. However, Onsager and Fuoss equation (Eq. (4)) [8], has allowed the estimation of diffusion coefficients with a good approximation for dilute solutions and symmetrical electrolytes of the type 1:1. This equation may be expressed by

$$D = \left(1 + c\frac{\partial \ln y_{\pm}}{\partial c}\right)\left(D^0 + \Sigma \Delta n\right) \tag{4}$$

where D is the mutual diffusion coefficient of the electrolyte, the first term in parenthesis is the activity factor, $y_{\pm}$ is the mean molar activity coefficient, c is the concentration in mol dm^{-3}, D^0 is the Nernst limiting value of the diffusion coefficient [Eq. (4)], and Δ_n are the electrophoretic terms given by

$$\Delta n = k_B T \, An \frac{\left(z_1^n t_2^0 + z_2^n t_1^0\right)^2}{|z_1 z_2| a^n} \tag{5}$$

where k_B is the Boltzmann's constant; T is the absolute temperature; A_n are functions of the dielectric constant, of the solvent viscosity, of the temperature, and of the dimensionless concentration-dependent quantity (ka), k being the reciprocal of average radius of the ionic atmosphere; t_1^0 and t_2^0 are the limiting transport numbers of the cation and anion, respectively.

Since the expression for the electrophoretic effect has been derived on the basis of the expansion of the exponential Boltzmann function, because that function is consistent with the Poisson equation, we only would have to take into account the electrophoretic term of the first and the second order ($n = 1$ and $n = 2$). Thus, the experimental data D_{exp} can be compared with the calculated D_{OF} on the basis of Eq. (6)

$$D = \left(1 + c\,\frac{\partial \ln y_\pm}{\partial c}\right)\left(D^0 + \Delta_1 + \Delta_2\right) \qquad (6)$$

It was observed and reported in the literature a good agreement between the observed values and those calculated by Eq. (6) for potassium chloride [9] and magnesium nitrate [10], but for cobalt chloride [11] and beryllium sulfate [12] they are definitely higher than theory predicts. This is not surprising, if we take into account the change with concentration of parameters, such as, viscosity, hydration and hydrolysis, factors not taken into account in Onsager-Fuoss model [8]. In fact, those differences may possibly be due to hydrolysis of the cations (Eq. (7)) [13–15], which would be more pronounced in cobalt chloride and beryllium sulfate than potassium chloride and magnesium nitrate.

$$M(H_2O)_x^{n+}(aq) \;\overset{\longleftarrow}{\rightarrow}\; M(H_2O)_{x-1}(OH)^{(n-1)+}(aq) + H^+(aq) \qquad (7)$$

We are proposing a model in order to allow the estimation of the amount of ion $H^+(aq)$ produced in each of the systems studied, using experimental diffusion coefficients determined by our condutimetric method and the respective predictive by that theory.

That is, having in mind the acidic character of some cations [14, 15], the percentages ion $H^+(aq)$ produced (or the amount of acid that would be necessary to add a given electrolyte solution in the absence of hydrolysis in order to simulate a more real system) are determined by the following system (Eqs. (8) and (9)),

$$D_{acid}z + D_{electrolyte}y = D_{exp} \qquad (8)$$

$$z + y = 1 \qquad (9)$$

being $z \times 100$ and $y \times 100$ the percentages of acid and electrolyte, respectively, and $D_{exp,}$ our values of the diffusion coefficients of electrolytes in aqueous solutions, and D_{acid} and $D_{electrolyte}$, the Onsager and Fuoss values of the diffusion coefficients for the acids and electrolytes, respectively, in aqueous solutions.

20.2   EXPERIMENTAL ASPECTS

20.2.1   CONDUCTIMETRIC TECHNIQUE

An open-ended capillary cell (Fig. 20.1) [5], which has been used to obtain mutual diffusion coefficients for a wide variety of electrolytes [5, 9–13, 16–19], has been described in great detail [5]). Basically, it consists of two vertical capillaries, each closed at one end by a platinum electrode, and positioned one above the other with the open ends separated by a distance of about 14 mm. The upper and lower tubes, initially filled with solutions of concentrations 0.75 c and 1.25 c, respectively, are surrounded with a solution of concentration c. This ambient solution is contained in a glass tank $(200 \times 140 \times 60)$ mm immersed in a thermostat bath at 25°C. Perspex sheets divide the tank internally and a glass stirrer creates a slow lateral flow of ambient solution across the open ends of the capillaries. Experimental conditions are such that the concentration at each of the open ends is equal to the ambient solution value c, that is, the physical length of the capillary tube coincides with the diffusion path. This means that the required boundary conditions described in the literature [5] to solve Fick's second law of diffusion are applicable. Therefore, the so-called Δl effect [5] is reduced to negligible proportions. In our manually operated apparatus, diffusion is followed by measuring the ratio w=R_t/R_b of resistances R_t and R_b of the upper and lower tubes by an alternating current transformer bridge. In our automatic apparatus, w is measured by a Solartron digital voltmeter (DVM) 7061 with 6 1/2 digits. A power source (Bradley Electronic Model 232) supplies a 30 V sinusoidal signal at 4 kHz (stable to

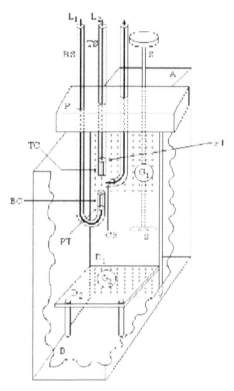

FIGURE 20.1 Conductimetric technique (TS, BS: support capillaries; TC, BC: top and bottom diffusion capillaries; CE: central electrode; PT: platinum electrodes; D1, D2: perspex sheets; S: glass stirrer; P: perspex block; G1, G2: perforations in perspex sheets; A, B: sections of the tank; L1, L2: small diameter coaxial leads [5]).

within 0.1 mV) to a potential divider that applies a 250 mV signal to the platinum electrodes in the top and bottom capillaries. By measuring the voltages V' and V" from top and bottom electrodes to a central electrode at ground potential in a fraction of a second, the DVM calculates w.

In order to measure the differential diffusion coefficient D at a given concentration c, the bulk solution of concentration c is prepared by mixing 1 L of "top" solution with 1 L of "bottom" solution, accurately measured. The glass tank and the two capillaries are filled with c solution, immersed in the thermostat, and allowed to come to thermal equilibrium. The resistance ratio $w = w_\infty$ measured under these conditions (with solutions in both capillaries at concentration c accurately gives the quantity $t_\infty = 10^4/(1 + w_\infty)$.

The capillaries are filled with the "top" and "bottom" solutions, which are then allowed to diffuse into the "bulk" solution. Resistance ratio readings are taken at various recorded times, beginning 1000 min after the start of the experiment, to determine the quantity $t = 10^4/(1+w)$ as t approaches t_∞. The diffusion coefficient is evaluated using a linear least-squares procedure to fit the data and, finally, an iterative process is applied using 20 terms of the expansion series of Fick's second law for the present boundary conditions. The theory developed for the cell has been described previously [5].

Inter-diffusion coefficients of potassium chloride [9], magnesium nitrate [10], cobalt chloride [11] and beryllium sulfate [12] in water at 298.15 K, and at concentrations from 0.001 to 0.01 mol dm^{-3}, have been measured using a conductimetric cell and an automatic apparatus to follow diffusion. The cell uses an open-ended capillary method and a conductimetric technique is used to follow the diffusion process by measuring the resistance of a solution inside the capillaries, at recorded times.

20.3   EXPERIMENTAL RESULTS AND DISCUSSION

Table 20.1 shows the estimated percentage of hydrogen ions, z, resulting from the hydrolysis of different ions at different concentrations at 298.15 K, by using the Eqs. (8) and (9).

From the analysis of Table 20.1, we see that the diffusion of cobalt ion and mainly beryllium ion are the most clearly affected by their hydrolysis.

TABLE 20.1   Estimated Percentage of Hydrogen Ions, z, Resulting From the Hydrolysis of Some Cations in Aqueous Solutions at Different Concentrations, c, and at 298.15 K, Using Eqs. (8) and (9)

$c/$ mol dm^{-3}	$z\%$ (KCl)	$z\%$ (Mg(NO$_3$)$_2$)	$z\%$ (CoCl$_2$)	$z\%$ (BeSO$_4$)
1×10^{-3}	2	–	5	–
3×10^{-3}	3	1	6	69
5×10^{-3}	2	2	8	48
8×10^{-3}	3	1	6	26
1×10^{-2}	3	1	6	34

At the lowest Be^{2+} ion concentration (3×10^{-3} mol dm^{-3}) the effect of the hydrogen ions on the whole diffusion process has an important and main role ($z = 69\%$, Table 20.1). However, we can, to a good approximation, describe these systems as binary if we consider certain facts. For example, in the case of solutions of beryllium sulfate, several studies [12] indicated that the predominant species present are, the solvent, Be^{2+}, SO_4^{2-} and H^+ ions, as well as indicated in the following scheme resulting from hydrolysis of the Be^{2+} ion.

$$[Be(H_2O)_4]^{2+} \leftrightarrow [Be(H_2O)_3(OH)]^- + H^+$$

$$[Be(H_2O)_3(OH)]^- + [Be(H_2O)_4]^{2+} \leftrightarrow [(H_2O)_3\,Be\text{-}OH\text{-}Be\,(H_2O)_3]^{3+}$$

However, based on pH measurements (i.e., $2.9 \leq pH \leq 3.9$) for various $BeSO_4$ solutions at concentrations ranging from 3×10^{-3} mol dm^{-3} and 1×10^{-2} mol dm^{-3}, and from the low equilibrium constants relative to the hydrolysis of the Be^{2+} ion [15], we can then treat this system as being a binary system. Moreover, given the uniformity of the concentration gradient imposed by our method, it is considered that the value of the diffusion coefficient, which is given to the scientific community, is a quantitative measure of all existing species that diffuse.

Concerning the other electrolytes (KCl and $Mg(NO_3)_2$), [9, 10] the reasonable agreement obtained between the calculated and the experimental values of diffusion coefficient, should result in negligible percentages for the hydronium ion (Table 20.1).

20.4   CONCLUSIONS

We have estimated the percentages of H_3O^+ (aq) resulting from the hydrolysis of some ions (i.e., beryllium, cobalt, potassium and magnesium ions), using a new model and the experimental and calculated mutual diffusion coefficients for potassium chloride, magnesium nitrate, cobalt chloride and beryllium sulfate in aqueous solutions, contributing in this way for a better knowledge of the structure of these systems.

Among the cited systems, we concluded that the diffusion of beryllium sulfate is clearly the most affected by the beryllium ion hydrolysis.

ACKNOWLEDGMENTS

Financial support of the Coimbra Chemistry Centre from the FCT through project Pest-OE/QUI/UI0313/2014 is gratefully acknowledged.

KEYWORDS

- aqueous solutions
- diffusion coefficients
- electrolytes
- hydration

REFERENCES

1. Robinson, R. A., Stokes, R. H., Electrolyte Solutions, second ed., Butterworths, London, 1959.
2. Harned, H. S., Owen, B. B., The Physical Chemistry of Electrolytic Solutions, third ed., Reinhold Pub. Corp., New York, 1964.
3. Tyrrell, H. J. V., Harris, K. R., Diffusion in Liquids: A Theoretical and Experimental Study, Butterworths, London, 1984.
4. Lobo, V. M. M., Handbook of Electrolyte Solutions, Elsevier, Amsterdam, 1990.
5. Agar, J. N., Lobo, V. M. M. (*Measurement of diffusion coefficients of electrolytes by a modified Open-Ended Capillary Method*), J. Chem. Soc., Faraday Trans. I, 71, 1659–1666 (1975).
6. Callendar, R., Leaist D. G. (*Diffusion coefficients for binary, ternary, and polydisperse solutions from peak-width analysis of Taylor dispersion profiles*), J. Sol. Chem. 35, 353–379 (2006).
7. Barthel, J., Feuerlein, F., Neuder, R., Wachter, R. (*Calibration of conductance cells at various temperatures*), J. Sol. Chem. 9, 209–212 (1980).
8. Onsager, L., Fuoss, R. M. (*Irreversible processes in electrolytes. Diffusion. conductance and viscous flow in arbitrary mixtures of strong electrolytes*), J. Phys. Chem. 36, 2689–2778 (1932).
9. Lobo, V. M. M., Ribeiro, A. C. F., Verissimo, L. M. P. (*Diffusion coefficients in aqueous solutions of potassium chloride at high and low concentrations*), J. Mol. Liq. 94, 61–66 (2001).
10. Lobo, V. M. M., Ribeiro, A. C. F., Veríssimo, L. M. P. (*Diffusion coefficients in aqueous solutions of magnesium nitrate*), Ber. Bunsen Gesells. Phys. Chem. Chem. PhYS. 98, 205–208 (1994).

11. Ribeiro, A. C. F., Lobo, V. M. M., Natividade, J. J. S. (*Diffusion coefficients in aqueous solutions of cobalt chloride at 298.15 K*), J. Chem. Eng. Data 47, 539–541 (2002).
12. Lobo, V. M. M., Ribeiro, A. C. F., Veríssimo, L. M. P. (*Diffusion coefficients in aqueous solutions of beryllium sulfate at 298 K*), J. Chem. Eng. Data 39, 726–728 (1994).
13. Valente, A. J. M., Ribeiro, A. C. F., Lobo, V. M. M., Jiménez, A. (*Diffusion coefficients of lead(II) nitrate in nitric acid aqueous solutions at 298.15 K*), J. Mol. Liq. 111, 33–38 (2004).
14. Baes, C. F., Mesmer, R. E., The Hydrolysis of Cations, John Wiley & Sons, New York, 1976.
15. Burgess, J. Metal Ions in Solution, John Wiley & Sons, Chichester, Sussex, England, 1978.
16. Ribeiro, A. C. F., Lobo, V. M. M., Oliveira, L. R. C., Burrows, H. D., Azevedo, E. F. G., S. Fangaia, I. G., Nicolau, P. M. G., Fernando, A. D. R. A. Guerra (*Diffusion coefficients of chromium chloride in aqueous solutions at 298.15 and 303.15 K*), J. Chem. Eng. Data 50, 1014–1017 (2005).
17. Ribeiro, A. C. F., **Esteso, M. A.**, Lobo, V. M. M., Valente, A. J. M., Sobral, A. J. F. N., Burrows, H. D. (*Diffusion coefficients of aluminum chloride in aqueous solutions at 298.15 K, 303.15 K and 310.15 K*), Electrochim. Acta.52, 6450–6455 (2007).
18. Veríssimo, A. L. M. P., Ribeiro, A. C. F., Lobo, V. M. M., Esteso, M. A. (*Effect of hydrolysis on diffusion of ferric sulfate in* aqueous solutions**),** J. Chem. Thermodyn. 55, 56–59 (2012).
19. Ribeiro, A., Gomes, A. C. F., Veríssimo, J. C. S., Romero, L. M. P. C., Blanco, L. H., Esteso, M. A. (*Diffusion of cadmium chloride in aqueous solutions at physiological temperature 310.15 K*), J. Chem. Thermodyn. 57, 404–407 (2013).

CHAPTER 21

MAGNETIC NANOPARTICLES' EFFECT ON PHOTOELECTRIC SENSITIVITY AND PHOTOCONDUCTION OF POLYMER COMPOSITES

B. M. RUMYANYZEV,[1] S. B. BIBIKOV,[1] V. I. BERENDYAEV,[2]
A. L. KOVARSKY,[1] O. N. SOROKINA,[1] A. V. BYCHKOVA,[1]
V. G. LEONTIEV,[1] and L. N. NIKITIN[1]

[1]*Russian Academy of Sciences, Moscow, Russia*

[2]*L.Ya.Karpov Institute of Physical Chemistry, Moscow, Russia*

CONTENTS

ABSTRACT

It has been found the influence of magnetic nanoparticles (Ni, Cu-Ni alloy, Fe in carbon nanotubes (Fe-CNT), Fe_3O_4) stabilized by polyalkanetherimide

(PAEI) on photoelectric sensitivity (PES), carrier photogeneration quantum yield β and the residual potential of composite films based on soluble photoconductive polyimides (PI) with PAEI, and also on p – and n – transport layers of photovoltaic cells based on polymer p-n heterostructures. It has been shown that the magnitude and sign of the effect is determined by the structure of PI, type of charge transfer interaction (interchain or intrachain) and also by the degree of surface coating of MNP with adsorbed PAEI macromolecules. At zero and low coverage (N < 50 macromolecules per MNP) the changing of positive effect to negative is observed. For N=100–200 there is the maximum positive effect (growth of PES by 4.7–5.0 times and the photogeneration quantum yield by 1.6–2.5 times), and for N > 200 effect decreases. It is suggested that the effect is connected with the redistribution of PAEI in the volume caused by effective adsorption of PAEI macromolecules on the surface of MNP. This redistribution significantly affects PES, the magnitude of β and carrier drift length l_D both because of the interchain nature of EDA interaction in PI and because of the strong (exponential) dependences of these values on the distance between transport centers for carrier hopping conductivity and photogeneration mechanisms. Alternative explanations (magnetic-spin effects in the field of MNP, the scattering of light by particles) is also considered. The magnitude of the magnetic field of Ni-Cu MNP has been estimated using studied earlier low field magnetic effect on the luminescence of PAEI composites with rubrene. Although the chaotic orientation fields of individual particles the magnetic spin effect from MNP, depending on square of local field strength, is not zero. Measurements showed that the luminescence intensity change in the MNP field (–3%) corresponds to a magnetic effect in homogeneous external field with the strength H_0 > 10 Oe. This value has been taken for the effective average MNP field strength within the composite film at particles content of 6 mass.%.

21.1 INTRODUCTION

Recent investigations demonstrate that single domain magnetic nanoparticles (MNP) possess by the numerous unusual properties as compared with bulk samples: increased magnetization, blocking temperature T_b, magneto resistivity, magnetic nanocomposites- anomalous high magneto caloric effect [1]. Within the temperature range $T_b < T < T_C$ (Curie temperature) these particles have nonzero magnetic momentum which change easily its orientation in external magnetic field. It is shown that magnetic properties of nanoparticles are determined not only by their chemical composition, type of crystal lattice, their size and shape, but to the great extent by their interaction with host matrix and neighboring particles [1]. As in the last time it can possible the preparing of MNP incorporated into various polymer matrices (polymer magnetic nanocomposites) the possibility appears to control their properties by varying the interactions with matrix. And if this interaction effect on MNP is studying intensively [2] the reverse MNP effect on matrix itself is not investigated so widely, its optical (luminescent) and photoelectrical characteristics (charge carrier photogeneration quantum yield, carrier transport, their capture and recombination) which determine their application in optoelectronics (photovoltaics, photodetectors, electroluminescent cells). This paper is devoted to investigation of MNP effect on photoelectric characteristics of photoactive polymers (original photoconducting soluble polyimides (PI)) and their composites with inert polymers (polyalkanetherimides (PAEI) and low molecular mass photoconductors-dopants [3] with the purpose of developing the approaches to synthesis of smart magnetic nanocomposites which photoelectrical and photovoltaic properties can be governed by MNP. If one takes into account that to the present time the numerous optoelectronic devices (photovoltaic (PV) and electroluminescent cells, photodetectors, polymer transistors [4]) are developed on base of polymer composites, the problem of control their characteristics becomes very actual. So, in patent claim [5] for increasing the polymer PV cell efficiency based on composites of conjugated polymers with fullerene the magnetite Fe_3O_4 MNP are added to active layer. During the cell preparing the mixture is undergone to external magnetic field action, which orients MNP magnetic moments forming numerous conducting particle chains. These chains induce built-in electric field, which promotes charge separation in electron-hole pairs, and improve carrier transport.

In Ref. [6] for the same purpose the magnetite MNP covered with fatty acid surface active layer are used. In this case the increase of cell efficiency is explained by the growth of long-lived triplet electron-hole pair population in MNP magnetic field (magnetic spin effect). In Ref. [7] replacing of organic hole transport layer PEDOT-PSS in PV cell by thin MNP Fe_3O_4 layer cast from solution results in not only increase of energy conversion efficiency on 12% of its value but also the improvement of temporal stability.

Adding of CoFe MNP into organic electroluminescent cells based on conjugated polymers results in the increase of emission quantum efficiency on 27% [8]. Its further growth is achieved by applying external magnetic field causing MNP spin orientation that results in increase of singlet electron-hole pairs population which recombination gives additional photon emission. So, use of MNP in various optoelectronic devices allows to regulate effectively their characteristics.

In this chapter, it is shown also that besides of MNP magnetic properties the significant role in photoelectric characteristics determination plays adsorption of PAEI macromolecules, which are used as nanoparticle stabilizer on the MNP surface. The adsorption effect of various polymers both synthetic and biopolymers on magnetic properties and electron magnetic resonance (EMR) spectra for magnetite MNP is investigated in details in Ref. [9] in connection with problem of vectorial drug delivery in human organism. It is shown in Ref. [9] that in some cases the adsorption efficiency is very high and the methods of polymer chemical modification for its improving are developed. It is found that adsorption is accompanied by MNP magnetic clusters formation.

21.2 EXPERIMENTAL PART

The nanoparticles of Ni and Cu-Ni alloy with 28 atomic % Cu content are used for investigation of the MNP effect on photoelectric sensitivity (PES) and charge carrier photogeneration quantum yield in composite films. MNP are synthesized by codeposition of proper salts with following reduction by hydrogen and thermal treating [10]. The MNP size is within the range from 5 to 50 nm. Their ferromagnetic properties are investigated by EMR and magnetostatic methods, which reveal that Curie temperature for these MNP $T_C = 40–60°C$ [10]. The original soluble photoconducting PIs [11] and their composites with PAEI or dopant-sensitizers are used as photoactive polymer matrices. For obtaining of magnetic liquid dispersion based on MNP 1% PAEI solution in chlorinated solvent is used. Magnetic dispersion with known MNP concentration (of about 20% mass.) is treated in ultrasonic bath at maximum power until formation of homogeneously colored solution. The solution remains stable during some hours. Magnetic solutions are introduced into PI or PEPC solution (or PI mixture with organic dopant-sensitizer perylenediimide derivative (PDID) 1:1) in tetrachloroethane (1% mass.) in amount of 30% volume. The obtained mixture cast onto glass ITO conducting plates 2×2 cm^2 with following drying at the air under room temperature in a magnetic field H=1,5 kOe. In such a way the magnetooriented homogeneous composite films of good quality are obtained with mass ratio of components PI: PAEI: MNP 10:2.5:0.6.

Fe nanoparticles incorporated into carbon nanotubes (FeCNT) and magnetite Fe$_3$O$_4$ MNP with average size 17 nm obtained in colloidal solutions as magnetic liquid by the method of alkaline deposition of Fe(II) and (Fe III) salts mixture in aqua media [12] are used also. MNP size and their distribution are determined by correlative spectrometry using single plate correlator of real time "Photocor-SP" as well as dynamic light scattering and electron microscopy methods. Film magnetization is measured by EMR method. This method is used also for observation of MNP aggregate formation in polymer matrix, determination of MNP part involved in aggregates and alignment degree in linear aggregates.

Photoelectric sensitivity (PES) and charge carrier photogeneration quantum yield (CCPGQY) for composite films (of 2–3 μm thickness) are measured by conventional electrophotographic method. This method consists of studying the kinetics of dark and photo induced surface potential decay for polymer films charged in positive (or negative) corona discharge [13]. Maximum surface charge potential V and its dark decay rate (dV/dt)$_d$ are determined by film conductivity. Photoinduced potential decay rate (1/I) (dV/dt)$_{ph}$ (I–photon flux, cm^{-2} sec^{-1}) is determined by photodischarge rate of capacitor with one transparent electrode formed by the ionic contact (air ions on polymer surface) and the second one –glass ITO plate. Its capacity equals to:

$$A\varepsilon_0/d,$$

where A – film surface area, ε_0 – dielectric constant, d- film thickness. Photodischarge rate depends on efficient CCPGQY (β_{ef}), carrier collection efficiency on electrodes C(Z) and part of absorbed light Π:

$$(1/I) \, [(dV/dt)_{ph} - (dV/dt)_d] = (ed)/(\varepsilon\varepsilon_0 \, \beta_{ef} \, \Pi) \tag{1}$$

where $\beta_{ef} = \beta C(Z)$, $Z = (\mu V\tau/d^2)$ – ratio of carrier drift length ($l_D = \mu V\tau/d$) to film thickness d, μ – carrier drift mobility, τ – their lifetime, e – electron charge. The expression of C(Z) function for strong and weak absorption is given in [13] from which it follows that for Z<1 C(Z) = Z and for Z>1 C(Z) =1 (strong absorption) while for weak absorption C(Z) = 1/2. For polymers β and C(Z) values usually strongly dependent on field strength E = V/d. PES value S is determined as the inverse exposition of initial charge potential half decay:

$$S = (Wt_{1/2})^{-1} = (\beta_{ef} \, \Pi de)/(E(h\nu) \, V\varepsilon\varepsilon_0) \, [m^2/J] \tag{2}$$

where E(hν) – excitation photon energy, W – excitation intensity, J/m^2 sec). So this method allows to obtain the next photoelectric characteristics of polymer films: PES, CCPGQY from Eq. (1) and to evaluate carrier drift length l_D from β_{eff} (E) field dependence which involves also C(Z) field dependence. For the latter transition from C(Z)=Z dependence to C(Z)=1 one occurs at field strength E_0 for which drift length is equal to film thickness: $l_D (E_0) = \mu E_0\tau = d$. For Z<1 the residue potential is observed on decay kinetic curve V(t) under inhomogeneous excitation (by UV light) which is caused by carriers trapped in film bulk.

The setup enables also to measure sample optical density D(λ) at monochromatic or integral excitation. This allows to evaluate the part of absorbed light:

$$\Pi = 1 - \exp\,[-(D-D_0)],$$

where D_0 – equivalent optical density caused by light reflection from front and back sample sides and light scattering.

The sign of major carriers can be elucidate via comparison of PES values under positive (S^+) and negative (S^-) corona charge of free surface for inhomogeneous excitation by UV light: if $S^+ > S^-$ the major carriers are holes, in contrary case – electrons.

The setup enables also to measure the integral luminescence intensity of samples and its external magnetic field effect at H < 1 kOe within T = 290–350 K range.

21.3   RESULTS AND DISCUSSION

21.3.1   STABILIZED MAGNETIC NANOPARTICLES EFFECT ON PHOTOELECTRIC AND DARK CHARACTERISTICS OF COMPOSITE FILMS BASED ON PI WITH PAEI

The influence of the Ni, Cu-Ni alloy, Fe in carbon nanotubes (Fe-CNT) and magnetite MNP stabilized with PAEI on the following dark and photoelectric characteristics of the PI+PAEI+MNP composite films obtained from organic solvents (chloroform + tetrachloroethane), photoelectric sensitivity S (PES), maximum surface charge potential V, the rate of surface potential dark decay $(dV/dt)_d$, residual potential U is investigated.

The same parameters of the films PI+PAEI without MNP have been measured (reference samples). The latter includes PAEI used as MNP stabilizer because of its lowering the PI film PES value. The influence of unstabilized MNP on PI film PES (without PAEI) is investigated below in subdivision 21.3.2. Three types of Cu-Ni alloy MNP with the same composition have been studied (MNP4, MNP5, MNP6). These MNP have slightly different particle size. The original soluble photoconductive polyimides on the base of N, N'-bis(n-aminophenyl)-N, N'-diphenylbenzidine [11] have been used in the work:

where X = O(PI1); – (PI2); CO – (PI3); S (PI4).

Synthesis and their properties are presented in Ref. [11], MM = 30000–60000. Polyimides PI1-PI4 differs from each other by the affinity of diimine (acceptor) chain fragment Ea and its steric factor. This steric factor defines the PI ability to crystallization and the strength of the (inter and intrachain) electron donor-acceptor (EDA) interaction with the charge-transfer complex (CTC) formation [11]. The last parameter defines photoelectric and optic properties of PI [11]. The most intensive absorption band of the interchain CTC in visible area (500–600 nm) is observed for the PI3 films (Ea =1.46 eV). Less intensive and more short wavelength band of the interchain CTC (400–500 nm) is observed for the PI1 (Ea = 1.12 eV). For the PI2 and PI4 demonstrating high enough Ea values (1,46 eV) the interchain CTC band observed in visible area has a very low intensity. Such a low value is due to the steric factor for the interchain interaction. For PI2 and PI4 the short wavelength band (350–450 nm) of intrachain CTC is much more intensive. Thus the PES (S) values and the optical density for the composites (D) and reference samples (S$_0$) have been measured in four spectral areas: UV (l<400 nm), blue (400–500 nm), green (500–600) and long wavelength (600–800 nm). The results of measured influence of the MNP on PES of the composites in four spectral areas are collected in Table 21.1. The main parameter used to characterize this effect (x) is calculated as:

$$x= S/S_0 [V_0/V]^{0.5}[\Pi_0/\Pi] \qquad (3)$$

where V, V$_0$, Π, Π_0 are the charge potential and the fraction of absorbed light energy for the composite films with MNP and reference sample (without MNP). The ξ values should be corrected to the same V and Π values because of the S dependence on them: S $\sim$ V$^{0.5}$ Π.

As it can be seen from the Table 21.1 the addition of MNP to the composites on the base of PI1 and PI3 with the PAEI leads to the sufficient PES increasing (ξ = 1.1–7.0) in each excitation area meanwhile for PI2 and PI4 the MNP influence is weak or does not observe at all (PI4). The insignificant influence on the PES is typical for Fe$_3$O$_4$ MNP. The influence of MNP on the PES is higher in the long wavelength excitation area than in other areas for each sample including PI2.

The most significant influence of MNP is observed for the PI composites with the most intensive absorption bands of interchain CTC (PI1 and PI3): UV, blue areas for PI1 and green, long wavelength area for PI3. It has been shown that in these cases (PI1 and PI3) the relative value of the residual potential U/V of the composite films with MNP decreases while for PI4, for which ξ < 1 (Table 21.1), U/V increases.

The addition of Ni and Ni-Cu MNP does not influence on dark parameters of the composite films (V and (dV/dt)$_d$), which can change both in positive and negative directions. This fact evidences the absence of sufficient

TABLE 21.1   The Influence of Stabilized MNP Addition (6 wt.%) on PES of Composites PI1-PI4 with PAEI (ξ)

Polymer	MH Ni	MH Ni-Cu (4)	Ni-Cu (5)	Ni-Cu (6)	FeCNT	Fe$_3$O$_4$
UV excitation (l < 400 nm)						
PI1	1.9	1.6	1.7	1.4	2.2	
PI2	1.0	0.9	0.9	0.9	0.5	
PI3	2.1	2.1	1.5	2.0	2.8	1.2
PI4	0.7		1.0	0.6		
Excitation in blue area (l= 400–500 nm)						
PI1	2.0	1.5	1.2	1.1	1.4	
PI2	1.1	1.1	1.2	1.2	<0.2	
PI3	2.6	1.9	1.6	1.4	1.5	1.1
PI4	0.6		0.7	0.5		
Excitation in green area (l= 500–600 nm)						
PI1	1.4	1.6	1.2	0.8	0.8	
PI2	1.3	1.2	1.2	1.1	<0.1	
PI3	3.0	1.6	1.7	1.3	1.5	0.6
PI4	0.7		0.6	0.4		
Excitation in long wavelength area (l= 600–800 nm)						
PI1	1.3	1.3	1.5	1.5	1.0	
PI2	2.2	1.6	2.2	2.8	<0.1	
PI3	2.6	7.0	5.7	3.4	4.3	0.5
PI4	0.5		0.5	0.9		

conductivity of percolation chains and their aggregates in magnetoanisotropic composites. Some influence on the dark parameters can be observed for FeCNT MNP (charge potential V decrease).

The same dependence demonstrates optical density D (the fraction of absorbed light energy Π): in some cases it increases for the films with MNP, in other ones it does not change or even decreases. There is no correlation between D value variations with ξ value. The Rayleigh scattering on MNP in the films volume influences on the D value first of all. It can also influence on the PES parameters due to the increase in the probability of the absorption and photogeneration of the carriers in CTC bands due to the photon optical path increase. Although the scattering coefficient $k \sim \lambda^{-4}$ its influence on the PES should be stronger in the green and long wavelength areas (the area of the weak absorption of interchain CTC where $D<1$ and $\Pi<1$) and should be weak (or absent) in the blue and UV areas where Π value is close to 1. The influence of MNP reaches its maximum in long wavelength area but it is observed as well in blue and UV areas. Thus it is impossible to explain the MNP influence on the PES only by light scattering on MNP. Moreover the correct estimation of MNP influence on the PES (ξ value) takes into account the fraction of the absorbed energy, which the light scattering can only effect on.

It is checked that the optical density of the films with the Ni MNP in transparent polymer matrix (PAEI) caused by the light scattering and self absorption by the particles is maximum in blue area, less in green area, and absent in red one. Since the MNP influence on PES is maximum in long wavelength area (Table 21.1) their participation in the carrier photogeneration (photoemission, photoconductivity of the chains) cannot be significant. Moreover such an explanation does not allow one to clarify the ξ value difference for various PI (Table 21.1).

The measurement of CCPGQY β for the composite films PI3 with MNP4 (Cu-Ni), for which the effect of influence is maximum (Table 21.1), for each spectral area has shown that its field dependences ($\beta(E) \sim E^n$ with $n = 1.0–1.65$ depending on spectral area of excitation) and spectral dependences are typical for PI with interchain EDA interaction [11]. In this case the carrier formation occurs due to the field assisted thermodissociation of spin-correlated electron-hole (ion-radical) pairs with rather large spatial separation ($r_0 = 4.2–5.0$ nm, 3.2–3.6 nm, 3.0 nm and 2.5–2.7 nm for UV, blue, green and long wavelength areas respectively) and initial yield $\Phi_0 = 0.07;\ 0.20;\ 0.20$ and 0.33 for the same areas). These parameters of photogeneration are significantly different for PI with the pronounced interchain (PI1 and PI3) and intrachain (PI2 and PI4) EDA interaction and also for PI with different intensity of this interaction. The influence of MNP on the CCPGQY β ($\omega = \beta(MNP)/\beta(0)$) has been determined. These results have been compared with the MNP influence on photoelectric sensitivity. Since the last includes both the β value and carrier collection rate defined as the ratio of drift length to the sample thickness l_D/d (see Eq. (2)), for the cases when this value < 1 it is possible to distinguish the MNP influence on each of these values. It has been found that when $l_D/d > 1$ (collection efficiency $C = 1$) the ratio $\xi/\omega = 1$, that means that MNP influence on photoelectric sensitivity is due to variation of photogeneration quantum yield. In cases when l_D /d < 1 (the content of PAEI in composite is rather high and level of the coating is high N>200) ξ/ω > 1, that means that the MNP influence on the drift length and collection rate is observed also.

Indeed, as it has been noted earlier for the composites PI1+PAEI and PI3+PAEI with MNP besides of the PES increasing the relative residual potential U/V decrease from 10–17% to 3–7% is observed under inhomogeneous excitation in short wavelength area (UV, blue one). As in pure PI films the U/V < 3%, the l_D decreasing for PI+PAEI composite films probably results from the PAEI addition. Since the PES-value depends on the collection rate of the carriers S ~ C(l_D/d) (2), the increase in l_D for the composite films with MNP can lead to the increase in S, especially in UV and blue areas of the spectrum. Thus the influence of MNP on the drift length of the charge carriers as well as on the photogeneration yield is the mechanism of composite photoelectric properties control.

The same positive influence of Cu-Ni particles stabilized by PAEI on the PES have been detected for the composite films of PI+PAEI+PDID and PEPC+PAEI+PDID contained sensitizer-dopant which is the perylene-diimide derivative (PDID) with the adsorption area nearby 400–600 nm (Fig. 21.1). The films of these composites are the *p*- and *n*-transport layers of organic *p-n* heterostructures, which are the base of photovoltaic cells [14]. As it is seen from the Fig. 21.1 (curve 3) for the *p*-layers the ξ value is maximum in blue and red areas of the spectrum and minimum in sensitizing area (480–550 nm). For the *n*-layers (PEPC+PAEI+PDI) the ξ value is vice versa maximum in sensitizing area and minimum in short wavelength and red areas. This difference is associated probably with the differences in transport centers for holes (donor fragment of PI in p-layers) and electrons (PDI particles in n-layers) and the mechanism of carrier photogeneration.

21.3.2 THE INFLUENCE OF UNSTABILIZED MAGNETIC NANOPARTICLES AND DEGREE OF THEIR COVERAGE BY ADSORBED POLYALKANETHERIMIDE MACROMOLECULES ON THE PES OF POLYIMIDE FILMS

The influence of MNP4 (Cu-Ni) on the photoelectric sensitivity *S* of pure PI1, PI2 and PI3 films without addition of polyalkanetherimide (PAEI) used for MNPs stabilization during preparation of the composite films

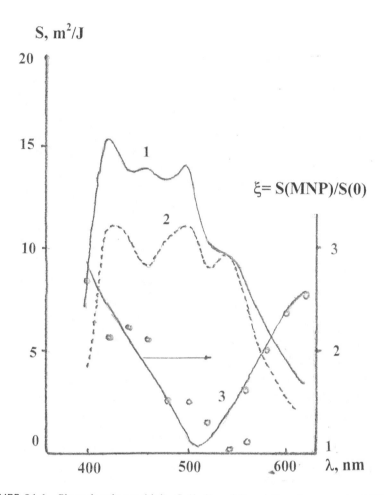

FIGURE 21.1 Photoelectric sensitivity S (1, 2) and its relative change under MNP 6 addition (with 9 mass% content) ξ (3) vs. excitation wavelength λ for composite films PI 3+ PAEI + PDID (sensitizer-dopant, perylenediimide derivative) (p-transport layer): 1 – with MNP additive; 2 – without MNP additive; 3 – ξ = S(MNP)/S(0) (with taking into account charge potential V and part of absorbed light flux Π).

has been studied. To prevent aggregation MNPs suspended in PI solution (12–17 wt.% content of MNPs in relation to the polymer) has been treated in the ultrasonic bath with the maximum output power for not less than 25 min until the homogeneous solution is obtained. The solution has been immediately used for the casting onto conducting glass templates with following drying at the air under room temperature in the magnetic field of $H=1.5$ kOe. In such a way the magneto-oriented samples have been obtained. Below in Table 21.2 the ratios of PES of the samples with MNP4 (S) to S_0 (PES for the pure PI films) $x=S/S_0$ corrected to the equivalent potential values V and the fraction of absorbed light energy Π (similar to Table 21.1) are given.

As it can be seen from Table 21.2, the addition of MNPs (without PAEI) to the pure PI films leads to the PES decrease ($\xi<1$) depending on the spectral area of the excitation and PI structure, the minimum effect being observed in UV area and for PI3 sample. It should be noted that according to Table 21.1 the highest ξ value for the composite films based on PI and MNPs stabilized by PAEI is observed for PI3. Since PES reduction is found not to be accompanied by a significant growth of the residual potential V in this case, it can be concluded that the unstabilized MNPs are not deep trapping centers for generated charge carriers but rather shallow traps that could reduce the effective drift mobility and carrier collection. The absence of the carrier capture by the stabilized MNPs is connected certainly with coating the particle surface by the adsorbed PAEI macromolecule chains, which leads to the neutralization of active sites on the surface serving as trapping centers due to the interaction with the active functional polymer groups. The addition of pure PAEI to the PI films (reference samples) is proved to lead to a decrease in PES, which is the most significant in the case of PI3. This phenomenon is likely due to the evident interchain CTC formation for this PI (Fig. 21.1). Therefore, a slight increase in the distance between the chains due to the addition of PAEI reduces the efficiency of

TABLE 21.2 ξ Values for Various Spectral Areas

UV excitation				
	PI1	**PI2**	**PI3+12% MNP**	**PI3+17%MNP**
ξ	0.60	0.66	0.78	0.86
Excitation in blue area (400–500 nm)				
ξ	0.21	0.40	0.93	0.65
Excitation in green area (500–600 nm)				
ξ	0.16	0.17	0.62	0.45
Excitation in long wavelength area (600–800 nm)				
ξ	0.42	0.20	0.50	0.45

electron transfer that depends exponentially on this distance. This reduction depends on the excitation wavelength and reaches its maximum within the interchain CTC band (blue and green light, Fig. 21.1). The observed PES decrease resulting from the PAEI addition is weaker for PI with intrachain CTC (PI2 and PI4). Sensitized p- and n-transport layers PI3+PDID and PEPC+PDID with the PAEI additive of the same content are characterized with the more strong PES decrease in the absorption band of PDIP than this value for the pure PI3 (Fig. 21.2). This means that in this case the additive affects not only the distance between the chains, but also the process of sensitization of carrier photogeneration by PDIP particles.

So, one who performs analysis of the MNP impact on PES value should take into account also the negative role of their surface (carrier trapping and capture).

According to the calculation the data given in Table 21.1 have been obtained for the ratio of N=100 macromolecules per one MNP (3 mg of polymer per 1 mg of MNPs). In this chapter, the dependence of ξ on N in the range from N=0 to 300 has been revealed for the composites based on PI and MNP4 or MNP6 (Fig. 21.3). It has been found that for N=0–30 (in the absence of coating and for low coverage of MNP surface) the value of $\xi \leq 1$, and for N > 50 observed $\xi > 1$, and a significant increase has been observed with N increasing in the range N = 75–200, but for N > 200 ξ value is constant or decreases. It follows that for N = 200, apparently, there is a complete coverage of MNPs surface by PAEI macromolecules and a maximum value of ξ=2.5–4.7 (for composites with MNP6) is observed. Further increase in N leads to a significant reduction of the drift length of carriers l_D and PES due to PI dilution with excess PAEI macromolecules. Last two experimental observations (PES reduction when PI is diluted with neutral PAEI and significant dependence of the magnitude and sign of ξ on a degree of MNP surface coverage by PAEI macromolecules) suggest that the effect of MNP on the quantum yield of carrier photogeneration β and their drift length l_D in composite films PI+ PAEI+MNP is due to PAEI redistribution in volume caused by adsorption of PAIE macromolecules on the MNP surface. The adsorption leads to PAEI globule formation around the MNP and as consequence PAEI leaves the volume in this way. This rearrangement significantly affects the PES, the β and l_D values both because of the interchain nature of CTC

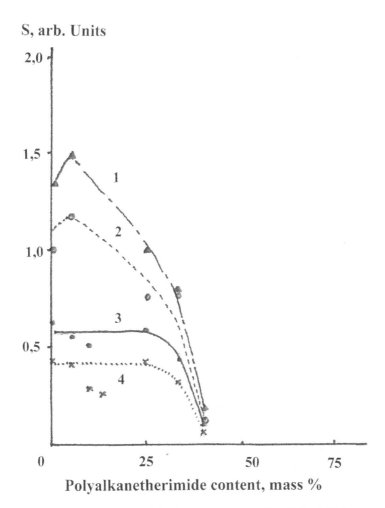

FIGURE 21.2   Photoelectric sensitivity S vs. PAEI content for (PI 3 + PAEI) composite films: 1- excitation by green light, 2- blue, 3- red (600–800 nm), 4- UV light; positive charge.

and because of the strong dependence of these values on the distance between the transport centers in carrier hopping conductivity and photogeneration model [15]. Increased efficiency of PAEI macromolecule adsorption can be associated with magnetic properties of Cu-Ni MNP, possessing significant permanent magnetization (upto 100–150 Oe [10])

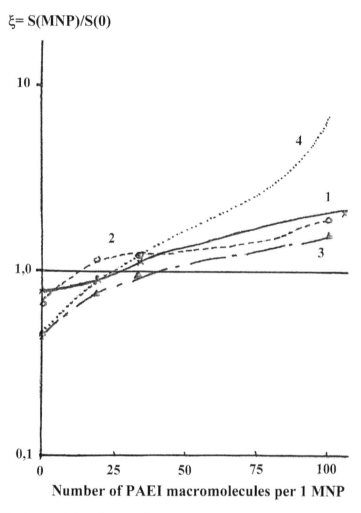

FIGURE 21.3 Relative photoelectric sensitivity change under MNP addition (9 mass %) ξ = S(MNP)/S(0) (with taking into account charge potential V and part of absorbed light flux Π) vs. PAEI macromolecules number N per 1 MNP4 (CuNi) at their constant concentration in composite PI 3 + PAEI + MNP: 1- UV excitation, 2- blue, 3- green, 4- in long-wave area (600–800 nm), positive charge.

at room temperature. The formation of MNP magnetic clusters in composite film prepared from the solution while drying leads to a capture and efficient adsorption of macromolecules in the clusters [9].

Another possible explanation is related to the magnetic properties of MNPs. Previously it has been shown that the quantum yield of photogeneration in the case of field assisted thermodissociation of spin-correlated radical ion pairs in the layers of pure and sensitized PI can vary in weak magnetic fields in the scale of hyperfine spin-nuclear interaction ($H \sim 100$ Oe) due to changing the rate of singlet-triplet transitions k_{ST} (H), induced by hyperfine interaction [16]. Magnetic fields of this range within the composite film can be created by magnetized magnetic particles, thus, despite the chaotic orientation of particle fields, there may be a nonzero magnetic effect because it is determined by the local field strength and independent of its direction. However, the magnitude of the effect in this case does not exceed 10%, which is much lower than the observed effect of MNPs.

21.3.3 MNP IMPACT ON MAGNETIC SPIN EFFECTS IN PHOTOPROCESSES (LUMINESCENCE AND PHOTOCONDUCTIVITY) IN POLYMER COMPOSITES PAEI WITH RUBRENE

This section demonstrates the presence of efficient build-up magnetic field of MNP Cu-Ni and its action on photoprocesses in polymer composites.

Films and crystals of the rubrene (5,6,11,12-tetraphenyltetracene) are one of the most known organic photoconductors. Photoelectric and luminescence characteristics of the rubrene can vary widely due to the high efficiency of its photo oxidation. The products of photooxidation (transannular peroxides and quinoid products) actively participate in the processes of photogeneration of the charged carriers (donor-acceptor complexes with charge-transfer) and luminescence (exciplexes formation).One of the unique features of the rubrene and its photoproduct is is a high sensitivity of the main photo processes to the external magnetic field (H< 300 Oe) at room temperature (magnetic-spin effects), which opens up the possibility of magnetic regulation of the material characteristics. Sensitivity in high fields (H > 1 kOe) is determined by two triplet fission of the singlet excitations S [17]:

$$S + S_0 \rightarrow T + T \tag{4}$$

In the weak fields (H < 100 Oe) the sensitivity of the rubrene samples to magnetic field is caused by the formation of the excited charge transfer complex (exciplex) between the singlet exciton (donor) and acceptor impurity on the surface of microcrystals A [18]:

$$S + A \rightarrow (S^+A^-) \qquad (5)$$

In this chapter, the polymer composites of polyalkanesterimides (PAEI) containing 70–95 mass % of microcrystals of rubrene with photoconductivity in the absorption area 400–600 nm have been obtained and studied.

For composites the exciplex luminescence quenching (upto 10%) and decrease of the photocurrent i_{ph} by 2–3% in the weak magnetic field (H < 100–200 Oe) at room temperature associated with magnetic modulation of the hyperfine interaction (HFI) within the ion-radical pairs has been found (Fig. 21.4). An important feature is the elucidation of the key role of rubrene microcrystals of a certain size in all photoprocesses in the composite films. Microcrystallization leads to a significant growth of photosensitivity, as well as to appearance of electrical and weak magnetic fields influence on exciplex luminescence. It was found significant influence of stable MNP6 (Cu-Ni) (with content 6 mass %) on the value of a photocurrent, its volt-ampere characteristics and spectral distribution. These changes (the long wavelength shift and the appearance of a new broad band in the red spectral area) are the evidence of adsorption of macromolecules PAEI on the surface of MNP and decrease in their content in the volume of the film, which leads to a slight increase in the concentration of the rubrene and the additional appearance of the rubrene phase and microcrystals.

It has been found also that the introduction of MNP additive decreases the maximum of absolute value of low field magnetic effect both on the luminescence intensity (from L/L = −10% to −7%) in composites with MNP), and on the photocurrent (from −2.3% to –(1.8–2.3)%, respectively) (Fig. 21.4). This reduction can be explained by the influence of the own magnetic field of MNP. Although these fields are oriented chaotically, the local field strength, which determines the magnetic spin effect, is different from zero. In addition in the magnetically ordered films there is directed field component of linear aggregates of MNP. When the polymer film contains MNP the luminescence relative intensity change in the magnetic field is expressed as follows:

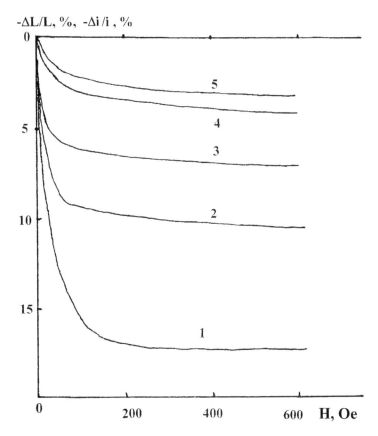

FIGURE 21.4   Relative change of luminescence intensity $\Delta L/L$ (H)= [L(H) – L(0)]/L(0) (1–3) and photocurrent $\Delta i_{ph}/i_{ph}$ (H) = [i_{ph}(H) – i_{ph}(0)]/i_{ph}(0) (4–5) vs. magnetic field strength H: 1 – for microcrystal rubrene film obtained by vacuum sublimation; 2,4 – composite film PAEI + rubrene (80 mass %); 3,5 – composite film PAEI + rubrene (80 mass %) with MNP 6 additive (6 vol %). Excitation 400–500 nm, luminescence registration at λ> 600 nm, photocurrent: planar cell with aquadag electrodes (2 mm distance), V=320 Volts (4) and V= 80 Volts (5), measurement in vacuum chamber at room temperature.

$$DL/L(H) = [L(H+H_0) - L(H_0)]/L(H_0), \qquad (6)$$

where H_0 is the averaged of local values of the field in the volume of the film induced by the MNP. The corresponding value for the sample without MNP is equal to:

$$DL/L(H) = [L(H) - L(0)]/L(0) \qquad (7)$$

Taking the values of the variables from Fig. 21.4 and inserting them into Eqs. (9) and (10), and also considering that $L(H + H_0) \approx L(H)$ (because $H >> H_0$), we find that

$$[L(H_0) - L(0)]/L(0) = -3.3\%.$$

From the curve 2 in Fig. 21.4 (for sample without MNP) we find that this value corresponds to the strength value $H = H_0 = 10$ Oe. The similar value of H_0 can be found from the values of magnetic effect for the photocurrent, although in this case the accuracy of H_0 determining is much lower (0.5%).

Thus, the polymeric composites on the basis of organic photoconductors (polyacene series) allow creation of new materials with magnetic and electric control of the main photoprocesses.

21.4 CONCLUSIONS

1. It has been found the influence of magnetic nanoparticles (Ni, Cu-Ni (nonconductive particles), Fe-CNT, Fe_3O_4) on photoelectric sensitivity, carrier photogeneration quantum yield and the residual potential of films based on soluble photoconductive polyimides (PI) and their composites with polyalkanetherimide (PAEI), and also for p- and n-transport layers of photovoltaic cells based on polymer p–n heterostructures. The magnitude of this effect – change of sensitivity and the quantum efficiency by some times.

2. It has been shown that the magnitude and sign of the effect is determined by the structure of PI and by the type of charge transfer interaction (interchain interaction gives positive effect and intrachain gives no effect or negative one), and also by the degree of surface coating of MNP with adsorbed PAEI macromolecules. The latters prevent aggregation of MNP and protect carriers against capture by the surface traps. At zero and low coverage (N < 50 macromolecules on MNP) the changing of positive effect to negative is observed. For N=100–200 there is

the maximum positive effect (growth of PES by 4.7–5.0 times and the photogeneration quantum yield by 1.6–2.5 times), and for N > 200 effect decreases.

3. It is suggested that the influence of MNP on the carrier photogeneration quantum yield β and on their drift length l_D in photoconductive composite films of PI with neutral PAEI is connected with the redistribution of the latter in the volume caused by effective adsorption of PAEI macromolecules on the surface of MNP. Adsorption causes PAEI folding in globule around MNP and so the macromolecules go out of the volume. This PAEI redistribution significantly affects PES, the magnitude of β and l_D both because of the inter-chain nature EDA interaction in PI, and because of the strong (exponential) dependences of these values on the distance between transport centers for carrier hopping conductivity and photogeneration mechanism. Alternative explanations (magnetic-spin effects in the field of MNP, the scattering of light by particles) is also considered.

4. The magnitude of the magnetic field of Ni-Cu MNP has been estimated using studied earlier low field magnetic effect on the luminescence of PAEI composites with rubrene. Although the average magnetic field of MNP in the film is zero because of the chaotic orientation fields of individual particles the magnetic spin effect from MNP, depending on the local magnetic field and not depending on its direction, is not zero. Measurements showed that the luminescence intensity change in the MNP field (–3%) corresponds to a magnetic effect in homogeneous external field with the strength $H_0 > 10$ Oe. This value has been taken for the effective MNP field strength within the composite film at particles content of 6 mass.%.

ACKNOWLEDGEMENTS

The work is supported by Russian Fund of Basic Researches (grant № 12-03-00564).

KEYWORDS

- adsorption
- charge carrier
- magnetic nanoparticles
- magnetic spin effect
- photoelectric sensitivity
- photogeneration
- polyalkanetherimide
- polymer composite
- quantum yield
- rubrene

REFERENCES

1. Gubin, S. P., Koksharov, Yu. A., Khomutov, G. B., Yurkov, G. Yu. *Uspekhi Khimii.* 2005, v. 74, №6, 539–574.
2. Pomogajlo, A. D., Rosenberg, A. S., Uflyand, I. E. *Metal Nanoparticles in Polymers.* Khimia publ. – Moscow, 2000.
3. Rumyantsev, B. M., Berendyaev, V. I., Tsegel'skaya, A. Yu., Kotov, B. V. Mol. Cryst. Liq. Cryst., 2002, v. 384, 61–67.
4. Hoppe, H., Sariciftsi, N. S. *J. Mater. Res.* 2004, v. 19, № 7, 1924–1945.
5. Xiong Gong. *Enhanced Efficiency Polymer Solar Cells Using Aligned Magnetic Nanoparticles.* Patent Application № 20130247 993. Publ. date 26. 09. 2013.
6. Zhang, W., Xu, Y., Wang, H., Xu, C., Yang, S. *Solar Energy Mater. & Solar Cells.* 2011. v. 95. Is. 10, 2850.
7. Wang, K., Ren, H., Yi, C., Liu, C., Wang, H., Huang, L., Zhang, H., Karim, A., Gong, X. *ACS Appl. Mater. Interfaces, Article ASAP. DOI*: 10. 1021/am4033179. Publ. date 24. 09. 2013.
8. Sun, C.-J., Wu, Y., Xu Zh., Hu, B., Bai, J., Wang, J.-P., Shen, J. *Appl. Phys. Lett.* 2007. v. 90, Is. 23. 232110.
9. Bychkova, A. V. *Synthesis of stable macromolecular coverings on magnetic nanoparticle surfaces for medical and biological applications. Ph. D dissertation*, Moscow, 2012, 186.
10. Kuznetsov, O. A., Sorokina, O. N., Leontiev, V. G., Shlyakhtin, O. A., Kovarski, A. L., Kuznetsov, A. A. *Journ. of Magnetism and Magnetic Materials*, 2007, v. 311, 204–207.
11. Kotov, B. V., Berendyaev, V. I., Rumyantsev, B. M. et al. *Doklady RAS, phys. khim.* 1999, v. 367, 183–187.

12. Sorokina, O. N., Bychkova, A. V., Kovarsky, A. L. *Khimicheskaya Fizika, khimicheskaya fizika nanomaterialov* 2009, v. 28, № 4, 92–97.
13. Grenishin, S. G. *Electrofotograficheskii process,* 1970, Nauka publ. Moscow, 374.
14. Rumyantsev, B. M., Berendyaev, V. I., Pebalk, D. V. *Khimicheskaya Fizika.* 2014, v. 33, № 4, 42.
15. Pope, M., Swenberg, C. *Elektronnye processy v organicheskikh kristallakh.* V. 1. Transl. from English, Mir Publ. Moscow, 1985, 347.
16. Rumyantsev, B. M., Berendyaev, V. I., Kotov, B. V. *Zh. Fiz. Khimii, fotokhimiya i magnetokhimiya.* 1999, v. 73, № 3, 538–547.
17. Rumyantsev, B. M., Lesin, V. I., Frankevich, E. L. *Optika i Spektroskopiya,* 1975, v. 38, 89–93.
18. Rumyantsev, B. M., Lesin, V. I., Frankevich, E. L. *Zh. Fiz. Khimii.* 1979, v. 53, № 6, 1509–1514.

CHAPTER 22

THE NEW EQUIPMENT FOR MODERNIZATION OF SYSTEM FOR CLEARING THE FLUE GASES

R. R. USMANOVA[1] and G. E. ZAIKOV[2]

[1]*Ufa State Technical University of Aviation; 12 Karl Marks str., Ufa 450000, Bashkortostan, Russia; E-mail: Usmanovarr@mail.ru*

[2]*N.M.Emanuel Institute of Biochemical Physics, Russian Academy of Sciences, 4 Kosygin str., Moscow 119334, Russia; E-mail: chembio@sky.chph.ras.ru*

CONTENTS

ABSTRACT

The various principles observed for the gas-cleaning installations are rotational, impact-sluggish and centrifugal. The experiment by definition efficiency of clearing of gas emissions is executed. The characteristic for optimization of hydrodynamic conditions of its work is given for each aspect of scrubbers. It will allow equipment designers to make a sampling of the necessary equipment. The modified row of apparatuses for wet clearing of gas emissions is devised. Designs on modernization of system of clearing of smoke fumes of refire kilns are devised. Builds of apparatuses confirm the patent for an invention.

22.1 INTRODUCTION

The rapid development of the industry, which was embraced in the second half of the twentieth-century in many countries of the world, has now led to a serious decline in the ecological situation. One of the burning issues is the pollution of air basin by gas emissions of the industrial factories. Growth of industrial outputs has served as the reason of increase in volumes of emissions in a circumambient. Working out of a considerable quantity of new processes promoted increase in quantity of the toxic substances arriving in an aerosphere. The problem of protection of a circumambient can be solved at the expense of a heading of the without waste, self-contained production engineering. However, now this direction yet has not had sufficient development, and therefore, the problem of creation of the perfect and effective equipment for clearing of gas emissions of the industrial factories is actual.

The problem essentially becomes complicated that volumes of gas emissions of the industrial factories make tens, and sometimes and hundreds, thousand m^3/h that does inconvenient application of the traditional clearing equipment. The majority of the apparatuses used now for clearing of gases from gaseous, liquid and firm impurity, are characterized by the low carrying capacity caused in the small maximum permissible speeds of gas in apparatuses. It serves as the reason of that high efficiency apparatuses have the big overall dimensions (for example, diameter of tower absorbers can attain $10 \div 12$), and expenses for their manufacturing, installation and transportation are unreasonably great. Besides, in apparatuses of the big diameter it is impossible to achieve a liquid phase uniform distribution on their cross-section that leads to sharp decrease in efficiency of clearing.

The specified problems have served as the reason of that many industrial gas emissions at all are not exposed to clearing. As an example it is possible to result smoke gases of the factories of metallurgy, power engineering, chemical, petrochemical and other industries, tank and scavenging gases of the various factories, emissions of a dust and steams of organic dissolvent in production areas of the factories.

The problem of clearing of great volumes of gas emissions of the industrial factories in an aerosphere can be solved for the account of application for these purposes of apparatuses of whirlwind type. Use in whirlwind apparatuses of centrifugal separation of phases removes restriction on maximum permissible speed of gas and allows to spend processes at the flow rate speeds of gas attaining 20–30 mps. High carrying capacity of whirlwind apparatuses causes their low metal consumption, rather small specific power expenses, simplicity of manufacturing. Design features of whirlwind apparatuses allow to spend to them complex clearing of gas emissions of the industrial factories as from harmful gaseous impurity, and/or small liquid and firm corpuscles. Apparatuses also are rather convenient for conducting in them of process of vapor cooling of high-temperature gas emissions at a stage of preparation of gases to clearing.

In spite of the fact that a principle of a design of apparatuses of whirlwind type are developed for a long time, their wide use in the industry is restrained by an insufficient level of scrutiny hydro- and aerodynamic regularity of work and absence of reliable and well-founded methods of calculation of efficiency of processes of clearing of gas.

22.2 ENGINEERING DESIGN AND EXPERIMENTAL RESEARCHES OF NEW APPARATUSES FOR GAS CLEARING

22.2.1 APPARATUSES OF IMPACT-SLUGGISH ACT

Mechanically each of such apparatuses consists of contact channel partially entrained in a liquid and the drop catcher merged in one body. The principle of act of apparatuses is based on a way of intensive washing of gases in contact channels of a various configuration with the subsequent separation of a gas-liquid stream in the drop catcher.

The liquid, which has thus reacted and separated from gas is not drained at once from the apparatus, and circulates in it and is repeatedly used in dust removal process. Circulation of a liquid in the wet-type collector occurs at the expense of a kinetic energy of a gas stream. Each apparatus is supplied by the device for maintenance of a fixed level of a liquid, and also the device for removal of sludge from the modular loading pocket of a scrubber.

Distinctive features of apparatuses:

1. Irrigation of gas by a liquid without use of injectors that allows to use for an irrigation a liquid with the high maintenance of suspended matters (to 250 mg/m^3);

2. The closed circulation of a liquid in apparatuses which allows to reuse a liquid in rotary connections of scrubbers and to scale down its charge on gas clearing in 10 and more times in comparison with other types of wet-type collectors;

3. Removal of the trapped dust from apparatuses in the form of dense sludges with low humidity that allows to simplify dust salvaging to reduce loading by water purification systems;

4. Configuration of the drop catcher in the body of the apparatus, which allows to reduce sizes of dedusters to secure with their compactness.

The specified features and advantages of such scrubbers have led to wide popularity of these apparatuses, active working out of various builds, research and implementation of wet-type collectors, as in Russia, and abroad.

The investigated rotoklon (Figs. 22.1 and 22.2) had 3 slot-hole channels speed of gas in which made to 15 mps. In the capacity of modeling system air and a talc powder have been used. The apparatus body was filled with water on level h$_\text{ж}$=0.175÷0.350м.

Rotoklon works as follows:

Depending on a dust content dust gas a stream the top blades by means of screw lifts, and the bottom blades 6 by means of flywheels 8 are established on a corner defined by an operating mode of the device.

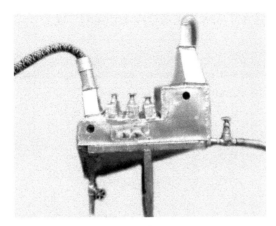

FIGURE 22.1   Experimental assembly "Rotoklon."

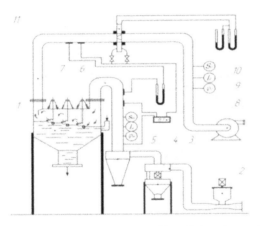

FIGURE 22.2   The Circuit design of experimental installation: 1 – rotoklon; the dust 2 – batcher; 3 – qualifier; 4 – collector of a coarse dust; 5 – cyclone; 6,7 – gas pipeline; 8 – the ventilating fan; 9 – potentiometer; 10 – differential pressure gauge; 11 – diaphragm.

Dusty gas acts in an entrance branch pipe in the top part of the case of 1 device. Hitting about a surface of a liquid, it changes the direction. Owing to high speed of the movement, cleared gas grasps the top layer of a liquid and splits up him in the smallest drops and foam with an advanced surface.

The caught dust settles in the bunker to rotoklon and through a branch pipe for plum slime waters, together with a liquid, is periodically removed from the device.

22.2.2 APPARATUSES OF ROTATIONAL ACT

Dynamic gas washer, according to Figs. 22.3 and 22.4, contains the vertical cylindrical case with the bunker gathering slime, branch pipes of input and an output gas streams.

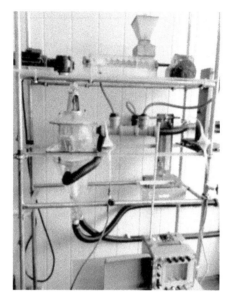

FIGURE 22.3   Experimental assembly "Dynamic gas washer."

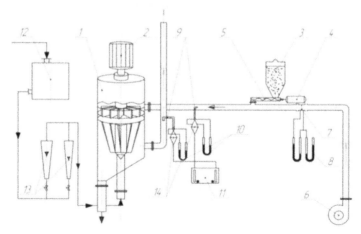

FIGURE 22.4   The Circuit design of experimental installation: 1 – scrubber; 2 – the drive; 3 – the dust loading pocket; 4 – the electric motor; 5 – the batcher; 6 – the fan; 7 – a diaphragm; 8,10 – differentia 12 – the pressure tank; 13 – rotametr; 14 – sampling instruments.

Inside of the case it is installed conic vortex generator. Dynamic gas washer works as follows:

The Gas stream containing mechanical or gaseous impurity, acts on a tangential branch pipe in the ring space formed by the case and rotor. The liquid acts in the device by means of an axial branch pipe. at dispersion liquids the zone of contact of phases increases and, hence, the effective utilization of working volume of the device takes place more.

The Invention is directed on increase of efficiency of clearing of gas from mechanical and gaseous impurity due to more effective utilization of action of centrifugal forces and increase in a surface of contact of phases. The Centrifugal forces arising at rotation of a rotor provide crushing a liquid on fine drops that causes intensive contact of gases and caught particles to a liquid.

Owing to action of centrifugal forces, intensive hashing of gas and a liquid and presence of the big interphase surface of contact, there is an effective clearing of gas in a foamy layer. The water resistance of the irrigated apparatus at change of loadings on phases has been designed.

Considered angular speed of twirl of a rotor and veering of twirl of guide vanes of an air swirler.

22.2.3 BUBBLING – THE VORTICAL DEVICE

Experimental installation (Figs. 22.5 and 22.6) has been devised. Experimental researches are directed on definition of effect of design data of an air swirler, a specific irrigation, apparatus operating modes on aerodynamics of a gas-liquid stream and efficiency of clearing of gas emissions.

The Optimum mode of clearing of gas emissions can be achieved at application of the developed design bubbling – the vortical device providing a mode evaporation – condensation cooling and centrifugal wet dust-separation.

In bubbling – the vortical device before vortex generator a gas stream the central atomizer is installed, and in each flowing section after vortex generator peripheral atomizers are located.

Bubbling – the vortical device works as follows. Dusty gas moves in the cylindrical chamber on an entrance pipe where vortex generator by

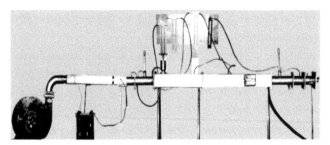

FIGURE 22.5　Experimental assembly "Bubbling – the vortical device."

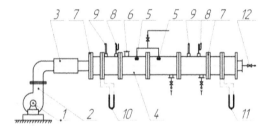

FIGURE 22.6　The Circuit design of experimental installation: 1 – electric motor; 2 – ventilating fan; 3 – hot-air heater; 4 – the whirlpool chamber; 5-atomizers; 6 – loading pocket; 7 – diaphragm; 8,10 – differential pressure gauges; 9 – the thermometer; 11 – the fine gauge strainer; 12 – the coincidence gate.

means of the blades forming flowing section, rejects a stream and gives to him rotary movement. Under action of centrifugal force arising at this disperse particles move to walls of the cylindrical chamber. For improvement of conditions of clearing of gases, before and after vortex generator are installed one central and four peripheral atomizers in which the irrigating liquid moves. The central atomizer installed before vortex generator, creates a volumetric torch atomization an irrigating liquid. At contact of the polluted gas and a liquid there is a partial evaporation of last and cooling of gas. The formed suspension is divided under action of the centrifugal force arising at rotation of a stream. The torch atomization the cooling liquid, formed by the central atomizer, (alongside with action of centrifugal forces) promotes outflow of disperse particles from the central zone of the cylindrical chamber that reduces a way of a particle up to a wall and reduces time of separation.

Technical and economic efficiency of use offered bubbling – the vortical device for clearing and cooling of smoke gases consists that he allows:

1. To increase efficiency dust separation due to installation in the device vortex generator a gas stream, the central and peripheral atomizers;
2. To lower hydraulic resistance of the device owing to a choice of optimum geometry of blades vortex generator;
3. To save material means and the areas of industrial premises due to an opportunity of installation bubbling – the vortical device in vent dust removal system.

22.3   CLEARING OF GASES OF DUST IN THE INDUSTRY

The results have been almost implemented in manufacture of roasting of refractory materials at conducting of redesign of system of an aspiration of smoke gases. The devised scrubber is applied to clearing of smoke gases of baking ovens of limestone in the capacity of a closing stage of clearing.

Temperature of gases of baking ovens in main flue gas breeching before the exhaust-heat boiler 500–600°C, and after exhaust-heat boiler 250°C. An average chemical compound of smoke gases (by volume): 17%CO_2; 16%N_2; 67% CO. Besides, in gas contains to 70 mg/m³ SO_2; 30 mg/m³ H_2S; 200 mg/m³ F and 20 mg/m³ CI. The gas dustiness on an exit from the converter reaches to 200/m3 the Dust, as well as at a fume extraction with carbonic oxide after-burning, consists of the same components, but has the different maintenance of oxides of iron. In it than 1 micron, than in the dusty gas formed at after-burning of carbonic oxide contains less corpuscles a size less. It is possible to explain it to that at after-burning CO raises temperatures of gas and there is an additional excess in steam of oxides. Carbonic oxide before a gas heading on clearing burn in the special chamber. The dustiness of the cleared blast-furnace gas should be no more than 4 mg/m³. The following circuit design (Fig. 22.7) is applied to clearing of the blast-furnace gas of a dust.

Gas from a furnace mouth of a baking oven 1 on gas pipes 3 and 4 is taken away in the gas-cleaning plant. In raiser and down taking duct gas is chilled, and the largest corpuscles of a dust, which in the form of sludge are

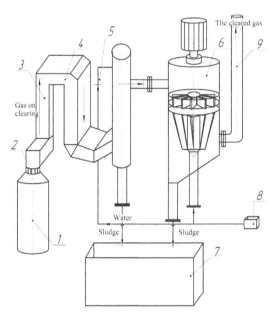

FIGURE 22.7 Process flow sheet of clearing of gas emissions: 1 – bake roasting; 2 – water block; 3 – raiser; 4 – downtaking duct, 5 – centrifugal scrubber; 6 – scrubber dynamic; 7 – forecastle of gathering of sludge, 8 – hydraulic hitch, 9 – chimney.

trapped in the inertia sludge remover, are inferred from it. In a centrifugal scrubber 5 blast-furnace gas is cleared of a coarse dust to final dust content 5–10/m³ the Dust drained from the deduster loading pocket periodically from a feeding system of water or steam for dust moistening. The final cleaning of the blast-furnace gas is carried out in a dynamic spray scrubber where there is an integration of a finely divided dust. Most the coarse dust and drops of liquid are inferred from gas in the inertia mist eliminator. The cleared gas is taken away in a collecting channel of pure gas 9, whence is fed in an aerosphere. The clarified sludge from a gravitation filter is fed again on irrigation of apparatuses. The closed cycle of supply of an irrigation water to what in the capacity of irrigations the lime milk close on the physical and chemical properties to composition of dusty gas is applied is implemented. As a result of implementation of trial installation clearings of gas emissions the maximum dustiness of the gases, which are thrown out in an aerosphere, has decreased with 3950 mg/m³ to 840 mg/m³, and total emissions of a dust from sources of limy manufacture were scaled down about 4800 to/a to 1300 to/a.

Such method gives the chance to make gas clearing in much smaller quantity, demands smaller capital and operational expenses, reduces an atmospheric pollution and allows to use water recycling system.

22.4 RESULTS OF INDUSTRIAL TESTS OF THE GAS-CLEANING PLANT ON THE BASIS OF DEVICE "ROTOKLON"

Let's consider system of wet clearing of the gases departing from the closed ferroalloy furnace 1. On this furnace comparative researches of the described system of wet dust separation (Fig. 22.8) have been conducted.

The slope breeching 2, actually is the hollow scrubber in diameter of 400 *mm* working in the evaporation cooling regime. At work gases arrive from it in a Venturi scrubber 3, consisting of two cylindrical columns in

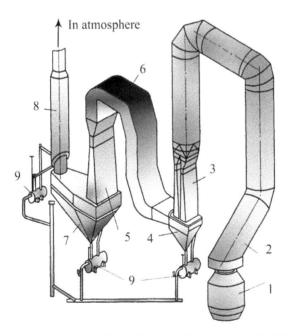

FIGURE 22.8 The scheme of clearing of flue gas with gas cooling in a Venturi scrubber and the subsequent clearing in a rotoklon: 1 – baking oven; 2 – gas exit branch; 3 Venturi scrubber; 4 – bunker – the drop catcher; 5 – a rotoklon; 6 – gas pipeline; 7 – inertial heat – and a mist eliminator; 8 – exhaust pipe; 9 – a hydraulic hitch.

diameter of 1000 *mm* with the general bunker. In each column of a scrubber it is established on three atomizers. The Venturi scrubber of the first step of clearing has a mouth in diameter of 100 *mm* and is irrigated with water from an atomizer established in front of the confusor.

Gases after a slope breeching go at first to the bunker the drop catcher 4, and then in a rotoklon of the another step, which consists from inertial a heat – and a mist eliminator 7. The exhaust of gases from the furnace is carried out by vacuum pump VVN-50 established behind devices of clearing of gas emissions. Purified gases are deduced in atmosphere.

Regulation of pressure of gases under a furnace roof and the expense of gases is carried out by a throttle in front of the vacuum pump. Slurry water from devices of clearing of gas emissions flows off a by gravity in a tank of a hydroshutter 9, whence also a by gravity arrives in a slurry tank. From a slurry tank water on two slurry clarifier is taken away on water purification. After clarification, chemical processing and cooling water is fed again by the pump on irrigating of gas-cleaning installations.

The dust containing in gases, differs high dispersion (to 80 weight. % of particles less than 5–6 microns). In Table 22.1 the compound of a dust of exhaust gases is resulted. In tests for furnaces almost constant electric regime that secured with identity of conditions at which parameters of systems of dust separation characterize was supported. The furnace worked on the fifth – the seventh steps of pressure at fluctuations of capacity 14.5–17.5 megawatt.

The quantity of dry gases departing from the furnace made 1500 2000 m^3/h. The temperature of gases before clearing of gas emissions equaled 750 850 ° C, and humidity did not exceed 4–5% (on volume).

Results of calculation of a payment for pollutant emission of system of dust separation are shown in Table 22.2.

TABLE 22.1　Results of Post-test Examination

Compound	Requisite concentration, g/m^3	Concentration after clearing, g/m^3
Dust	0.02	0.00355
NO_2	0.10	0.024
SO_2	0.03	0.0005
CO	0.01	0.0019

TABLE 22.2 Results of Calculation of a Payment for Pollutant Emission

The list of pollutants (the substance name)	It is thrown out for the accounting period, t/year	Including		The base specification of a payment within admissible specifications, a Russian rouble/t	The size of a payment for a maximum permissible emission, Russian rouble/year	The base specification of a payment within the established limits, a Russian rouble/t	Total a payment on the enterprise, a Russian rouble/year
	In total	VPE	MPE				
1	2	3	4	5	6	7	8
The inorganic Dust	19.710		—	21	228.65	105	228.65
Nitrogen dioxide	105.1		—	52	379.19	260	379.19
Carbon monoxide	288.2		—	0.6	2.99	3	2.99
Sulfurs dioxide	197.1		—	40	539.14	200	539.14
Total:					1149.97		1149.97

Thus, we have chosen the scheme of clearing of gases which allows to lower concentration of pollutants to preset values and consequently, and to lower payments by the enterprise for emissions.

22.5   INDUSTRIAL APPLICATION OF VORTICAL DEVICES

It is settlement – theoretical and design – design works were carried out with reference to operating conditions of manufacture hypochlorite calcium of Joint-Stock Company "Caustic."

On Joint-Stock Company "Caustic" the device is offered to be used for clearing smoke gases of furnaces of roasting 1 (Fig. 22.9).

According to the technological scheme, departing from the furnace of roasting 1 gases at temperature 550°C act in the bubbling-vortical device 2. Here on an irrigation 1–3% a solution of limy milk (pH = 11.5–12.5) move. Separated slime acts in a drum – slake 3; clarification and cooling of limy milk happens in the filter-sediment bowl 4: from which it feed on irrigation. The cleared gas stream smoke exhauster 7 is thrown out in an atmosphere.

Thus, introduction bubbling – the vortical device will allow to solve a problem of clearing of smoke gases with return of all caught dust in the

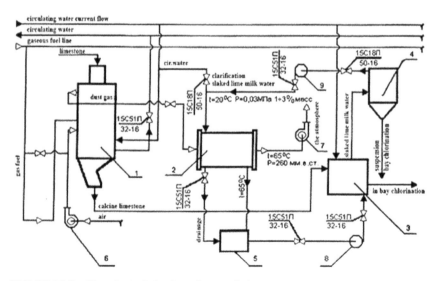

FIGURE 22.9   Flowchart of clearing of gas emissions.

form of slime in branches of slaking lime, that provides captives manufactures.

Bubbling – the vortical device is supposed to mount in vent dust
removal systems with the purpose of economy of the areas of industrial
premises, thus hydraulic resistance of system does not exceed 500 Pascal,
power inputs on clearing of gas in 3 times below, than in known devices.

Application bubbling – the vortical device allows to achieve an intensi-
fication of process of gas purification with reduction of a gassed condition
of air pool.

22.6   THE MODIFIED VARIETY OF APPARATUSES

By results of research the industrial modified rows of apparatuses in diameter from 0.8 to 1.5 m. Apparatuses are devised secure with clearing of
gases over the range productivity from 500 till 22000 m³/h. Apparatuses
are introduced in the capacity of the another echelon of wet clearing of
gas emissions in manufacture of hypochlorite of calcium for clearing of a
waste-heat of a limestone refire kiln; in burning department for clearing of
smoke fumes of a refire kiln of a barite; in commodity output department
(Fig. 22.10).

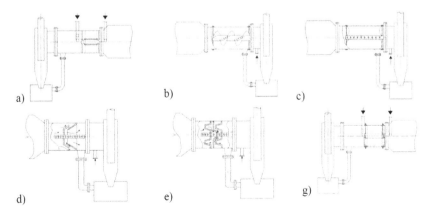

FIGURE 22.10 The modified variety of apparatuses under patents for inventions:
(a) №2182843; (b) № 2305457; (c) № 316383; (d) № 2382680; (e) № 2403951; (g) №
2234358.

Distinctive feature of offered systems is realization closed loop system smoke sucker – Bubbling – the vortical device – a cyclone and retrace of the trapped product to technological manufacture. It is necessary to note simplicity of a design and maintenance of apparatuses, and also betterment of work of the flue-gas pump (the vane wheel rotor and the body are not choked) along with high separation efficiency of gases. Besides, the apparatus has indisputable advantage when the deduster should place in gas pipes for the factories with the restricted floor spaces. The rational values of parameters of a dynamic spray scrubber had in-process are implemented in type apparatuses: a scrubber. Results of work have been used at designing of gas-cleaning plants of some industrial productions (Fig. 22.11).

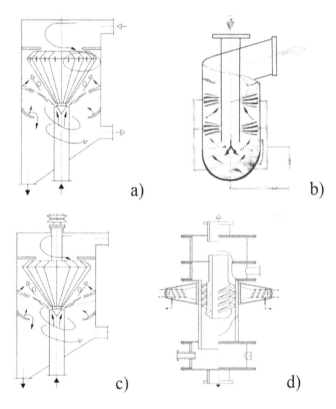

FIGURE 22.11 The modified variety of apparatuses scrubber under patents for inventions: (a) Patent decision № 2012153318/15; (b) Patent decision № 2012157446/15; (c) № 2339435; (d) № 2482923.

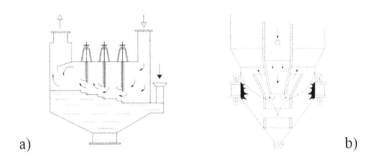

a)   b)

FIGURE 22.12   The modified variety of apparatuses "Rotoklon" under patents for inventions: (a) № 2317845; (b)The request №2007120000.

The complex of the spent researches has formed the basis for designing of system of air purification of production areas. The had results have been almost implemented in commodity output manufacture. The apparatus is resistant to against oscillations both the general loading, and a relationship of gas rates. One of the major advantages of configuration of systems clearings of gas emissions "Rotoklon" is possibility of the closed cycle of an irrigation thanks to system internal circulation of a liquid in the apparatus (Fig. 22.12).

Recommendations about rational designing of dust removal apparatuses are devised. By results of work it is introduced in various manufactures more than 10 dedusters. Economic benefit of implementation of systems and recommendations has made more than 3 million roubles/ years.

22.7   CONCLUSIONS

1. The solution of an actual problem on perfection of complex system of clearing of gas emissions and working out of measures on decrease in a dustiness of air medium of the industrial factories for the purpose of betterment of hygienic and sanitary conditions of work and decrease in negative affecting of dust emissions on a circumambient is in-process given.

2. Designs on modernization of system of an aspiration of smoke gases of baking ovens of limestone with use of the new scrubber

which novelty is confirmed with the patent for the invention are devised. Efficiency of clearing of gas emissions is raised. Power inputs of spent processes of clearing of gas emissions and power savings at the expense of modernization of a flowchart of installation of clearing of gas emissions are lowered.

3. The modified variety of apparatuses for wet clearing of the gas emissions, which have confirmed high separation efficiency both in laboratory, and in industrial conditions. The ecological result of implementation of systems and recommendations consists highly clearings of a waste-heat and betterment of ecological circumstances in a zone of the factories. Economic benefit of implementation has made more than 3 million rouble/year.

KEYWORDS

- **bubbling-swirling apparatus**
- **dynamic separator**
- **Rotoklon**
- **smoke fumes**
- **water recycling**

REFERENCES

1. Usmanova, R.R. Bubbling – the vortical device. The patent of the Russian Federation the invention №2182843. 27 May 2002. The bulletin. №15.
2. Usmanova, R.R. Bubbling – the vortical device with adjustable blades. The patent of the Russian Federation the invention №2234358. 20 August 2004. The bulletin. №23.
3. Usmanova, R.R. Bubbling – the vortical device with an axial sprinkler. The patent of the Russian Federation the invention №2316383. 10 February 2008. The bulletin. №4.
4. Usmanova, R.R. Rotoklon with adjustable sine wave blades. The patent of the Russian Federation the invention № 2317845 27 February 2008. The bulletin. №6.
5. Usmanova, R.R. Bubbling-swirling apparatus with parabolic swirler. The patent of the Russian Federation the invention № 2382680, 27 February 2010. The bulletin. №6.

6. Usmanova, R.R. Bubbling-swirling apparatus with conical swirler. The patent of the Russian Federation the invention №2403951, 20 November 2010. The bulletin. №32.

7. Usmanova, R.R. Dynamic gas washer. The patent of the Russian Federation the invention №2339435. 20 November 2008. The bulletin. №33.

8. Usmanova, R.R. The whirlwind apparatus with application of ultrasonic vibrations. The patent of the Russian Federation the invention №2482923. 27 May 2013 The bulletin. №15.

9. Usmanova, R.R. Scrubber. Patent decision № 2012153318/15 From 07.02.2014.

10. Usmanova, R.R. Hydrodynamic deduster. Patent decision № 2012157446/15 From 24.02.2014.

CHAPTER 23

RECENT ADVANCES ON MONTMORILLONITE-BASED FIRE RETARDANT COMPOSITES IN POLYMER MATRICES[1]

GUIPENG CAI and CHARLES A. WILKIE

Department of Chemistry and Fire Retardant Research Facility, Marquette University PO Box 1881, Milwaukee, WI 53201, USA, E-mail: charles.wilkie@marquette.edu

CONTENTS

[1]Dedicated to Gennady E. Zaikov on the occasion of his 80th birthday and in celebration of his 60 years in science.

ABSTRACT

Fire retardancy in the future will likely involve a combination of materials that all function as fire retardants but each offers something different. In this paper, it is suggested that a nano-dimensional material, possibly and organically modified montmorillonite, will be one of the components. Drawing examples from a number of different systems, this is elucidated in this chapter.

23.1 INTRODUCTION

Fire retardancy of polymeric composites has been extensively studied in the past decade and the halogen-based fire retardants, which have been the workhorse of the industry for many years, have been more and more questioned due to environmental concerns. Recent studies on human exposure to flame retardant chemicals reveal that the bromine-containing fire retardants (FRs) can degrade to form carcinogenic brominated dioxins and can even be absorbed by the human body [1–3]. Thus the need for green FRs, which can replace the halogens, is more and more necessary. While there are a large number of potential fire retardant, this review focuses on montmorillonite (MMT) and its combination with other materials. Since the fire retardancy due to MMT was first discovered in polymer matrices by Gilman et al. [4], this type of green, low-loading effective (usually 3–5 wt. %) material has been of continued interest globally. A substantially reduced peak heat release rate (PHRR) is the principal interesting feature of MMT-based polymer composites. Two major factors, which may be considered detrimental, are the shortened time to ignition (TTI) and the almost unchanged total heat release (THR) [5]. In addition, MMT alone in polymers does not improve the classification in the UL94 protocol. Thus, if MMT is to be useful, it must be used in combination with other fire retardants. We believe that combinations of fire retardants is, in general, the wave of the future, and, in principal, we think that MMT can be useful as one of the materials to be combined. In this review, combinations of clay (montmorillonite) with other fire retardants are examined.

First of all, the processes by which MMT enhances fire retardancy will be reviewed. This will be followed by a section on the effects of dispersion, then combinations with intumescent fire retardants, then with other fillers and finally the layer-by-layer technique to achieve good fire retardancy will be addressed.

23.2 PROCESSES BY WHICH FIRE RETARDANCY IS ACHIEVED USING MMT

There are two processes by which fire retardancy is achieved using MMT, the formation of a barrier, which reduces transport of material from the polymeric phase to the vapor phase and also reduces the ability of heat to transfer to the underlying polymer, and paramagnetic radical trapping, which traps degrading polymeric radicals so they cannot enter the vapor phase. Barrier formation was first proposed by Gilman [6] who suggested that the rising bubbles during burning may carry the clay to the surface where it acts as a barrier. This has been further studied in a series of papers from this laboratory in which it was shown that the reduction in the peak heat release rate is related to the stability of the radicals that are formed. When very stable radicals are produced, the reduction in the PHRR is low while for more unstable radicals, the reduction can be larger [7–11]. The notion that the material actually rises to the surface as opposed to simply acting as in *in situ* barrier was addressed in Ref. [12].

Paramagnetic radical trapping was first suggested in 2001 [13]. It was found that a significantly reduced PHRR was obtained when the amount of clay was clearly insufficient to produce the necessary barrier. By comparing iron-containing clays with those in which iron was absent, the necessity of the iron was shown. This has recently been experimentally shown to be true by Carvalho et al. [14].

23.3 DISPERSION OF MMT IN POLYMER MATRIX

In most case, the dispersion of an organically modified MMT is the most important factor in the fire performance – if the MMT is poorly dispersed, the fire performance will be poor while if there is good dispersion, whether intercalated or exfoliated, there will be a substantial reduction in the PHRR [13, 15]. Thus, the dispersion of organo-MMT in polymer matrices and the adaptation of new surfactants to enhance the dispersion of organoclay in polymer matrices is also described herein.

During the past two decades, the investigation of polymer/MMT composites reveals that dispersion of MMT plays an important role in the physical, mechanical and flammability properties [13, 16–18]. Polymers are generally covalent compounds and show weak or no polarity but pristine montmorillonite is an ionic compound and exhibits relatively strong polarity. Therefore, the compatibility between the polymer and clay must be enhanced and this is done using long chain surfactants. Surfactants play the role of an amphiphile, having two functional moieties, a polar end and a nonpolar end, which enhance the compatibility between polymers and clays. There are some problems with the conventional quaternary ammonium surfactants, primarily their low thermal stability. However, phosphonium surfactants are more thermally stable compared to the ammonium salts and can be used [19]. Pack et al. employed resorcinol diphenylphosphate (RDP) oligomers as a surfactant to modify pristine MMT [20]. RDP is a phosphorus-based fire retardant with good thermal stability, high efficiency, and low volatility [21]. If RDP can be intercalated into MMT, it can play a dual role both as a surfactant and as a fire retardant. First of all, one part (by weight) RDP was melted in a beaker with continuous stirring, then four parts of pristine MMT clay was added with stirring until the liquid was completely absorbed into the clay powder. The modified clay was then placed in a vacuum oven at 100°C for 24 h to remove moisture and unabsorbed liquid. Small-angle X-ray scattering (SAXS) and atomic force microscope (AFM) results supported the intercalation of RDP into MMT galleries. This method does not need solvent, so it is a green method to produce modified MMT. The RDP-MMT was incorporated into various polymers, including high-impact polystyrene (HIPS), poly(methyl methacrylate) (PMMA), acrylonitrile butadiene styrene (ABS) and polypropylene (PP), and binary polymer blends, like PS/PMMA and polycarbonate (PC)/ poly (styrene-coacrylonitrile) (SAN24). They also used Cloisite 20A to blend with the above polymers as a reference. RDP-MMT is more compatible with the styrene groups compared to the Cloisite clays, which could lead to exfoliation in HIPS and ABS. Although they did not use cone calorimetry, the gasification evaluation revealed that PC/SAN24/RDP-MMT exhibits a more remarkable PHRR and MLR reduction than the polymer blend with Cloisite 20A. In a similar study, Lai et al. also adopted RDP-modified MMT in a PP/IFR system [22]. They did not employ the direct melt blending method to intercalate RDP into MMT, rather they used solvent mixing in ethanol. The d-spacing of modified MMT is 2.15 nm, which is close to that of the melt blended system (2.23 nm). Due to the high thermal stability of RDP, the onset decomposition temperature and the maximum decomposition rate temperature of PP/IFR/RDP-MMT are higher than PP composites with conventional quaternary ammonium salt modified MMT (O-MMT). The amount of MMT is varied between 0 and 6% and PP/IFR/RDP-MMT shows a slightly increased LOI value compared to PP/IFR/O-MMT (less than or close to 1% increase) and the UL94 ratings of those two systems are also the same; in other words there is no improvement by these methods of evaluation. The cone results of PP/IFR/RDP-MMT show a more reduced PHRR value compared to the system without RDP, PP/IFR/O-MMT, and thicker chars are obtained after burning compared with the composites with or without O-MMT.

If the surfactant used to modify pristine MMT can also intumesce, this might make a nice method to prepare fire retardant materials. Huang et al., synthesized a phosphorus–nitrogen containing compound, 2-(2-(5, 5-dimethyl-1, 3, 2-dioxaphosphinyl-2-ylamino) ethyl-amino)-N, N, N-tri-ethyl-2-oxoethanaminium chloride, which was denoted as compound c [23]. It takes three steps for the synthesis and the final yield is about 30%. They used this compound to modify pristine MMT and combined it with polyurethane to form composites. No discernible XRD peaks of the composites indicated that a disordered structure was obtained, although they claimed exfoliation instead. The TEM images reported in this paper to our eyes do not suggest exfoliation. A major problem occurs when nano-dispersion is asserted based only on the absence of a peak in the XRD or poor quality TEM images. The PHRR reduction from the cone calorimetry is moderate (25%) even at 20% c-MMT loading, which also indicates a disordered structure rather than an exfoliated structure. The two advantages of this system are the prolonged time to ignition (TTI) and reduction of smoke. Later, they synthesized another phosphorus–nitrogen containing compound, 2, 4, 8, 10-Tetraoxa-3, 9-diphosphaspiro [5.5]-undecane-3,

9-dioxide-disubstitutio-acetamide-N,   N-dimethyl-N-hexadecyl-ammo-
nium bromide (PDHAB), which they combined with LDPE/EVA [24].
The synthesis is also a three-step process, but the final yield is higher
than previous one, around 40%. Moreover, supported by XRD and TEM,
this two-long-tail surfactant seems to enhance the compatibility between
LDPE/EVA and MMT. The two advantages of previous system in cone
calorimetry remain in this system. The PHRR reduction in this system is
still moderate (28%). Despite the nice idea of combination of surfactant
and intumescent effect, the complicated synthetic process and relatively
low yield may limit practical applications of the above systems.

In another study, a twice modified organoclay, which they call a dual
modified clay (DMC), was prepared by successive cationic exchange
reactions of pristine MMT with two intercalating reagents, hexadecylt-
riphenyl-phosphonium bromide and 2-methacryloxy-ethyl-hexadecyl-
dimethyl-ammonium bromide [25]. This method increases the d-spacing
of organo-modified MMT from 1.20 nm of pristine MMT to 2.49 nm.
The XRD patterns and TEM images supported an exfoliated structure of
PMMA/DMC-3% nanocomposites. Mechanical properties, such as, the
tensile strength, elongation at break, break stress, and modulus, increase
considerably compared with control PMMA. The only fire test conducted
in this paper was LOI. All samples' LOI (DMC at 1%, 3% and 5%) values
are lower or equal to 21, which means they have had no effect on the ease
of extinguishment, which is what is evaluated by LOI. The PMMA/DMC
nanocomposites are prepared through a casting method, which needs a
highly dilute solution in N-methyl-2-pyrrolidone (NMP) as a solvent.
Thus, this method is not appropriate for large scale production and the
process may not be green.

Based on a previous study, which suggested that ultrasonic treatment
enhances the intercalation of clay in HDPE, increasing the d-spacing up to
50%, Swain et al. further investigated PMMA/clay nanocomposites synthesized by an ultrasonic-assisted emulsifier-free emulsion polymerization
technique [26, 27]. In order to enhance the dispersion of the clay layers
in the polymeric matrix, ultrasonic waves of different power and frequencies were used. XRD and TEM results indicated enhanced dispersion of
MMT after ultrasonication. Among the mechanical properties, toughness
increased significantly compared with the composites without ultrasound.

LOI was the only fire evaluation in this work and it was insignificantly improved by ultrasound in PMMA/clay composites. PMMA is very difficult to fire retard so this may not be a good material to use to evaluate novel systems.

23.4 COMBINATION OF ORGANOCLAY AND INTUMESCENT FIRE RETARDANTS (IFRS)

On one hand, intumescent flame retardants (IFRs) have been widely studied due to their advantages, such as, low smoke, low release of toxic gasses and antidripping, etc., on the other hand, there are also disadvantages, including cost-ineffectiveness, low thermal stability and the weakness of the char structure. The barrier effect of nano-clay may help improve the char strength and fire retardant efficiency as well. Ribeiro et al. investigated the synergy between various commercial MMTs and a standard intumescent formulation containing ammonium polyphosphate (APP) and pentaerythritol (PER), in the matrix of a poly (ethylene (30%)–butyl acrylate) copolymer. They found a relation between the d-spacing of MMTs and fire retardancy performance [28]. For the five commercial organo-MMTs (Cloisite 10A, 15A, 20A, 93A, 30B), the LOIs were all the same when only the clay was added, indicating that clay alone does not have an effect. On the other hand, when both clay and the intumescent system were present, the LOIs increased from 20 to almost 30. This is an increase but it is very doubtful if that will have any effect on the fire retardancy. They also selected the lowest d-spacing clay (30B) and highest d-spacing clay (15A) in the same polymer matrix and conducted the UL94 evaluation. The results show that the former obtains a V0 rating while the latter obtains only a V2 rating; however, the polymer/Cloisite Na$^+$ also obtains a V0 rating, which indicates that it is the intumescent, and not the clay, that dominate.

In a paper from the group of Wang et al., IFR composites based on ethylene-propylene-diene terpolymer (EPDM) with organo-MMT and organo-modified MgAl-LDH were studied. The IFR was composed of APP and a nitrogen-containing carbonization agent (CA) poly (1, 3, 5-triazin-2-aminoethanol diethylenetriamine), in a 3:1 ratio (by weight) [29].

They keep the total loading of filler(s) at 32 wt.% and the respective fillers (IFR, organo-MMT or LDH) vary. The combination of high IFR loading (30%) and loading of organo-MMT or LDH (2%) is more effective in fire performance than the combination of low loading of IFR (27%) and high loading of organo-MMT or LDH (5%). The same amount of organo-MMT is more effective in PHRR reduction (54% and 69%, respectively) than that of LDH (51% and 60%, respectively). This result is consistent with previous research from the Wilkie group, which also compared MMT and MgAl-LDH in PMMA matrix, but without IFR [30].

23.5   COMBINATION WITH OTHER FILLERS (CNT, ZRP, ETC.)

The combination of MMT and other fillers may result in enhanced fire retardancy properties, however, some questions still need to be addressed. Is there always synergy between MMT and other FR fillers? How important is compatibility; is there an optimum ratio? The combination of MMT and metal hydroxides are studied initially by Beyer et al. in EVA matrix in order to reduce the high loading of metal hydroxides, normally 60 wt% or higher [31]. Instead of alumina trihydrate (ATH), Lenza et al., did a thorough investigation of magnesium hydroxide (MH) in the high-density polyethylene (HDPE) matrix in combination with MMT [32]. The MMT loading is in a range of 1–5 wt% and MH 10–55 wt%. In this study, the MH-containing composites show an increased TTI compared to control HDPE. MH at 45 wt% loading in HDPE resulted in remarkable reduction in PHRR (86%); the addition of 5 wt% MMT did not produce a substantial increase in PHRR reduction (88%). The LOI value increases from 25.6 (HDPE/MH45%) to 28.3 (HDPE/MH45%/EVA/MMT5%), which is the same value obtained in HDPE/MH50%. In another study, in order to enhance the compatibility of hydrophilic MH with hydrophobic polymers, the incorporation of MH, with and without salinization, was investigated by Suihkonen et al., in epoxy-based composites [33]. The silanized MH shows better compatibility with epoxy matrix by scanning electron microscope (SEM) observation. However, when the fire performance was evaluated by cone calorimetry neither nano-scale nor microscale MH, whether silanized or not, exhibited a discernible difference. This is probably due to the mechanism of MH during the burning process, which decomposes and produces water vapor to absorb the heat of combustion and nano-dispersion is not necessary. This result also reminds us that, although the term "nano" has been an oft-used concept, it is not necessary that every material be a nano-material. Different situation may need different consideration.

Carbon nanotubes (CNT) have found various applications due to their desirable mechanical and thermal properties [34, 35]. Since there are innumerable papers addressing CNT's many properties, here the focus is only on the fire retardancy behavior, especially in combination with MMT. Beyer and Kashiwagi did some pioneering work in this area [36–38]. They found that CNT shows superior or equivalent efficiency in FR relative to that with same amount of MMT. Despite the reductions in HRR/MLR of polymer/CNT nanocomposites (using a conventional cone calorimeter), in most cases, the polymer was completely burnt leaving behind only a layer of CNTs. So, it is believed that CNTs may be addressed as an adjuvant in certain systems and scenarios rather than as a general fire retardant [39]. Moreover, due to relatively high cost of CNT, combination of CNT with cheaper MMT may result in better fire performance and be more cost effective. Cai et al. investigated the combination of multiwall carbon nanotubes (MWCNT) and organoclay modified with two-chain benzimidazolium surfactants with melamine as an additional FR in polyamide 6 (PA6) [40]. The higher steric hindrance in the presence of benzimidazolium modified silicate layers resulted in poor dispersion in a PA 6 matrix. Furthermore, this had a negative effect on the dispersion of CNTs in the ternary nanocomposite. The additional presence of melamine also had a negative effect on the thermal stability of CNT reinforced composites. Also due to poor dispersion of organoclay, no positive interaction between organoclay and CNT in FR performance was observed. This must remind the reader that simply combining two or more materials does not lead to a synergistic combination. Synergy is a badly misused term in the field of fire retardancy and it must be mathematically proven in each instance.

The combination of MMT and other materials, such as, a Lewis acid type transition metal has been investigated in the past decade [41, 42]. The purpose of this combination is to enhance formation of more homogeneous char without or with fewer cracks. Recently, Kashiwagi, et al., studied MMT-based poly (acrylonitrile-co-styrene) (SAN) nanocomposites with zinc chloride as a catalyst system, which were prepared by the solvent casting method to explore possible synergistic flame retardant effects [43]. The presence of zinc chloride did increase the amount of char of the composites, but with larger cracks as well. They explained that the char formation of zinc chloride-containing composites occurs in the later stage of the gasification process, while, SAN-clay nanocomposites forms small cracks in the surface layer at the early stage. Thus, each additive acts independently, without any synergy. At the end of their work, they suggested that catalysts forming char at earlier times are needed to obtain such synergy. In subsequent work from the same group, a series of phosphomolybdates (PMo) with different cations, including ammonium phosphomolybdate hydrate (NHPMo), melamine phosphomolybdate hydrate (MEPMo), zinc phosphomolybdate hydrate (ZnPMo) and sodium phosphomolybdate hydrate (NaPMo), were used as catalysts, combined with MMT clay, to flame retard SAN [44]. Among the four PMo salts, some synergistic effects were seen in the combination of NHPMo with Cloisite 20A which is probably due to the overlapping degradation temperature range of the combination of 20A/NHPMo and SAN40, which probably enhances char formation. This result flows directly from suggestions raised in the previous paper[43]. Another possible reason is that more NHPMo may be around or on the clay stacks and less in the gallery space, which probably bridges the clay stacks with char formed around clay and forms a stronger network during the degradation process. The other three PMo salts also form network structures, but they break down at higher temperatures.

In another study, nano-dispersed cerium (IV) dioxide (CeO_2) has been found to exhibit significant FR enhancement at low loadings (2.3 wt% or 4.6 wt%) when incorporated in PMMA through a precursor via emulsion polymerization [45]. It was suggested that this performance might be due to the ability of Ce (III) to act as a reluctant at elevated temperatures. Because it is not convenient to add FR fillers homogeneously through emulsion polymerization and with an intention to study potential synergy with MMT, a melt blending process was employed for the mixing of PMMA and CeO_2 in combination with MMT. No nano-dispersion of the cerium oxide was observed, while that is observed in emulsion polymerization. Thus, with microdispersion of CeO_2, MMT is dominant in fire retardancy when evaluated by cone calorimetry [46]. Hence, there is no synergy between microdispersed CeO_2 and MMT. Even when CeO_2 is nano-dispersed, the combination with MMT is ineffective; this is likely due to the clay layers blocking the transport of Ce (III) particles to the surface.

23.6   LAYER-BY-LAYER TECHNIQUE (LBL)

LbL technique was first developed by Decher et al. [47, 48]. Since this is a relatively simple technique to fabricate multifunctional thin films that are typically less than one micrometer thick, it has aroused wide interest worldwide during the past decade [49–51]. The advantage of LbL in fire performance of the materials with which it is coated is that it makes full use of the barrier properties of clay at the top surface. For other well-dispersed nano-composites, either intercalated or exfoliated, the rising bubbles of decomposition products of polymers during burning push the clay layers to the top surface. Due to some factors, such as, viscosity, clay loading, etc., it is not easy to form a homogeneous char; cracks are usually seen after burning. Since MMT layers are negatively charged, it is possible to combine a positively charged substance to employ LbL technique for FR purpose. Grunlan's group did some interesting work in this area with the LbL technique [52–56]. They first employed this technique with an assembly of branched polyethylenimine (BPEI) and Laponite clay. Laponite clay is a synthetic clay whose aspect ratio is lower than MMT. By adjusting the pH values, the LbL film coated fabric samples exhibit moderately enhanced FR behavior compared to the control fabric by the vertical flame test [51]. Since the aspect ratio of MMT is larger than Laponite and is more cost effective, in subsequent work, they investigated cotton fabric coated with BPEI and MMT clay, prepared *via* LbL [52]. In this study, they tried two different pH values (7 and 10) and two clay suspensions (0.2 wt% and 1 wt%). During actual burning in the vertical flame test, compared with control samples, afterglow time of cotton fabric coated with LbL technique was significantly reduced. The woven structure of the fabric, as observed in SEM images, is well preserved relative to the chars from coated fabrics, whereas the few ashes from the control fabric showed little structure. To obtain more details for FR behavior, they carried out microscale combustion calorimeter (MCC). The results indicate there is no evidence that more bi-layers (bL) (20bLs) bring more heat release capacity (HRC) reductions than less bL coatings (5bLs). However, the samples prepared at pH=10 and 1wt% MMT with 5bLs coating shows lowest HRC and total heat release (THR). This paper did not give an explanation on the FR mechanism of this system, thus, further studies are needed to interpret the above phenomena. Following the same procedure, another study from their group investigated an environmentally benign LbL coating composed of positively charged chitosan and anionic MMT[56]. In this work, they also tried two different pH values (3 and 6) with 1 wt% MMT suspension. The 10 bL coated PU prepared at pH=6 exhibits a 52% PHRR reduction compared with control PU through cone calorimetry. While the samples prepared at pH=3 only bring about less than 40% PHRR reduction. The first result is equivalent to 5% clay loading in PU matrix if conventional polymer nanocomposites are employed. But in their cone data, no TTI and smoke value are provided. From this work, it can be seen that pH value of the solution is an important factor, which affects the final FR results. It seems that the LbL technique is a sound and practical method to achieve FR. However, can one scratch the LbL coatings off and thus expose the underlying fabric, in which case burning is assured? This, and similar questions on the durability of these coatings, must be addressed.

23.7 CONCLUSION AND PROSPECTS

During the past several years, research on FR polymer-MMT based materials is still a hot area worldwide. The dispersion of MMT clay in polymer matrices is a major concern in FR behavior, as without good dispersion, there is no FR enhancement. There have been numerous updates in combination of various FR fillers such as CNT, LDH and IFR etc. For the future, we expect to see the increased use of combinations with the green FR materials, that is, halogen-free materials. Another factor that must be considered is the ease of fabrication. As can be seen in this review, there have been some attempts to explore novel surfactants to modify MMT, however, some involve a complicated synthetic route, and sometimes, with low yields. As this review was being considered, the Nobel Prize in Chemistry for 2013 was awarded *"for the development of multiscale models for complex chemical systems" [57]*. This award may indicate a trend in chemical and materials science research in future. As described in this paper, fire retardancy of polymer-MMT nanocomposites, when in junction with other FR components, are sometimes very complicated. The trial and error method, which is the current situation, is expensive both in time and in money and this must be improved. The expense may be reduced substantially if new FR formulations are initiated by first by computer modeling. Thus the requirements for FR researchers in future may include not only synthesis, fabrication, processing and evaluation of polymeric materials, but also proficiency with computer modeling. FR researchers must also broaden their view to find some effective means from related areas to better enhance FR performance of polymer composites. There is an old saying in China that "there are other hills whose stones are good for working jade." Hopefully, there will come up some "jades" in FR area to make our life safe and green.

KEYWORDS

- clay–fire retardant combinations
- fire retardancy
- montmorillonite
- nanocomposites

REFERENCES

1. Chevrier, J., Harley, K. G., Bradman, A., Gharbi, M., Sjödin, A., Eskenazi, B., Polybrominated Diphenyl Ether (PBDE) Flame Retardants and Thyroid Hormone during Pregnancy, *Environ Health Perspect* 2010, 118, 1444–1449.
2. Herbstman, J. B., Sjödin, A., Kurzon, M., Lederman, S. A., Jones, R. S., Rauh, V., Needham, L. L., Tang D Niedzwiecki, M., Wang, R. Y., Perera, F., Prenatal Exposure to PBDEs and Neurodevelopment, *Environ Health Perspect* 2010, 118, 712–719.
3. DiGangi, J., Blum, A., Bergman, Å., de Wit, C. A., Lucas, D., Mortimer, D., Schecter, A., Scheringer M Shaw, S. D., Webster, T. F., 2010. San Antonio Statement on

Brominated and Chlorinated Flame Retardants, *Environ Health Perspect* 2010, 118, A516–A518.

4. Gilman, J., Kashiwagi, T., Lichtenhan, J. Nanocomposites: a revolutionary new flame retardant approach. *SAMPE Journal* 1997, 33, 40–46.

5. Kiliaris, P., Papaspyrides, C. D., Polymer/layered silicate (clay) nanocomposites: An overview of flame retardancy. *Progress in Polymer Science* 2010, 35, 902–958.

6. Gilman, J. W., Kashiwagi, T., in *Polymer-Clay nanocomposites*, T.J. Pinnavaia and, G.W. Beall, Eds, John Wiley and Sons, New York, 2000, pp 193–206.

7. Jang, B. N., Wilkie, C. A., The Thermal Degradation of Polystyrene Nanocomposites, *Polymer*, 2005, 46, 2933–2942.

8. Jang, B. N., Wilkie, C. A., The Effect of Clay on the Thermal Degradation of Polyamide 6 in Polyamide 6/Clay Nanocomposites, *Polymer*, 2005, 46, 3264–3274.

9. Costache, M. C., Jiang, D. D., Wilkie CA Thermal Degradation of Ethylene-Vinyl Acetate Copolymer Nanocomposites, *Polymer*, 2005, 46, 6947–6958.

10. Costache, M. C., Wang, D. Y., Heidecker, M. J., Manias, E., Wilkie, C. A., The Thermal Degradation of Poly(Methyl Methacrylate) Nanocomposites with Montmorillonite, Layered Double Hydroxides and Carbon Nanotubes, *Polymers Adv. Tech.*, 2006, 17, 272–280.

11. Jang, B. N., Costache, M. C., Wilkie, C. A., The Relationship Between Thermal Degradation Behavior of Polymer and the Fire Retardancy of Polymer/Clay Nanocomposites, *Polymer*, 2005, 46, 10678–10687.

12. Chen, K., Wilkie, C. A., Vyazovkin, S., Revealing Nano-confinement in Degradation and Relaxation Studies of Two Structurally Different Polystyrene-Clay Systems, *J. Phys. Chem. B*, 2007, 111, 12685–12692.

13. Zhu, J., Uhl, F., Morgan, A. B., Wilkie, C. A., Studies on the mechanism by which the formation of nanocomposites enhance thermal stability, *Chem. Mater.*, 2001, 13, 4649–4654.

14. Carvalho, H. W. P., Santilli, C. V., Briois, V., Pulcinelli, S. H., Polymer-clay nanocomposites thermal stability: experimental evidence of the radical trapping effect, *RSC Advances*, 2013, 3, 22830–22833.

15. Shi, Y., Kashiwagi, T., Walters, R. N., Gilman, J. W., Lyon, R. E., Sogah DY. Ethylene vinyl acetate/layered silicate nanocomposites prepared by a surfactant-free method: enhanced flame retardant and mechanical properties. *Polymer* 2009, 50, 3478–3487.

16. Ray, S. S., Okamoto, M. Polymer/layered silicate nanocomposites: a review from preparation to processing, *Progress in Polymer Science* 2003, 28, 1539–1641. .

17. Gilman, J. W., Jackson, C. L., Morgan, A. B., Harris, R., Manias, E., Giannelis, E. P., et al. Flammability properties of polymer-layered-silicate nanocomposites. Polypropylene and polystyrene nanocomposites, *Chem. Mater.* 2000, 12, 1866–1873.

18. Morgan AB. Flame retarded polymer layered silicate nanocomposites: a review of commercial and open literature systems, *Polymers Adv. Tech.*, 2006, 17, 206–217.

19. Zhu, J., Morgan, A. B., Lamelas, F. J., Wilkie CA. Fire properties of polystyrene–clay nanocomposites. *Chem Mater* 2001, 13, 3774–3780.

20. Pack, S., Kashiwagi, T., Cao, C. H., Korach, C. S., Lewin, M., Rafailovich, M. H., Role of surface interactions in the synergizing polymer/clay flame retardant properties, *Macromolecules* 2010, 43, 5338–5351.

21. Levchik, S. V., Weil, E. D., A review of recent progress in phosphorus-based flame retardants, *J Fire Sci*, 2006, 24, 345–364.
22. Lai, X. J., Zeng, X. R., Li HQ , Liao, F., Yin, C. Y., Zhang, H. L., Synergistic Effect of Phosphorus-Containing Montmorillonite with Intumescent Flame Retardant in Polypropylene, *Journal of Macromolecular Science, Part B: Physics*, 2012, 51, 1186–1198.
23. Huang, G. B., Gao, J. R., Li, Y. J., Han, L., Wang, X., Functionalizing nano-montmorillonites by modified with intumescent flame retardant: preparation and application in polyurethane, *Polym Degrad Stab* 2010, 95, 245–253.
24. Huang, G. B., Li, Y. J., Han, L. A., Gao, J. R., Wang, X. A novel intumescent flame retardant-functionalized montmorillonite: preparation, characterization, and flammability properties, *Appl Clay Sci* 2011, 51, 360–365.
25. Wang, W. S., Liang, C. K., Chen, Y. C., Su, Y. L., Tsai, T. Y., Chen-Yang, Y. W., Transparent and Flame Retardant PMMA/Clay Nanocomposites Prepared with Dual Modified Organoclay, *Polym Adv Technol* 2012, 23, 625–631.
26. Swain, S. K., Isayev, A. I., Effect of ultrasound on HDPE/clay nanocomposites: Rheology, structure and properties, *Polymer* 2007, 48, 281–289.
27. Patra, S. K., Prusty G and Swain, S. K., Ultrasound assisted synthesis of PMMA/clay nanocomposites: Study of oxygen permeation and flame retardant properties, *Bull Mater Sci* 2012, 35, 27–32.
28. Ribeiro, S. P. S., Estevao, L. R. M., Nascimento, R. S. V., Effect of clays on the fire-retardant properties of a polyethylenic copolymer containing intumescent formulation, *Sci. Technol. Adv. Mater* 2008, 9, 024408 (7pp).
29. Shen, Z. Q., Chen, L., Lin, L., Deng, C. L., Zhao, J., and Wang, Y. Z., Synergistic effect of layered nanofillers in intumescent flame- retardant EPDM: montmorillonite versus layered double hydroxides, *Ind. Eng. Chem. Res.* 2013, 52, 8454−8463.
30. Wang, L., Xie, X., Su, S., Feng, J., Wilkie, C. A., A comparison of the fire retardancy of poly (methyl methacrylate) using montmorillonite, layered double hydroxide and kaolinite. *Polymer Degradation and Stability* 2010, 95, 572–578.
31. Beyer, G., Flame retardant properties of EVA-nanocomposites and improvements by combination of nanofillers with aluminum trihydrate, *Fire and Materials* 2001, 25, 193–197.
32. Lenza, Joanna, Merkel, Katarzyna, Rydarowski, Henryk, Comparison of the effect of montmorillonite, magnesium hydroxide and a mixture of both on the flammability properties and mechanism of char formation of HDPE composites, *Polymer Degradation and Stability* 2012, 97, 2581–2593.
33. Suihkonen, R., Nevalainen, K., Orell, O., Honkanen, M., Tang, L. C., Zhang, H., Zhang, Z., Vuorinen, J., Performance of epoxy filled with nano- and micro-sized magnesium hydroxide, J Mater Sci 2012, 47,1480–1488.
34. Iijima, S. *Nature* (London) 1991, 354(6348), 56–58.
35. Xie, X. L., Mai, Y. W., Zhou XP. *Mat Sci Eng* 2005, 49(4), 89–112.
36. Beyer, G., Short communication: carbon nanotubes as flame retardants for polymers, *Fire Mater.* 2002, 26, 291–293.
37. Kashiwagi, T., Grulke, E., Hilding, J., Groth, K., Harris, R., Butler, K., Shields, J., Kharchenko, S., Douglas, J., Thermal and flammability properties of polypropylene/carbon nanotube nanocomposites. Polymer 2004, 45, 4227–4239.

38. Takashi Kashiwagi, Fangming Du, Jack, F. Douglas, Karen, I. Winey, Richard, H. Harris Jr, John R Shields, Nanoparticle networks reduce the flammability of polymer nanocomposites, Nature Materials, 2005, 4, 928–933.

39. Schartel, B., Potschke, P., Knoll, U., Abdel-Goad, M. Fire behavior of polyamide 6/ multiwall carbon nanotube nanocomposites, *Eur Polymer J* 2005, 41, 1061–1070.

40. Cai, G. P., Dasari, A., Yu, Z. Z., Du, X. S., Dai, S. C., Mai, Y. W., Wang, J. Y., Fire response of polyamide 6 with layered and fibrillar nanofillers, *Polym Degrad Stab* 2010, 95(5), 845–851.

41. Jang, B. N., Wilkie, C. A., The effects of clay on the thermal degradation behavior of poly(styrene-co-acrylonitirile), *Polymer* 2005, 46, 9702–9713.

42. Cai, Y., Hu, Y., Song, L., Xuan, S., Zhang, Y., Chen, Z., Fan, W., Catalyzing carbonization function of ferric chloride based on acrylonitrile-butadiene-styrene copolymer/organophilic montmorillonite nanocomposites, *Polym Degrad Stab* 2007, 92, 490–496.

43. Kashiwagi, T., Danyus, R., Liu, M., Zammarano, M., Shields, J. R., Enhancement of char formation of polymer nanocomposites using a catalyst, *Polymer Degradation and Stability* 2009, 94, 2028–2035.

44. Meifang Liu, Xin Zhang, Mauro Zammarano, Jeffrey, W. Gilman, Takashi Kashiwagi, Flame retardancy of poly(styrene-co-acrylonitrile) by the synergistic interaction between clay and phosphomolybdate hydrates, *Polymer Degradation and Stability* 2011, 96, 1000–1008.

45. Cai, G. P., Lu, H. D., Zhou, Y., Hao, J. W., Wilkie, C. A., Fire retardancy of emulsion polymerized PMMA/Cerium (IV) dioxide and PS/ Cerium (IV) dioxide nanocomposites, *Thermochimica Acta* 2012, 549, 124–131.

46. Guipeng Cai, Charles A. Wilkie, Shaorong Xu, Zhengzhou Wang, Further studies on polystyrene/ cerium (IV) oxide system: melt blending and interaction with montmorillonite, *Polym Adv Technol* 2014, 25, 217–222.

47. Decher, G., Hong, J. D. European patent, application number 91113 464.1.

48. Decher, G., Hong, J. D. *Makromol. Chem.,Macromol. Symp.* 1991, 46, 321–327.

49. Ariga, K., Hill, J. P., Ji, Q. Layer-by-layer assembly as a versatile bottom-up nanofabrication technique for exploratory research and realistic application, *Phys. Chem. Chem. Phys.* 2007, *9*, 2319–2340.

50. Bertrand, P., Jonas, A., Laschewsky, A., Legras, R. Ultrathin polymer coatings by complexation of polyelectrolytes at interfaces: suitable materials, structure and properties, *Macromol. Rapid Commun.* 2000, *21*, 319–348.

51. Podsiadlo, P., Shim, B. S., Kotov, N. A., Polymer/clay and polymer/carbon nanotube hybrid organic-inorganic multilayered composites made by sequential layering of nanometer scale films, *Coord. Chem. Rev.* 2009, *253*, 2835–2851.

52. Li, Y. C., Schulz, J., Grunlan JC. Polyelectrolyte/nanosilicate thin-film assemblies: influence of pH on growth, mechanical behavior, and flammability. ACS Applied Materials & Interfaces 2009, 1(10), 2338–2347.

53. Li, Y. C., Schulz, J., Mannen, S., Delhom, C., Condon, B., Chang, S., Zammarano, M., Grunlan, J. C., Flame retardant behavior of polyelectrolyte-clay thin film assemblies on cotton fabric, *ACS Nano* 2010, 4, 3325–3337.

54. Laufer, G., Carosio, F., Martinez, R., Camino, J., Grunlan, J. C., Growth and fire resistance of colloidal silica-polyelectrolyte thin film assemblies, *J. Colloid Interface Sci.* 2011, 356, 69–77.

55. Li, Y. C., Mannen, S., Morgan, A. B., Chang, S., Yang, Y. H., Condon, B., Grunlan, J. C., Intumescent all-polymer multilayer nanocoating capable of extinguishing flame on fabric, *Adv. Mater.* 2011, 23, 3926–3931.

56. Laufer, G., Kirkland, C., Cain, A. A., Grunlan, J. C., Clay-chitosan nanobrick walls: completely renewable gas barrier and flame-retardant nanocoatings, *ACS Applied Materials & Interfaces* 2012, 4, 1643–1649.

57. http://www.nobelprize.org/.

CHAPTER 24

SYNTHESIS OF O-INCLUDING COMPOUNDS BY CATALYTIC CONVERSATION OF OLEFINS

G. Z. RASKILDINA,[1] N. G. GRIGOR'EVA,[2] B. I. KUTEPOV,[2]
S. S. ZLOTSKY,[1] and G. E. ZAIKOV[3]

[1]*Ufa State Petroleum Technological University, 1 Kosmonavtov Str., 450062 Ufa, Russia; Phone (347) 2420854, E-mail: nocturne@mail.ru*

[2]*Institute of Petrochemistry and Catalysis of RAS, 141 pr. Oktyabria, 450075, Ufa, Russia*

[3]*N. M. Emanuel Institute of Biochemical Physics, Russian Academy of Sciences, 4 Kosygin str., 119334 Moscow, Russia; E-mail: Chembio@sky.chph.ras.ru*

CONTENTS

ABSTRACT

By studying the reactions of hydration of norbornene and 2-vinyl-2-me-thil-gem-dichlorocyclopropane and the reactions of styrene and norborn-ene with different alcohols and carbonic acids in presence heterogenic catalyst, it was found that the selected zeolite H-Beta is active and selec-tive catalyst for these reactions. Alcohol, ethers, esters and diesters of nor-bornene have exo-configuration. It has been established that reaction of norbornene with diols, catalyzing by zeolite Beta, leads to the formation of esters, which did not find before.

24.1 INTRODUCTION

We have been investigated the reactions of commercially available alkenes (norbornene and 2-vinyl-2-methil-*gem*-dichlorocyclopropane) with water in the presence of zeolite catalyst H-Beta, which successfully is used in petrochemical processes alkylation, isomerization and so on [1, 2]. The results of this reactions are formation of correspond alcohols. It was deter-mined that norbornene is hydrated easier (T=10 °C, under atmospheric pressure) than 2-vinyl-2-methil-*gem*-dichlorocyclopropane (T=150 °C, 6 bar gauge autoclave pressure). The results have reported that the hetero-geneously catalytic joining of water to olefins is interest as an effective and cheap method of obtaining of correspond alcohols. Furthermore, we investigated addition reaction of alcohols and ethers to norbornene and sty-rene. Addition of *O*-including compounds to the multiple carbon-carbon bonds in presence homogeneous and heterogeneous catalysts finds a wide application in synthesis of ethers and esters [3]. The use of homogeneous (mineral acids) and heterogeneous (cationite) catalysts has some draw-backs and do not provide required high yields and selectivity of products.

24.2 RESULTS AND DISCUSSION

The obtaining of alcohols by hydratation of olefins widespread in petro-chemical synthesis [4–6]. Recently was shown hydratation over mineral (sulfuric, phosphoric) and organic (toluene sulfonic) acids and over a cat-ionite KU-2–8 in H$^+$-form [7–10].

The hydratation of model alkenes (norbornene **2** and 2-vinyl-2-methil-*gem*-dichlorocyclopropane **3**) over zeolite catalyst H-Beta, which is widely used in industrial scale [8–10], we investigated.

The corresponding alcohols (**4, 5**) is obtained at 10°C (under atmospheric pressure) in water solvent. Conversion of more active norbornene **2** is 90% for 2 h (yield of alcohol is 50%), while less active 2-vinyl-2-methil-*gem*-dichlorocyclopropane **3** gives alcohol **5** (yield 24%) in an autoclave at 150°C, 6 bar gauge autoclave pressure for 4 h is shown in (Table 24.1).

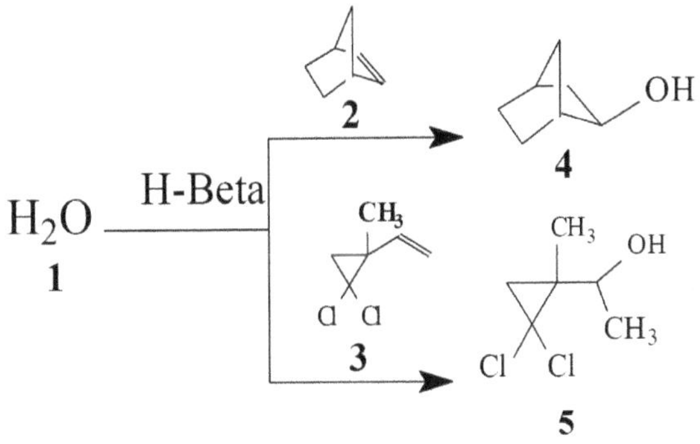

SCHEME 1   Hydratation of olefins.

In this chapter, we found (scheme 2) that monohydric alcohols (**7–9**) of various structures (butyl, allylic, benzyl) are connected to styrene (**6**) and norbornene (**2**) selectivity with formation corresponding ethers (**10–15**).

In the studied conditions (Table 24.2) there is total conversion of olefin. And selectivity of formation of target ethers is more 85%, which weakly dependent on the structure of reagents.

TABLE 24.1   The Interaction of Olefins **2, 3** With Water **1** (Molar Ratio Olefins: Water = 1: 24; 20 wt% Catalyst)

Reagents		Temperature (°C)	Time (h)	K^a, %	Selectivity (%)
	2	10	2	90	**4** (50)
1		80	24	26	**5** (24)
	3	150*	2	48	**5** (25)

K^a – conversion of olefin.

*Autoclave.

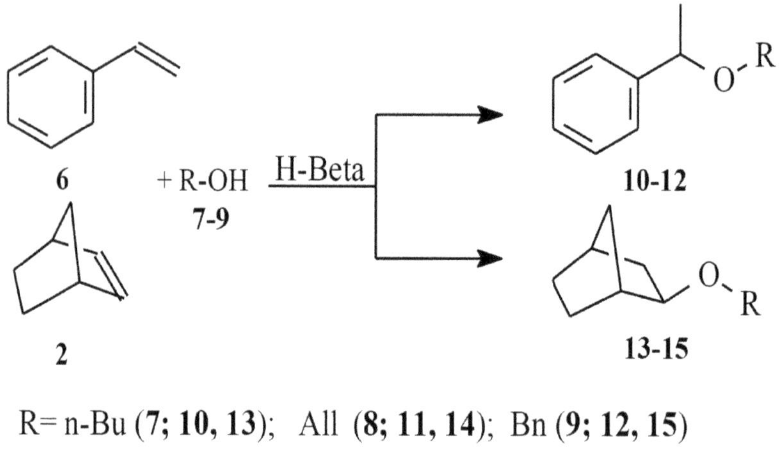

R= n-Bu (**7; 10, 13**); All (**8; 11, 14**); Bn (**9; 12, 15**)

SCHEME 2 Formation of ethers (**10–15**) by reaction of alcohols and olefins **2, 6**.

TABLE 24.2 Reaction of Olefins **6, 2** With Alcohols **7–9** in Presence Zeolite H-Beta (Molar Ratio Olefin: Alcohol = 1: 3, 20 wt% Catalyst, T=80°C, 5 h)

Olefin	Alcohol	Selectivity, %
6	7	10 (92)
2	7	13 (93)
6	8	11 (89)
2	8	14 (90)
6	9	12 (92)
2	9	15 (95)

In the case reaction of diols (**16, 17**) with norbornene the reaction proceeds by consistent formation of mono- (**18, 19**) and diethers (**20, 21**) in presence of zeolite H-Beta (Scheme 3).

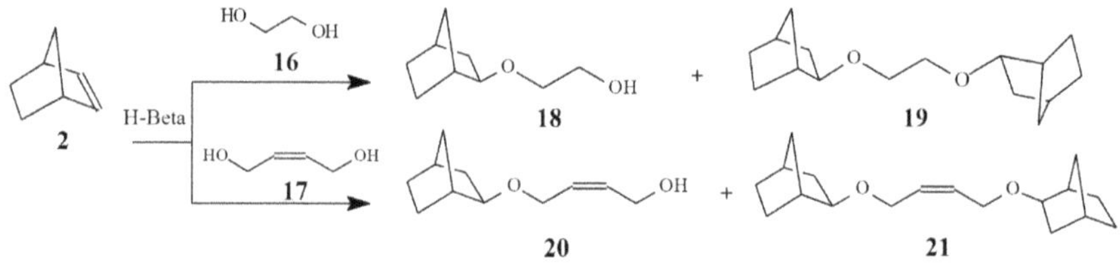

SCHEME 3 Formation of mono- and diethers of norbornene over zeolite H-Beta.

Increasing the temperature from 50°C to 80°C conversion of norbornene **2** is changes slightly. However selectivity of diethers' formation increases from 4% to 32%. Moreover increasing the molar ratio olefin: diol from 1:3 to 3:1 selectivity of diethers' formation increases by more than twice (Table 24.3).

In the studied conditions activities of ethylene glycol (**16**) and *cis*-2-butene-1,4-diol (**17**) are similar. Bicyclic olefin **2** reacts quantitatively with monohydric acids (**22–25**) producing appropriate esters. These yields are 80–99% (a four-fold molar excess of acid) and depend on identity of acids.

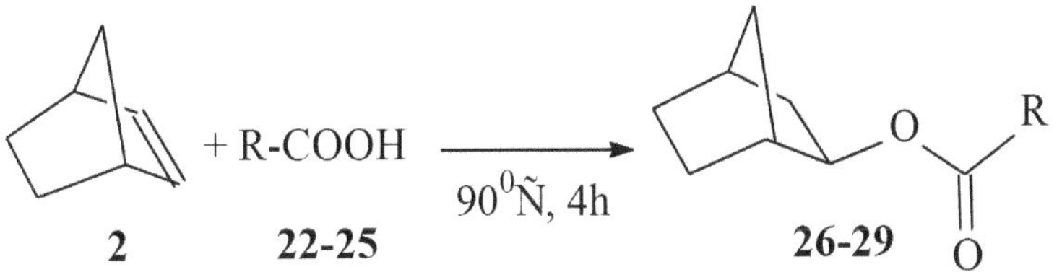

$$R= -CH_3 \ (22, 28); \ -n-C_3H_7 \ (23, 27); \ -CH_2Cl \ (24, 28); \ -C_3H_5(25, 29)$$

SCHEME 4   Formation of esters (**26–29**) by reacting monobasic carboxylic acids (**22–25**) with norbornene **2**.

At the same time, judging from yield of monoethers, monochloracetic **24** and methacrylic **25** acids are six times low active than acetic acid **22**.

The reaction of olefin **2** and dicarbonic acids is going to with the formation mono- and diethers appropriatly (Scheme 5).

TABLE 24.3   Reaction of Norbornene with Diols **16, 17** (Molar Ratio **2**: Diol = **A:B**, 20% wt. Catalyst H-Beta, 5 h)

Diol	A:B	T, °C	Selectivity, %	
			Monoether	Diether
16	1:3	50	**18** (92)	**19** (4)
		60	**18** (84)	**19** (12)
		80	**18** (58)	**19** (32)
17	3:1	80	**20** (25)	**21** (68)
	1:1		**20** (46)	**21** (41)
	1:3		**20** (60)	**21** (30)

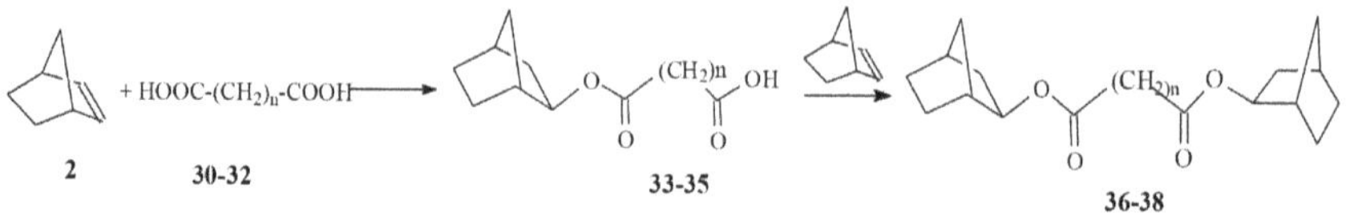

2   30-32

33-35

36-38

n = 0 (30, 33, 36); 1 (31, 34, 37); 2 (32, 35, 38)

SCHEME 5 Formation esters (**33–35**) by react monocarbonic acids (**30–32**) with norbornene **2**.

Spatial structure of diesters of norbornene (**36–38**) identified by methods of homo- (COSY, NOESY) and heteronuclear (HSQC, HMBC) two-dimensional ^{1}H, ^{13}C NMR spectroscopy. So, in ^{13}C NMR spectra of target esters (**36–38**), signals of atoms C-7 of bicyclic fragment are situated in the range of 35.6…36.2 ppm area., indicating that *exo*-configuration is, because signals *endo*-isomers located in the weaker field (~40 ppm area).

Exo-, exo- configuration of dibicyclo[2.2.1]hept-2-yl diester of malonic acid **37** is confirmed by interaction of protons of atoms C-2 and C-12 with protons of atoms C-6 and C-16. In case of *endo-, endo*-isomer protos of atoms C-2 and C-12 correlation with protons of atoms C-7 and C-17, that is not found in our case. Esters by *exo-, endo*-configuration are not have in the products of reactions from NMR-spectrums.

24.3 EXPERIMENTAL PART

An HRGS 5300 Mega Series "Carlo Erba" chromatograph with a flame ionization detector was used for the qualitative and quantitative analysis of starting material and reaction products. The chromatograph was equipped with a thermo-conductivity were registered using the "Bruker AVANCE-400" spectrometer (400.13 and 100.62 MHz, respectively) in CDCl$_3$ solvent, where benzene-d$_6$, toluene-d$_8$ were used as internal standards. High resolution mass spectra were measured on a Fisons Trio 1000 instrument, whose chromatograph was equipped with a DB-560 quarts column (50 m); the temperature of the column was increased from 50 to 320 °C with a programed heating rate of 4 °C min^{-1}; the electron impact (70 eV).

In order to carry out the reactions, zeolite BEA (Beta) (mole ratio SiO$_2$/Al$_2$O$_3$ = 18.0), synthesized in the JSC "Angarsk Catalysts and Organic Synthesis" in NH$_4$-form, was used as catalyst. Zeolite Beta was formed to H-form by heated in air at 540°C for 3 h. Before experiments the catalyst sample was dried in air for 4 h at 350°C.

24.3.1   CATALYST

Zeolite NH_4-Beta was produced by the public corporation Angarsk Factory of Catalysts and Organic Synthesis. Zeolite NH_4-Beta was transferred into H-Beta form by calcinations at 540 °C for 4h before all experiments.

24.3.2   THE INTERACTION OF NORBORNENE 2 WITH WATER OVER H-BETA ZEOLITE

The zeolite H-Beta (0.28 g) was added to a mixture of norbornene **2** (1.22 g, 0.013 mmol) and water **1** (50 g, 3.13 mmol) at 10 °C and stirred for 2h. After the catalyst was filtered and the reaction mass was extraction with ether. The latter was removed at the reduced pressure. The alcohol **4** (72°C/20 mm Hg) were isolated under vacuum. The mixture of *exo*-2-norborneol **4** and di-norbornyl ether of this mixture was (% w/w): 50/50% accordingly. Compounds were identified by NMR-spectroscopy.

24.3.3   EXO-2-NORBORNEOL (4)

^{1}H NMR, δ: 1.02–1.05 (m, 3H, C^6H_a, C^7H_a, C^5H_a), 1.12–1.18 (m, 2 H, C^3H_a, C^5H_b), 1.36–1.51 (m, 2H, C^7H_b, C^6H_b), 1.62–1.67 (m, 1H, C^3H_b), 2.12–2.35 (m, 2H, C^4H, C^1H), 3.68 (d, 1H, C^2H). ^{13}C NMR, δ: 24.39 C^6, 29.27 C^5, 34.38 C^7, 35.40 C^4, 42.37 C^3, 44.34 C^1, 74.95 C^2. IR-spectrum: 2851–2954 (C-H, CH_2), 3437 (-OH); m/z: 122 M$^+$ (30), 107 (100), 79 (93), 77 (52), 43 (30), 51 (21), 105 (10), 50 (10), 80 (7); Kovaė index I_k 1065.

24.3.4   THE INTERACTION OF 2-VINYL-2-METHIL-GEM-DICHLOROCYCLOPROPANE 3 WITH WATER OVER H-BETA ZEOLITE

The zeolite H-Beta (0.43 g) was added to a mixture of 2-vinyl-2-methil-*gem*-dichlorocyclopropane **3** (1.96 g, 0.013 mmol) and water **1** (50 g, 3.13 mmol) at 80 °C/150 °C* and stirred for 24/2*h. After the catalyst was filtered and the reaction mass was extraction with ether. The latter was removed at the reduced pressure. The alcohol **5** (95°C/25 mm Hg) were isolated under vacuum. The mixture of 1,1-dichloro-2-methyl-2-(hydroxyethyl-1)cyclopropane **5** of this mixture was 24%. Compound was identified by NMR-spectroscopy.

After the completion of the reaction mass was separated from the catalyst by filtration. The conversion of initial olefins and the quantitative composition of the alcohol fraction were determined using gas-liquid chromatography (GLC). The chemical stricter of alcohols (**4, 5**) was established by means of GC/MS spectrometry and NMR spectroscopy.

24.3.5   1,1-DICHLORO-2-METHYL-2-(HYDROXYETHYL-1) CYCLOPROPANE (5)

^{1}H NMR, δ: 1.23–1.24 2H (d, cyclC^{3}H$_2$), 1.42 3H (s, C^{5}H$_3$), 1.70 1H (s, OH), 1.92 3H (s, C^{6}H$_3$), 4.01–4.06 1H (m, C^{4}H ^2J=12.8, ^3J=6). ^{13}C NMR, δ: 20.71 C^6, 23.39 C^5, 34.66 C^3, 44.70 C^2, 66.39 C^4, 77.26 C^1. IR-spectrum: 2877–2972 (C-H, CH$_2$), 3418 (-OH); m/z: 169 M$^+$ (0.6), 45 (100), 124/126/128 (35/22/4), 87/89/91 (6/35/11), 53 (14), 53 (14), 43 (13), 51 (11). Kovaė index I$_k$=1123. *Autoclave.

24.3.6   METHOD OF REACTION OF OLEFINS (2) WITH ALCOHOLS (7–9)

A mixture of 0.255 M alcohol **7** (or 0.255 M alcohol **8,** or 0.255 M alcohols **9**) and 0.085 M norbornene **2,** 20% wt. catalyst H-Beta was carried out 80°C and mixed intensively for 5 h. The reaction mass was separated from the catalyst by filtering after the reaction termination and unreacted alcohol was removed at a low pressure. Ethers were isolated by vacuum distillation for calibration.

1-n-butyl-1-phenylethan (10): b.p. 95–96°C (10 mm Hg). ^{1}H-NMR (CDCl$_3$, δ ppm, JHz): 0.86 (t, 3H, CH$_3$), 1.33 (m, 2H, CH$_2$), 1.35 (m, 2H, CH$_2$), 1.45 (d, 3H, CH$_3$), 3.42 (q, 2H, OCH$_2$), 4.30 (q, 1H, CH), 7.3 (m, 5H, ArH). ^{13}C-NMR (CDCl$_3$, δ ppm): 13.78 C^6, 18.90 C^5, 21.15 C^2, 32.65 C^4, 69.75 C^3, 72.86 C^1, 124.97–127.04 Ar, 143.03 C$^{1'}$.

1-allyloxy-1-phenylethan (11): b.p. 68°C (5 mm Hg): ^{1}H-NMR (CDCl$_3$, δ ppm, JHz): 1.5 (d, 3H, CH$_3$), 3.85 (dddd., 1H, CH$_a$, ^2J 12.8, ^3J 5.8), 3.94

(dddd., 1H, CH$_b$, ^2J 12.8, ^3J 5.2), 4.51 (dd., 1H, CH, ^3J 6.4), 5.19 (dd., 1H, CH$_a$, ^2J 1.6, ^3J 10.4), 5.29 (dd., 1H, CH$_b$, ^2J 1.6, ^3J 17.2), 5.95 (dddd., 1H, CH, ^3J 5.2, ^3J 5.8, ^3J 10.4, ^3J 17.2), 7.28–7.40 (m., 5H, Ar).

1-cyclohexyloxy-1-phenylethan (12): b.p. 91–92°C (9 mm Hg): ^{1}H-NMR (CDCl$_3$, δ ppm, JHz): 1.46 (d, 3H, CH$_3$), 1.12–1.20 (m, 6H, CH$_2$), 1.78–1.94 (m, 4H, CH$_2$), 3.40 (m, 1H, CH), 4.64 (q, 1H, ArCH), 7.23–7.33 (m, 5H, ArH).

Exo-2-(butoxy)bicyclo[2.2.1]heptane (13): b.p. 77 °C (27 mm Hg). ^{1}H-NMR (CDCl$_3$, δ ppm, JHz): 0.91 (m, 3H, CH$_3$), 0.96–1.07 (m, 3H, CH$_2$), 1.32–1.41 (m, 3H, CH$_2$), 1.46–1.55 (m, 4H, CH$_2$), 2.20 (s, 1H, CH), 2.29 (d, 1H, CH, ^2J=4), 3.24–3.40 (m, 1H, CH), 3.24–3.40 (m, 2H, CH$_2$). MS (70eV), m/z (J, %): 168 [M–1]$^+$ (≤1), 94 (100); 66 (74); 79 (65); 67 (56), 41 (49), 95 (39), 57 (28), 83 (19), 55 (17), 68 (16), 56 (13), 112 (12).

2-(alliloxy)bicyclo[2.2.1]heptane (14): b.p. 78 °C (10 mm Hg). ^{1}H-NMR (CDCl$_3$, δ ppm, JHz): 0.95–1.12 (m, 3H, CH$_2$), 1.37–1.60 (m, 5H, CH$_2$), 2.25 (s, 1H, CH), 2.33 (d, 1H, CH), 3.89–4.00 (m, 2H, CH$_2$), 5.10–5.18 (d, 1H, CH$_a$, ^2J=1.6, ^3J=10.4), 5.22–5.32 (d, 1H, CH$_b$, ^2J=1.6, ^3J=17.2), 5.83–5.99 (m, 1H, CH). MS (70eV), m/z (J, %): 152 [M–1]$^+$ (1), 67 (100); 41 (56); 94 (51); 95 (41), 66 (34), 79 (29), 55 (27), 93 (23), 91 (11), 81 (11), 77 (11).

2-(benzyloxy)bicyclo[2.2.1]heptane (15): b.p. 70 °C (20 mm Hg). ^{1}H-NMR (CDCl$_3$, δ ppm, JHz): 0.92–1.05 (m, 3H, CH$_2$), 1.25–1.40 (m, 2H, CH$_2$), 1.51–1.56 (m, 3H, CH$_2$), 2.19 (s, 1H, CH), 2.28 (s, 1H, CH), 3.64–4.51 (m, 1H, CH), 3.64–4.51 (m, 2H, CH$_2$), 7.00–7.05 (m, 5H, Ar).

2-(bicyclo[2.2.1]heptyl-2-oxy)ethanol (18): b.p. 130°C (10 mm Hg). ^{1}H-NMR (CDCl$_3$, δ ppm, JHz): 0.95–1.10 (m, 3H, CH$_2$), 1.38–1.59 (m, 5H, CH$_2$), 2.23 (m, 1H, CH), 2.33 (m, 1H, CH), 2.55 (1H, OH), 3.39 (d, 1H, CH), 3.49–3.57 (m, 2H, CH$_2$), 3.70–3.77 (m, 2H, CH$_2$). ^{13}C-NMR (CDCl$_3$, δ ppm): 24.61 C^6, 28.55 C^5, 34.77 C^7, 35.17 C^4, 39.56 C^3, 40.36 C^1, 63.71 C^9, 67.66 C^8, 83.0 C^2. MS (70eV), m/z (J, %): 156 [M–1]$^+$ (2), 95 (100); 67 (49); 155 (19), 111 (18), 94 (16), 94 (15), 66 (15), 41 (15).

2,2′-[ethane-1,2-diylbis(oxy)]bicyclo[2.2.1]heptane (19): b.p. 158°C (5 mm Hg). ^{1}H-NMR (CDCl$_3$, δ ppm, JHz): 0.86–1.26 (m, 8H, CH$_2$), 1.32–1.40 (m, 8H, CH$_2$), 1.55–1.58 (m, 2H, CH$_2$), 1.63 (m, 1H, CH), 1.73 (m, 2H, CH$_2$), 2.04 (m, 2H, CH, CH), 3.20 (m, 2H, CH$_2$), 3.42 (m, 2H, CH$_2$), 3.57 (d, 1H, CH). MS (70eV), m/z (J, %): 250 [M–1]$^+$ (1), 95 (100);

94 (95); 66 (87), 79 (77), 67 (52), 45 (47), 41 (33), 55 (19), 57 (18); 83 (17); 44 (16); 77 (13); 65 (13); 43 (14); 53 (11).

Exo-4-(bicyclo[2.2.1]hept-2-yloxy)but-2-en-1-ol (20): b.p. 141 °C (2 mm Hg). ^{1}H-NMR (CDCl$_3$, δ ppm, JHz): 0.95–1.03 (m, 2H, C^{6}H$_b$, C^{3}H$_a$), 1.03–1.14 (m, 2H, C^{6}H$_a$, C^{3}H$_b$), 1.37–1.48 (m, 1H, C^{5}H$_b$), 1.50–1.60 (m, 1H, C^{5}H$_a$), 2.17 (s, 1H, C^{1}H), 2.25 (s, 1H, –OH), 2.34 (d, 1H, C^{4}H), 3.40 (d, 1H, C^{2}H), 3.95–4.09 (m, 1H, C^{8}H$_a$), 4.19(m, 1H, C^{8}H$_b$, ^2J=6.4, ^3J=18.8), 4.22 (d, 2H, C^{11}H$_a$, C^{11}H$_b$, ^2J=4.4 ^3J=16.8), 5.67–5.75 (m, 1H, C^{9}H), 5.76–5.85 (m, 1H, C^{10}H). ^{13}C-NMR (CDCl$_3$, δ ppm): 24.58 C^6, 28.40 C^5, 35.14 C^4, 39.51 C^3, 40.28 C^1, 58.53 C^{11}, 64.03 C^8, 82.67 C^2, 128.82 C^{10}, 131.82 C^9. MS (70eV), m/z ($J.$, %): 182 [M–1]$^+$ (2), 164 (7), 138 (12), 109 (12), 95 (100), 94 (22), 81(18), 79 (46), 77 (10), 71 (17), 70 (27), 67 (90), 66 (40), 57 (10), 55 (27), 53 (18), 43 (40).

Exo-exo-[(2Z)-but-2-en-1,4-diylbis(oxy)]bisbicyclo[2.2.1]heptane (21): b.p. 183°C (2 mm Hg). ^{1}H-NMR (CDCl$_3$, δ ppm, JHz): 0.97–1.12 (m, 6H, C^{6}H$_a$, C^{14}H$_a$, C^{5}H$_a$, C^{17}H$_a$, C^{7}H$_a$, C^{18}H$_a$), 1.34–1.48 (m, 6H, C^{3}H$_a$, C^{14}H$_a$, C^{5}H$_b$, C^{16}H$_b$, C^{6}H$_b$, C^{17}H$_b$), 1.49–1.58 (m, 4H, C^{7}H$_b$, C^{18}H$_b$, C^{3}H$_b$, C^{14}H$_b$), 2.23 (s, 2H, C^{4}H, C^{15}H), 2.30–2.34 (d, 2H, C^{1}H, C^{12}H), 3.35–3.40 (dd, 2H, C^{2}H, C^{13}H), 3.95–4.05 (m, 2H, C^{8}H$_a$, C^{8}H$_b$), 4.09 (dd, 1H, C^{11}H$_a$, 4.19 (dd, 1H, C^{11}H$_b$), 5.63–5.74 (m, 1H, C^{10}H), 5.75–5.88 (m, 1H, C^{9}H). ^{13}C-NMR (CDCl$_3$, δ ppm): 23.75 C^6, 27.62 C^5, 33.93 C^7, 34.27 C^4, 38.71 C^3, 39.42 C^1, 57.54 C^{11}, 63.04 C^8, 81.25 C^2, 129.41C^{10}, 131.96 C^9. MS (70eV), m/z ($J.$, %): 276 [M–1]$^+$ (<1), 164 (6), 96 (10), 95 (100), 93 (6), 79 (6), 70 (10), 67 (42), 41 (13).

24.3.7 METHOD OF REACTION OF NORBORNENE (2) WITH MONOCARBOXYLIC ACIDS (22–25)

A mixture of 0.34 M acetic acid **22** (or 0.34 M *n*-butyric acid **23**, or 0.34 M chloracetic acid **24**, or 0.34 M methacrylic acid **25**), 0.085 M of norbornene, 20% wt. catalyst H-Beta was carried out 90°C and mixed intensively for 4 h. For homogenization of initial compounds (**24, 25**) nonane was used as a solvent. The reaction mass was separated from the catalyst by filtering after the reaction termination and unreacted acid was removed at a low pressure. Esters were isolated by vacuum distillation for calibration.

Exo-bicyclo[2.2.1]hept-2-yl ester of acetic acid (26): b.p. 95°C (20 mm Hg). ^{1}H-NMR (CDCl$_3$, δ ppm, JHz): 1.08–1.18 (m, 4H, C^{3}H$_a$, C^{6}H$_a$, C^{6}H$_b$, C^{3}H$_b$), 1.42–1.54 (m, 3H, C^{4}H$_b$, C^{7}H$_a$, C^{7}H$_b$), 1.73 (m, 1H, C^{4}H$_a$), 2.02 (s, 3H, C^{9}H$_3$), 2.30 (m, 2H, C^{2}H, C^{5}H), 4.61 (d, 1H, C^{1}H). ^{13}C-NMR (CDCl$_3$, δ ppm): 21.42 C^9, 24.33 C^3, 28.13 C^4, 35.24 C^5, 35.37 C^7, 39.60 C^6, 41.40 C^2, 77.60 C^1, 170.82 C^8. MS (70eV), m/z (J, %): [M–1]$^+$ 154 (6), 43 (100), 66/67 (70/68), 94/95 (68/52), 79 (65), 111/112 (64/51), 41 (53), 71 (52).

Exo-bicyclo[2.2.1]hept-2-yl ester of n-butyric acid (27): b.p. 106°C (10 mm Hg). ^{1}H-NMR (CDCl$_3$, δ ppm, JHz): 0.96 3 (t, 3H, C^{11}H$_3$), 1.17–1.20 (m, 3H, C^{3}H$_a$, C^{7}H$_a$, C^{4}H$_a$), 1.39–1.49 (m, 2 H, C^{6}H$_a$, C^{4}H$_b$), 1.50–1.58 (m, 2H, C^{7}H$_b$, C^{3}H$_b$), 1.65 (m, 2H, C^{10}H$_2$), 1.69–1.78 (m, 1H, C^{6}H$_b$), 2.25 (t, 2H, C^{9}H$_a$, C^{9}H$_b$), 2.27–2.32 (m, 2H, C^{5}H, C^{2}H), 4.62 (d, 1H, C^{1}H). ^{13}C-NMR (CDCl$_3$, δ ppm): 13.66 C^{11}, 18.54 C^{10}, 24.31 C^3, 28.85 C^4, 35.26 C^7, 35.38 C^9, 36.59 C^5, 39.65 C^6, 41.45 C^2, 77.29 C^1, 173.40 C^8. MS (70eV), m/z (J, %): [M–1]$^+$ 182 (2), 71 (100), 95 (52), 43 (40), 139 (31), 111 (30), 154 (15), 79 (12).

Exo-bicyclo[2.2.1]hept-2-yl ester of chloracetic acid (28): b.p. 95°C (6 mm Hg). ^{1}H-NMR (CDCl$_3$, δ ppm, JHz): 1.09–1.22 (m, 3H, C^{3}H$_b$, C^{7}H$_b$, C^{4}H$_b$), 1.44–1.57 (m, 4H, C^{3}H$_a$, C^{4}H$_a$, C^{7}H$_a$, C^{6}H$_a$), 1.75–1.80 (m, 1H, C^{6}H$_b$), 2.32 (m, 1H, C^{5}H), 2.36 (d, 1H, C^{2}H), 4.04 (s, 2H, C^{9}H$_a$, C^{9}H$_b$), 4.72 (d, 2H, C^{1}H). ^{13}C-NMR (CDCl$_3$, δ ppm): 24.13 C^6, 28.03 C^5, 35.23 C^7, 35.35 C^4, 39.38 C^3, 41.20 C^9, 41.39 C^1, 79.71 C^2, 167.01 C^8. MS (70eV), m/z (J, %): [M–1]$^+$ 188 (<1), 66/67/68 (100/78/29), 94/95 (56/64), 77/79 (38/87), 41 (40), 49 (20), 55 (17), 42 (13), 53 (10).

Exo-bicyclo[2.2.1]hept-2-yl ester of methacrylic acid (29): b.p. 90°C (6 mm Hg). ^{1}H-NMR (CDCl$_3$, δ ppm, JHz): 1.12–1.20 (m, 3H, C^{3}H$_a$, C^{4}H$_a$, C^{7}H$_a$), 1.43–1.57 (m, 4H, C^{3}H$_b$, C^{4}H$_b$, C^{6}H$_a$, C^{7}H$_a$), 1.74–1.79 (m, 1H, C^{6}H$_b$), 1.93 (s, 3H, C^{11}H$_3$), 2.31 (m, 1H, C^{5}H), 2.35 (m, 1H, C^{2}H), 4.68 (m, 1H, C^{1}H), 5.52 (s, 1H, C^{10}H$_a$), 6.07 (s, 1H, C^{10}H$_b$). ^{13}C-NMR (CDCl$_3$, δ ppm): 18.26 C^{11}, 24.24 C^3, 28.17 C^4, 35.33 C^7, 35.37 C^5, 39.57 C^6, 41.45 C^2, 77.74 C^1, 124.81 C^{10}, 136.90 C^9, 167.09 C^8. MS (70eV), m/z (J, %): [M–1]$^+$ 180 (<1), 69 (100), 41 (84), 66 (71), 94 (56), 95 (40), 79 (19), 70 (19), 109 (11), 97 (11), 55 (10), 124 (10), 137 (10).

Method of reaction of norbornene (2) with dicarboxylic acids (30–32).

A mixture of 0,34 M oxalic acid **30** (or 0,34 M malonic acid **31**, or 0,34 M succinic acid **32**), 0,085 M of norbornene, 20% wt. catalyst H-Beta was carried out 90°C and mixed intensively for 4 h. For homogenization of initial compounds (**30–32**) nonane was used as a solvent. The reaction mass was separated from the catalyst by filtering after the reaction termination and unreacted acid was removed at a low pressure. Esters were isolated by vacuum distillation for calibration.

The resulting physic-chemical properties, NMR-spectra and mass-spectra of compounds **33–35** correspond to literature data [11].

Exo-dibicyclo[2.2.1]hept-2-yl ester of oxalic acid (36): b.p. 122°C (7 mm Hg). ^{1}H-NMR (CDCl$_3$, δ ppm, J/Hz): 1.12–1.23 (m, 6H, C^{6}H$_a$, C^{5}H$_a$, C^{7}H$_a$, C^{15}H$_a$, C^{14}H$_a$, C^{16}H$_a$), 1.44–1.62 (m, 8H, C^{6}H$_b$, C^{5}H$_b$, C^{3}H$_a$, C^{7}H$_b$, C^{15}H$_b$, C^{14}H$_b$, C^{12}H$_a$, C^{16}H$_b$), 1.76–1.82 (dddd, 2H, C^{3}H$_b$, C^{12}H$_b$), 2.33 (s, 2H, C^{4}H, C^{13}H), 2.43 (d, 2H, C^{1}H, C^{10}H), 4.75 (d, 2H, C^{2}H, C^{11}H). ^{13}C-NMR (CDCl$_3$, δ ppm): 24.11 C^3, C^{15}, 28.03 C^5, C^{14}, 35.29 C^7, C^{16}, 35.37 C^4, C^{13}, 39.26 C^3, C^{12}, 41.30 C^1, C^{10}, 80.46 C^2, C^{11}, 158.05 C^8, C^9. MS (70eV), m/z (J., %): [M–1]$^+$ 278 (0.1), 95/96 (100/8), 66/67/68 (8/24/2), 41 (9), 77/79 (3/4), 93 (4), 53/55 (2/4), 65 (3).

Exo-dibicyclo[2.2.1]hept-2-yl ester of malonic acid (37): b.p. 172°C (2 mm Hg). ^{1}H-NMR (CDCl$_3$, δ ppm, J/Hz): 1.08–1.19 (m, 6H, C^{5}H$_a$, C^{15}H$_a$, C^{6}H$_a$, C^{16}H$_a$, C^{7}H$_a$, C^{17}H$_a$), 1.43–1.59 (m, 8H, C^{3}H$_a$, C^{13}H$_a$, C^{7}H$_b$, C^{17}H$_b$, C^{5}H$_b$, C^{15}H$_b$, C^{6}H$_b$, C^{16}H$_b$), 1.72–1.77 (m, 2H, C^{3}H$_b$, C^{13}H$_b$), 2.30 (s, 2H, C^{4}H, C^{14}H), 2.35 (s, 2H, C^{1}H, C^{11}H), 3.29 (s, 2H, C^{9}H$_a$, C^{9}H$_b$), 4.67 (d, 2H, C^{2}H, C^{12}H). ^{13}C-NMR (CDCl$_3$, δ ppm): 24.18 C^6, C^{16}, 28.10 C^5, C^{15}, 35.24 C^7, C^{17}, 35.34 C^4, C^{14}, 39.34 C^3, C^{13}, 41.33 C^1, C^{11}, 42.26 C^9, 78.79 C^2, C^{12}, 166.34 C^8, C^{10}. MS (70eV), m/z (J., %): [M–1]$^+$ 292 (0.2), 95 (100), 67 (13), 111 (10), 199 (8).

Exo-dibicyclo[2.2.1]hept-2-yl ester of succinic acid (38)L: b.p. 180°C (1 mm Hg). ^{1}H-NMR (CDCl$_3$, δ ppm, J/Hz): 1.12–1.24 (m, 6H, C^{7}H$_a$, C^{5}H$_a$, C^{16}H$_a$, C^{6}H$_a$, C^{17}H$_a$, C^{18}H$_a$), 1.42–1.53 (m, 8H, C^{14}H$_a$, C^{3}H$_a$, C^{5}H$_b$, C^{16}H$_b$, C^{7}H$_a$, C^{18}H$_b$, C^{6}H$_b$, C^{17}H$_b$), 1.70–1.75 (m, 2H, C^{3}H$_b$, C^{14}H$_b$), 2.29 (m, 4H, C^{4}H, C^{15}H, C^{1}H, C^{12}H), 2.57 (m, 4H, C^{9}H$_a$, C^{9}H$_b$, C^{10}H$_a$, C^{10}H$_b$), 4.63 (d, 2H, C^{2}H, C^{13}H). ^{13}C-NMR (CDCl$_3$, δ ppm): 24.25 C^6, C^{17}, 28.13 C^5, C^{16}, 29.58 C^9, C^{10}, 35.25 C^7, C^{18}, 35.36 C^4, C^{15}, 39.52 C^3, C^{14}, 41.39 C^1, C^{12}, 77.89 C^2, C^{13}, 171.93 C^8, C^{11}. MS (70eV), m/z (J., %): [M–1]$^+$ 306 (0.1), 95 (100), 67 (24), 195 (18), 55 (9), 79 (9), 111 (9), 213(9), 41 (7), 162 (7).

24.4 CONCLUSION

The obtained results indicate that the zeolite H-Beta is active and selective catalyst for synthesis of alcohols, ethers and esters from olefins with acids and alcohols.

It should be noted that offered methods are simple compared to methods based on the use of traditional acid catalysts. In this case products of reactions are separated from the catalyst by filtering and zeolite H-Beta can be recovered for use later.

Besides high activity and selectivity, zeolite catalyst H-Beta makes it possible to obtain new structure compounds, which were not obtained based on homogeneous acidic catalysts [12].

KEYWORDS

- **2-vinyl-2-methil-gem-dichlorocyclopropane**
- **alcohols**
- **diesters**
- **esters**
- **ethers**
- **heterogenic catalyst**
- **norbornene**
- **styrene**
- **zeolite H-Beta**

REFERENCES

1. Valencia, S., Corma, A., Cambor, M. Microporous and Mesoporous Materials, 1998, 25, 59–74.
2. Minchayev, Ch. M., Kondratyev, D. A. Uspehi Khimii ("Successes of Chemistry," in Rus.), 1983, 52(12), 1921–1973.
3. Reutov, O. A., Kurts, A. L., Butin, K. P. Organic Chemistry (in Rus.). Part 2, 1999, 624.
4. Patent 2,345,573, US. Herman, A. Bruson, http://www.freepatentsonline.com/2345573. html. 1944. #4.

5. Xulai Yang, Sabornie Chatterjee, Zhijun Zhang, Xifeng Zhu, and Charles, U. Pittman. Jr., Ind. Eng. Chem. Res. 2010. V.49. P. 2003.
6. Brown, H., Kawakami, J., J. Amer. Chem. Soc. 1990. V.92. 1970.
7. Butlerov, A. M. Selected Works in Organic Chemistry. M.: Publishing House of the Academy of Sciences of the USSR, 1951. P. 333.
8. Azinger, Ph. Chemistry and Technology of monoolefins. M.: Gostoptehidat, 1960. 467.
9. Sumio, A. Chem. Engng. 1973. V.56. p. 80.
10. Menjajlo, A. T. Synthesis of alcohols and organic compounds from oils. Tr. NIISa. M.: Goschimizdat, 1960. pp.226.
11. Mamedov, M. K. Journal of Organic Chemistry (in Rus.), 2006, 42(8), 1159–1162.
12. Gasanov, A. G., Nagiev, A. V. Journal of Organic Chemistry (in Rus.), 1994, 30(5), 707–709.

CHAPTER 25

THE INTERACTION OF SUBSTITUTED *GEM*-DICHLOROCYCLOPROPANES WITH ALCOHOLS AND PHENOLS

A. N. KAZAKOVA,[1] A. A. BOGOMAZOVA,[2] N. N. MIKHAILOVA,[1] S. S. ZLOTSKY,[1] and G. E. ZAIKOV[3]

[1]*Ufa State Petroleum Technological University, 1 Kosmonavtov Str., 450062 Ufa, Russia; Tel: (347) 2420854, E-mail: nocturne@mail.ru*

[2]*Sterlitamak branch of Bashkir State University, 49 Lenina avenue, 453103 Sterlitamak, Russia*

[3]*N. M. Emanuel Institute of Biochemical Physics, Russian Academy of Sciences, 4 Kosygin str., 119334 Moscow, Russia; E-mail: chembio@sky.chph.ras.ru*

CONTENTS

ABSTRACT

This chapter discloses the synthesis of cyclopropanone ketals by the reaction of gem-dichlorocyclopropanes with alcohols and phenols. In these conditions brom containing gem-dichlorocyclopropane is destroyed with formation of propargyl aldehyde acetals.

25.1   INTRODUCTION

Contemporary and effective methods for generation of dichlorocarbene in aqueous-organic media allow quantitatively receive substituted *gem*-dichlorocyclopropanes and corresponding unsaturated hydrocarbons. Thereupon studying of functionalization and transformation of available substituted *gem*-dichlorocyclopropanes is significant interest because it allows obtaining valuable intermediate products and biological active compounds, which is not easily accessible by other methods.

25.2   RESULTS AND DISCUSSION

The chlorination of butadiene affords 3,4-dichlorobut-1-ene **I** in a high yield, which is widely used in organic synthesis [1].

We used compound **I** to obtain the stereoisomeric 1,2-dichloroethyl-*gem*-dichlorocyclopropanes **II** by the Makosza method [2].

The diastereoisomers **IIa** and **IIb** (1:1) were separated by vacuum distillation and characterized by the ^1H and ^{13}C NMR spectroscopy and mass spectrometry.

Thus, in the case of compound **IIa**, in the ^1H NMR spectrum the signal of the CH-proton of the cyclopropane ring is observed as a doublet of doublets of doublets at 2.5 ppm, with the spin-spin coupling constants $^3J_{3a-2} = 7.7$ Hz, $^3J_{3b-2} = 10.4$ Hz, $^3J_{2-1} = 9.4$ Hz, whereas the signal of the similar proton in the spectrum of compound **IIb** appears at 2.05 ppm as a doublet of doublets of doublets with the coupling constants $^3J_{3b-2} = 7.7$ Hz,

$^3J_{3a-2} = 10.4$ Hz, $^3J_{2-1} = 5$ Hz. According to the literature data [3], the values of the spin-spin coupling constants of the CH-proton of the cyclopropane ring for compounds **IIa** (9.4 Hz) and **IIb** (5 Hz) indicate the *threo*- and *erythro*- configuration of compounds **IIa** and **IIb**, respectively.

In the ^{13}C NMR spectra of **IIa** the signals of the carbon atoms of CHCl- and CH$_2$Cl-groups are in a weaker field at 60.54 and 47.96 ppm, than similar signals of the carbon atoms in the spectrum of compound **IIb** (60.27 and 47.48 ppm, respectively). The upfield shift of the signals of the carbon atoms in the spectrum of *erythro*-isomer indicates a stronger syn-interaction between the atoms of CHCl- and CCl$_2$-groups than in the *threo*-structure. On the contrary, the signals of the CH$_2$- and CH-carbon atoms of the cyclopropane ring in compound **IIb** are in a weaker field as compared with the signals of similar atoms in **IIa** (the chemical shifts of the CH$_2$- and CH-carbon atoms in the *erythro*- and *threo*-isomer are 28.07, 34.87 and 27.29, 34.45 ppm, respectively).

The resulting mixture of stereoisomeric *gem*-dichlorocyclopropanes **IIa** and **IIb** was we used in the *O*-alkylation of phenols **III** in dimethyl sulfoxide (DMSO) in the presence of solid NaOH.

$R^1 = R^2 = H$ (**IIIa, IVa, Va**);
$R^1 = CH_3$, $R^2 = H$ (**IIIb, IVb, Vb**);
$R^1 = H$, $R^2 = Cl$ (**IIIc, IVc, Vc**).

The reaction is complete at room temperature within 15–20 min and results in ketals with the exocyclic double (**IVa–IVc**) and triple (**Va–Vc**) bonds.

The total yield of compounds **IV** and **V** equals 70–80% and is independent of the nature of the aromatic reactant **IIIa–IIIc**. However, the nature and position of the substituent affect the ratio of the ethylene and acetylene structures (**IVa–IVc/Va–Vc**). In the reaction with phenol **IIIa** the yield of acetylene ketal **Va** is much higher, while in the reaction with *p*-chlorophenol **IIIc** the yield of the fully dehydrochlorinated product **Vc** is only 2-fold higher than that of the partially dehydrochlorinated product **IVc**. However, in the case of alcohols of linear structure [allyl (**VIa**) and butyl (**VIb**)] the total yield of similar ketals **VIIa, VIIb, VIIIa and VIIIb**) is much lower (20–30%).

R–OH +   **IIa, IIb** $\xrightarrow[\substack{-HCl}]{\substack{DMSO\\NaOH}}$ RO   OR $\xrightarrow{-HCl}$ RO   OR

VIa,VIb   **VIIa, VIIb**   **VIIIa, VIIIb**

R = CH$_2$=CH-CH$_2$ (**VIa, VIIa, VIIIa**);
R = C$_4$H$_9$ (**VIb, VIIb, VIIIb**).

In this case, the acetylene product **VIII** dominates in the reaction products.

Earlier, we mentioned the elimination of the hydrogen halides in phenylpolyhaloalkyl ethers in the study of the *O*-alkylation of phenols and alcohols with polyhaloalkanes. [4]

The individual compounds **IVa–IVc, Va–Vc**, as well as a mixture of isomers **VIa, VIb** and **VIIa, VIIb** were isolated by column chromatography from the reaction mixture of phenols **IIIa–IIIc** and alcohols **VIa, IVb** with the reagents **IIa, IIb.**

In the ^{1}H NMR spectrum of compound **IVc** the CH proton of the cyclopropane ring is observed at 2.40 ppm as a doublet of doublets with the spin-spin coupling constants $^3J_{2-3a}$ = 7.4 Hz, $^3J_{2-3b}$ = 10.2 Hz. The trans-positioned CH$_2$-proton of the cyclopropane ring appears as a doublet of doublets at 1.56 ppm (^2J = 6.9 Hz, $^3J_{3a-2}$ = 7.4 Hz), while the *cis*-positioned CH$_2$-proton appears as a doublet of doublets in a weak field at 1.75 ppm (^2J = 6.9 Hz, $^3J_{3b-2}$ = 10.2 Hz). The proton of the terminal double bond, which is *cis*-positioned relative to the chlorine atom, is observed as a doublet at 5.28 ppm (^2J = 1.4 Hz), while the *trans*positioned proton is observed as a doublet in a weak field (5.41 ppm, ^2J = 1.4 Hz). A multiplet in the range of 7.08–7.27 ppm belongs to the protons of the aromatic ring. The structures of compounds **VIa, IVb, VIIa, VIIb** were identified in a similar way.

The mass spectra of compounds **IVa–IVc** contains low-intensive peaks (<1%) of the molecular ions. The main decay directions are caused by eliminating the halogen ion and aryloxy radical. The most abundant ions are [ArOH]+×with m/z (83%), 108 (100%), 128/130 (100%) in the spectra of compounds **IVa–IVc**, respectively.

The ^{1}H NMR spectrum of compound **Vc** contains the CH-proton of the cyclopropane ring as a doublet of doublets of doublets at 2.11 ppm ($^3J_{2-3a}$ = 6.9 Hz, $^3J_{2-3b}$ = 10.5 Hz, $^4J_{2-2}$ = 2.2 Hz). The *trans* positioned CH_2-proton of the cyclopropane ring appears as a doublet of doublets at 1.49 ppm (^2J = 6.3 Hz, $^3J_{2-3a}$ = 6.9 Hz), while the cispositioned CH_2-proton is observed as a doublet of doublets in a weaker field at 1.70 ppm (^2J = 6.3 Hz, $^3J_{2-3b}$ = 10.5 Hz). A doublet signal at 1.98 ppm ($^4J_{2-2}$ = 2.2 Hz) belongs to the proton of the terminal triple bond. A multiplet at 7.05–7.27 ppm corresponds to the aromatic ring protons. The structures of compounds **Va, Vb, VIIIa, VIIIb** were identified in a similar way.

The mass spectra of compounds **Va–Vc** are characterized by the low-intensive peaks (≤1%). Acetylene ketals **Va–Vc** are capable of undergoing cleavage of the CH_2CO moiety from the ions produced with the release of aryloxy radical from the M$^+$× ion, and the peaks of these ions have a maximum intensity (100%) [m/z 115, 129, 149/151 for compounds **Va– Vc**, respectively]. Compounds **VIIa, VIIb, VIIIa, VIIIb** are characterized by the absence of the molecular peaks (<0.1%). The main decay direction is caused by the breaking of bonds with oxygen and the formation of oxonium ions, which then loose the olefin molecule from the second alkoxy group.

Thus, the diastereoisomeric 1,2-dichloroethyl-*gem*dichlorocyclopropanes were obtained via the dichlorocarbenation by the Makosza method. The reactions of the obtained 1,2-dichloroethyl-*gem*dichlorocyclopropanes with phenols and alcohols in DMSO are a convenient method of the synthesis of ketals of chlorovinyl- and ethynylcyclopropanes.

We have received, increasing the number of investigated objects, 2-bromo-2-phenyl-1,1-dichlorocyclopropane (**X**) by the dichlorocarbenation of α-bromostyrene and studied his interaction with phenol (**IIIa**), allyl and *p*-butyl alcohols (**VIa, b**).

It was interesting to study the transformations under these conditions of 2-bromo-2-phenyl-1,1-dichlorocyclopropane (**IX**), which readily forms via the reaction of available α-bromostyrene with dichlorocarbene [5].

We have found (Table 25.1) that compound **IX** reacts with phenols **IIIa–IIIc** under phase transfer catalysis conditions to yield acetals of phenylpropargyl aldehyde **Xa–Xc**.

$$\text{IX} \quad + \quad \text{ROH} \quad \longrightarrow \quad \text{Xa-Xc, XIa, XIb}$$

IX   IIIa-IIIc, VIa, VIb   Xa-Xc, XIa, XIb

R = Ph (**IIIa, Xa**); 2-CH$_3$C$_6$H$_4$ (**IIIb, Xb**); 4-ClC$_6$H$_4$ (**IIIc, Xc**); CH$_2$=ÑÍ -ÑÍ $_2$ (**VIa, XIa**); n-C$_4$H$_9$ (**VIb, XIb**).

Endocyclic elimination of HBr is likely to lead to intermediate cyclopropenes in which the chloroallyl fragments react rapidly with phenols **IIIa–IIIc**. The reaction is accompanied by ring opening to form the acetylene bond.

Allyl alcohol (**VIa**) and butyl alcohol (**VIb**) do not react under these conditions with compound **IX**. We managed to accomplish this reaction using as a catalyst sodium metal in excess of the initial alcohol. The yields of resultant acetals in the reaction with alcohols are much lower than in the reactions with phenols, while conditions are more severe (Table 25.1).

TABLE 25.1 Reaction of Phenols **IIIa–IIIc** and Alcohols **VIa** and **VIb** With 2-bromo-2-phenyl-1,1-dichlorocyclopropane **IX**

Reactant	Reaction conditions	Reaction product (yield, %)
IIIa		Xa (78)
IIIb	A	Xb (69)
IIIc		Xc (33)
VIa	Б	XIa (30)
VIb		XIb (11)

Note: System A: molar ratio NaOH:**IIIa–IIIc:IX** = 0.02:0.007:0.003, 3.5 mL of DMF, 20 °C, reaction time 4 h (2 h for phenol **IIIa**). System B: molar ratio Na:**VIa, VIb:IX** = 0.1:0.012:0.005, 100 °C, reaction time 8 h (12 h for alcohol **IVb**).

A nonselective formation of acetylene structures was noted previously in the condensation of polysubstituted *gem*-dibromocyclopropanes with alkanols [6]. We established for the first time that gemdichlorocyclopropanes containing bromine atom in the ring can participate in similar transformations.

25.3 EXPERIMENTAL PART

The ^{1}H and ^{13}C NMR spectra were recorded on a Bruker AM-300 spectrometer (300.13 and 75.47 MHz, respectively) in CDCl3 relative to internal TMS. The GC-MS analysis was performed on a Shimadzu GCMS-QP2010 Plus instrument (EI, 70 eV; an ion source temperature 200 °C, the direct input temperature 40–290 °C, the heating rate 12 deg min^{-1}). The GLC analysis was performed on a LKhM-8 MD chromatograph equipped with a thermal conductivity detector, carrier gas helium (flow rate 1.5 l h−1, the column length 2 m, 5% SE-30 on a Chromaton N-AW carrier). The TLC analysis was performed using Silufol plates (Merk) eluting with a hexane–AcOEt mixture, 98:2. The preparative separation was carried out by column chromatography on silica gel eluting with hexane with increasing ethyl acetate content of 5 to 100%.

General procedure of dihalocarbenation. To a mixture of 0.1 mol of 3,4-dichlorobut-1-ene **IX**, 0.2 g of a phase transfer catalyst (catamine AB in 300 mL of chloroform) was added drop wise 320 g of 50% NaOH solution at 40 °C under stirring within 6 h. Then the mixture was stirred for 2 h at the same temperature. The reaction mixture was washed with water until the neutral reaction. The extract was dried over calcined $MgSO_4$. The solvent was evaporated, and the residue was distilled in a vacuum.

(2R*)-1,1-Dichloro-2-[(1S*)-1,2-dichloroethyl]-cyclopropane (IIa, *threo*-diastereomer). Yield 40%, colorless liquid, bp 96°C (2 mm Hg). ^{1}H NMR spectrum (CDCl$_3$), δ, ppm (*J*, Hz): 1.50 d.d (1H, C^{3}H$_a$, ^2J 7.4, ^3J 7.7), 1.88 (d.d., 1H, C^{3}H$_b$, ^2J 7.4, ^3J 10.4), 2.05 (d.d.d., 1H, C^{2}H, ^3J 7.7, ^3J 10.4, ^3J 9.4), 3.76–3.96 (m., 3H, C^{1}HCl, C^{2}H$_2$Cl). ^{13}C NMR spectrum (CDCl$_3$, δ, ppm): 27.29 (C^{3}H$_2$), 34.45 (C^{2}H), 47.96 (C^{2}H$_2$Cl), 58.48 (C^{1}Cl$_2$), 60.54 (C^{1}HCl). Mass spectrum, m/e (I$_{rel}$, %): 206/208/210/212 (<1) [M]$^+$, 171/173/175 (1) [M-Cl]$^+$, 157/159/161 (5/4/1) [M-CH$_2$Cl]$^+$, 144

(12/11/3), 135 (11/8/1), 109/111/113 (100/61/10), 96/98/100 (15/10/3), 75/77 (48/15), 51 (15).

(2R*)-1,1-Dichloro-2-[(1R*)-1,2-dichloroethyl]-cyclopropane (IIb, *erythro*-diastereomer). Yield 40%, colorless liquid, bp 99°C (2 mm Hg). ¹H NMR spectrum (CDCl₃), δ, ppm (*J*, Hz): (1.50 d.d., 1H, C³H$_a$, ²*J* 7.4, ³*J* 7.7), 1.87 (d.d., 1H, C³H$_b$, ²*J* 7.4, ³*J* 10.4), 2.05 (d.d.d., 1H, C²H, ³*J* 7.7, ³*J* 10.4, ³*J* 5.0), 3.75–3.94 (m., 3H, C¹HCl, C²H₂Cl). ¹³C NMR spectrum (CDCl₃, δ, ppm): 28.07 (C³H₂), 34.87 (C²H), 47.48 (C²H₂Cl), 58.48 (C¹Cl₂), 60.27 (C¹HCl). Mass spectrum, m/e (I$_{rel}$, %): 206/208/210/212 (<1) [M]⁺, 171/173/175 (0.5) [M-Cl]⁺, 157/159/161 (2/1/0.3) [M-CH₂Cl]⁺, 144 (12/11/3), 135 (11/8/1), 109/111/113 (100/61/10), 96/98/100 (11/8/2), 75/77 (32/10), 51 (11).

Reaction of phenols **IIIa–IIIc** and alcohols **VIa** and **VIb** with a 1,2-dichloroethyl-*gem*-dichlorocyclopropanes **IIa** and **IIb** mixture. A mixture of 7.5 mmol of phenol **IIIa–IIIc** (or 12.5 mmol of alcohol **VIa** or **VIb**), 15 mmol of NaOH (or 25 mmol in the case of alcohols **VIa** or **VIb**) in 2.1 mL of DMSO was stirred at 55–60°C for 1 h. Then to the mixture was added a solution of 2.5 mmol of **IIa** or **IIb** in 1 mL of DMSO at 20°C. The reaction mixture was diluted with 30 mL of water and extracted with chloroform. The organic layer was washed with 20 mL of 20% NaOH solution, then with water to pH 7, and dried over MgSO₄. Chloroform was evaporated in a vacuum. The residue was chromatographed on silica gel (100–400 µm, hexane–AcOEt, 98:2).

1,1'-[(2-(1-Chlorovinyl)cyclopropane-1,1-diyl)bis-(oxy)]dibenzene (IVa). Yield 6%, colorless liquid, Rf 0.31 (hexane–AcOEt, 98:2). ¹H NMR spectrum (CDCl₃), δ, ppm (*J*, Hz): 1.57 (d.d., 1H, C³H$_a$, ²*J* 6.6, ³*J* 7.4), 1.77 (d.d., 1H, C³H$_b$, ²*J* 6.6, ³*J* 10.5), 2.40 (d.d., 1H, C²H, ³*J* 7.4, ³*J* 10.5), 5.30 (d., 1H, =C²H$_a$, ²*J* 1.5), 5.41 (d., 1H, = C²H$_b$, ²*J*), 6.97–7.28 (m., 10H, Ph). Mass spectrum, m/e (I$_{rel}$, %):286/288 M⁺ (<1), 251 [M-Cl]⁺ (2), 193/195 [M-PhO˙]⁺ (19/5), 157 (63), 151/153 (50/21), 129 (60), 115 (72), 94 (83), 77 (100), 65 (22), 51 (49).

1,1'-[(2-Ethynylcyclopropane-1,1-diyl)bis(oxy)]-dibenzene (Va). Yield 71%, white crystals, mp 63–65°C, Rf 0.26 (hexane–AcOEt, 98:2). ¹H NMR spectrum (CDCl₃), δ, ppm (*J*, Hz): 1.52 (d.d., 1H, C³H$_a$, ²*J* 6.2, ³*J* 7.0), 1.75 (d.d., 1H, C³H$_b$, ²*J* 6.2, ³*J* 10.2), 2.01 (d., 1H, ≡C²H, ⁴*J* 2.1), 2.15 (d.d.d., 1H, C²H, ³*J* 7.0, ³*J* 10.2, ⁴*J* 2.1), 7.02–7.35 (m., 10H, Ph). ¹³C NMR spectrum (CDCl$_3$, δ, ppm): 16.03 (C^{2}H), 21.48 (C^{3}H$_2$), 66.5 (≡C^{2}H), 80.26 (C^1≡), 87.42 (C^1), 116.77, 117.22 (ortho-Ph), 122.59 (para-Ph), 129.27, 129.48 (meta-Ph), 155.60, 155.93 (Ph). Mass spectrum, m/e(I$_{rel}$, %): 250 M$^+$ (<1), 157 [M-PhO˙]$^+$ (5), 156 [M-PhOH]$^+$ (3), 128 (26), 115 (100), 94 (28), 77 (29), 65 (9), 51 (8).

1,1'-[(2-(1-Chlorovinyl)cyclopropane-1,1-diyl)bis-(oxy)]bis(2-methylbenzene) (IVb). Yield 15%, white crystals, mp 65–67°C, Rf 0.37 (hexane–AcOEt, 98:2). ^{1}H NMR spectrum (CDCl$_3$), δ, ppm (*J*, Hz): 1.62 (d.d., 1H, C^{3}H$_a$, ^2J 6.8, ^3J 7.4), 1.87 (d.d., 1H, C^{3}H$_b$, ^2J 6.8, ^3J 10.2), 2.22 (d., 6H, CH$_3$), 2.43 (d.d., 1H, C^{2}H, ^3J 7.4, ^3J 10.2), 5.36 (d., 1H, =C^{2}H$_a$, ^2J 1.4), 5.44 (d., 1H, =C^{2}H$_b$, ^2J 1.4), 6.90–7.62 (m., 8H, Ph). ^{13}C NMR spectrum (CDCl$_3$, δ, м.д.): 16.21 (CH$_3$), 21.63 (C^{3}H$_2$), 32.65 (C^{2}H), 87.51 (C^1), 114.26 (=C^{2}H$_2$), 115.16 (орто-Ph), 122.08 (para-Ph), 126.57, 130.88 (meta-Ph), 127.65 (CH$_3$-C), 137.59 (=C^{1}Cl), 154.22 (Ph). Mass spectrum, m/e (I$_{rel}$, %): 314/316 M$^+$ (<1), 279 [M-Cl]$^+$ (6), 207/209 [M-PhO˙]$^+$ (25/8), 171 (26), 165/167 (63/21), 143 (26), 129 (42), 108 (100), 91 (59), 77 (15), 65 (31).

1,1'-[(2-Ethynylcyclopropane-1,1-diyl)bis(oxy)]-bis(2-methylbenzene) (Vb). Yield 58%, colorless liquid, Rf 0.32 (hexane–AcOEt, 98:2). ^{1}H NMR spectrum (CDCl$_3$), δ, ppm (J, Hz): 1.57 (d.d., 1H, C^{3}H$_a$, ^2J 6.3, ^3J 7.4), 1.85 (d.d., 1H, C^{3}H$_b$, ^2J 6.3, ^3J 10.3), 2.08 (d., 1H, ≡C^{2}H, ^4J 1.9), 2.20 (d.d.d., 1H, C^{2}H, ^3J 7.4, ^3J 10.3, ^4J 1.9), 2.28 (s., 3H, CH$_3$), 2.37 (s., 3H, CH$_3$), 6.98–7.57 (m., 8H, Ph). ^{13}C NMR spectrum (CDCl$_3$, δ, ppm): 16.36, 16.42 (CH$_3$), 20.10 (C^{3}H$_2$), 32.65 (C^{2}H), 66.52 (≡C^{2}H), 80.71 (C^1≡), 86.94 (C^1), 114.71, 114.89 (ortho-Ph), 122.02 (para-Ph), 126.78, 126.90, 130.88 (meta-Ph), 126.72, 127.23 (CH$_3$-C), 153.77 (Ph). Mass spectrum, m/e (I$_{rel}$, %): 278 M$^+$ (<1), 171 [M-PhO˙]$^+$ (6), 170 [M-PhOH]$^+$ (3), 142 (6), 129 (100), 108 (53), 91 (23), 77 (8), 65 (14).

1,1'-[(2-(1-Chlorovinyl)cyclopropane-1,1-diyl)bis-(oxy)]bis(4-chlorobenzene) (IVc). Yield 23%, colorless liquid, Rf 0.30 (hexane–AcOEt, 98:2). ^{1}H NMR spectrum (CDCl$_3$), δ, ppm (J, Hz): 1.56 (d.d., 1H, C^{3}H$_a$, ^2J 6.9, ^3J 7.4), 1.75 (d.d., 1H, C^{3}H$_b$, ^2J 6.9, ^3J 10.2), 2.40 (d.d., 1H, C^{2}H, ^3J 7.4, ^3J 10.2), 5.28 (d., 1H, =C^{2}H$_a$, ^2J 1.4), 5.41 (d., 1H, = C^{2}H$_b$, ^2J 1.4), 7.08–7.27 (m., 8H, Ph). Mass spectrum, m/e (I$_{rel}$, %): 354/356/358 M$^+$ (<1), 319/321/323 [M-Cl]$^+$ (5/3/0.6), 227/229/231 [M-PhO˙]$^+$ (13/9/2), 191/193 (31/11), 185/187/189 (92/61/11), 163/165 (17/6), 149/151 (41/14), 128/130 (100/30), 111/113 (43/14), 99/101 (20/7), 75 (26), 63 (8).

1,1'-[(2-Ethynylcyclopropane-1,1-diyl)bis(oxy)]-bis(4-chloroben-zene) (Vc). Yield 47%, colorless liquid, Rf 0.27 (hexane–AcOEt, 98:2). ^{1}H NMR spectrum (CDCl$_3$), δ, ppm (J, Hz): 1.49 (d.d., 1H, C^{3}H$_a$, ^2J 6.3, ^3J 6.9), 1.70 (d.d., 1H, C^{3}H$_b$, ^2J 6.3, ^3J 10.5), 1.98 (d., 1H, ≡C^{2}H, ^4J 2.2), 2.11 (d.d.d., 1H, C^{2}H, ^3J 6.9, ^3J 10.5, ^4J 2.2), 7.05–7.27 (m., 8H, Ph). Масс Mass spectrum, m/e (I$_{rel}$, %): 318/320/322 M$^+$ (1), 191/193 [M-PhO$^•$]$^+$ (2/0.7), 190/192 [M-PhOH]$^+$ (2/0.7), 162/164 (6/2), 149/151 (100/32), 128/130 (44/14), 111/113 (14/5), 99/101 (5/2), 75 (9), 63 (3).

A mixture of 1,1-bis(allyloxy)-2-(1-chlorovinyl)-cyclopropane (VIIa) and 1,1-bis(allyloxy)-2-ethynylcyclopropane (VIIIa). Yield 4 (**VIIa**) and 26% (**VIIIa**), colorless liquid, Rf 0.24 (hexane–AcOEt, 98:2). ^{1}H NMR spectrum (CDCl$_3$), δ, ppm (J, Hz): 0.88 (d.d., 1H, C^{3}H$_a$ (**VIIa**), ^2J 6.6, ^3J 7.2), 1.18 (d.d., 1H, C^{3}H$_a$ (**VIIIa**), ^2J 5.5, ^3J 6.4; 1H, C^{3}H$_b$ (**VIIa**), 1.33 (d.d., 1H, C^{3}H$_b$ (**VIIIa**), ^2J 5.5, ^3J 10.0), 1.76 (d.d.d., 1H, C^{2}H (**VIIIa**), ^3J 6.4, ^3J 10.0, ^4J 2.2), 1.91 (d., 1H, ≡C^{2}H (**VIIIa**), ^4J 2.2), 2.09 (d.d., 1H, C^{2}H (**VIIa**) ^3J 7.2, ^3J 10.0), 4.07–4.19 (m., 4H, CH$_2$–O (**VIIIa**)), 4.20–4.38 (m., 4H, CH$_2$–O (**VIIa**)), 5.10–5.35 (m., 4H, CH$_2$= (**VIIIa**); 4H, CH$_2$= (**VIIIa**), 2H, =C^{2}H$_a$, =C^{2}H$_b$ (**VIIIa**)), 5.82–6.04 (m., 2H, =CH (**VIIIa**); 2H, =CH (**VIIIa**)). Масс (VIIa) m/e, (I$_{отн}$, %): 214/216 (<0.1) M$^+$, 179 [M-Cl]$^+$ (3), 173/175 (3/1) [M-C$_3$H$_3$$^•$]$^+$, 157/159 [M-C$_3H_5O^•$]$^+$ (2/0.7), 155 (2), 151 (5), 137 (19), 131/133 (28/9), 127 (16), 121 (5), 117/119 (25/11), 115 (30), 113 (18), 109 (24), 97 (13), 95 (24), 93 (100), 91 (74), 88 (63), 81 (33), 79 (25), 77 (44), 75 (12), 67 (30), 65 (9), 55 (30), 53 (70), 42 (18). Mass spectrum (VIIa), m/e (I$_{rel}$, %): 178 (<1) M$^+$, 137 [M-C$_3$H$_5$$^•$]$^+$ (4), 121 [M-C$_3$H$_5$O$^•$]$^+$ (6), 119 (12), 109 (24), 95 (27), 93 (20), 91 (100), 81 (64), 79 (66), 77 (84), 67 (48), 65 (21), 57 (6), 55 (98), 52 (64), 42 (18).

A mixture of 1,1-dibuthoxy-2-(1-chlorovinyl)cyclopropane (VIIb) and 1,1-dibuthoxy-2-ethynylcyclopropane (VIIIb). Yield 6 (**VIIb**) and 14% (**VIIIb**), colorless liquid, Rf 0.27 (hexane–AcOEt, 98:2). ^{1}H NMR spectrum (CDCl$_3$), δ, ppm (J, Hz): 0.93 (t., 12H, CH$_3$ (**VIIb, VIIIb**)), 1.10 (d.d., 1H, C^{3}H$_a$ (**VIIIb**), ^2J 5.3, ^3J 6.3), 1.24 (m., 8H, CH$_2$–CH$_3$ (**VIIb, VIIIb**); 8H, CH$_2$–CH$_2$–CH$_3$ (**VIIb, VIIIb**); 1H, C^{3}H$_b$ (**VIIIb**); 2H, C^{3}H$_a$, C^{3}H$_b$ (**VIIb**); 1H, C^{2}H (**VIIIb**)), 1.87 (d., 1H, ≡C^{2}H (**VIIIb**), ^4J 2.3), 2.01 (d.d., 1H, C^{2}H (**VIIb**), ^3J 7.0, ^3J 10.0), 3.53–3.62 (m., 8H, CH$_2$–O (**VIIb, VIIIb**)), 5.12 (d., 1H, =C^{2}H$_a$ (**VIIb**), ^2J 1.3), 5.23 (d., 1H, =C^{2}H$_b$ (**VIIb**), ^2J 1.3). Mass spectrum (**VIIb**), m/e (I$_{rel}$, %): 246/248 M$^+$ (<0.1), 211 [M-Cl]$^+$

(<1), 210 [M-HCl]$^+$ (<1), 173/175 [M-C$_4$H$_9$O$^•$]$^+$ (<1), 155 (20), 134/136 (6/2), 117/119 (6/2), 99 (100), 89 (7), 57 (42). Mass spectrum (**VIIIb**), m/e (I$_{rel}$, %): 210 M$^+$ (<1), 153 [M-C$_4$H$_8$$^•$]$^+$ (3), 137 [M-C$_3$H$_5$O$^•$]$^+$ (4), 135 (15), 107 (100), 83 (38), 81 (30), 79 (65), 74 (3).

Synthesis of compound IX. A 50% solution of NaOH (320 g) was added dropwise for 2 h to a vigorously stirred mixture of 0.1 mol of αbromostyrene and 0.2 g of triethylbenzylammonium chloride as the phase transfer catalyst in 300 mL of chloroform at 40°C. After that, the mixture was stirred for another 1 h at 40°C. Then, the reaction mixture was washed with water, the solvent was evaporated, and the residue was distilled in a vacuum.

2-Bromo-2-phenyl-1,1-dichlorocyclopropane IX. Bp 125–127 °C (4 mmHg).

^{1}H NMR (CDCl$_3$, δ, ppm, J, Hz): 2.06 (d, 1H, C^{3}H$_a$, ^{2}J 9.0), 2.09 (d., 1H, C^{3}H$_b$, ^{2}J 9.0), 7.17–7.39 (m., 4H, Ph). ^{13}C NMR (CDCl$_3$; δ, ppm): 35.4 (CH$_2$), 43.0 (CBr), 62.9 (CCl$_2$), 128.7, 128.9, 129.3, 138.9 (Ph).

MS m/z (I$_{rel}$, %): 264/266/268/270 [M]$^{+•}$ (1), 192/194 (5/5), 185/186/188 [M-Br$^•$]$^+$ (10/6/1), 149/151 (100/30), 115 (47), 89 (28), 75 (18), 63 (22).

Synthesis of compounds Xa–Xc. A solution of 0.003 mol of 2-bromo-2-phenyl-1,1-dichlorocyclopropane (**IX**) in 1 mL of DMF was added dropwise to a vigorously stirred mixture of 0.007 mol of phenol **IIIa–IIIc**, 0.02 mol of NaOH, and 2.5 mL of DMF. After 2–4 h, the reaction mixture was diluted with water, extracted with chloroform, and washed with water, the solvent was removed, and the residue was chromatographed on silica gel (hexane–ethyl acetate, 9: 1, as an eluent).

(1-Phenylprop-1-yn-3,3-diyl)bis(hydroxybenzene), Xa. R$_f$ 0.22. ^{1}H NMR (CDCl$_3$; δ, ppm, J/Hz): 7.05–7.40 (m., 15H, Ph; 1H, CH). ^{13}C NMR (CDCl$_3$; δ, ppm): 85.32 (C^2), 95.10 (C^1), 103.39 (C^3), 126.58, 127.61, 128.10, 128.35, 129.00, 131.02, 158.37 (Ph). MS m/z (I$_{rel}$, %): 300 [M]$^{+•}$ (<0.1), 207 [M-$^•$OC$_6$H$_5$]$^+$ (2), 175 (100), 158 (91), 144 (13), 131 (15), 116 (28), 102 (19), 91 (12), 77 (10), 63 (11).

(1-Phenylprop-1-yn-3,3-diyl)bis(hydroxyl-2-methylbenzene), Xb. R$_f$ 0.27. ^{1}H NMR (CDCl$_3$; δ, ppm, J/Hz): 2.2 (s., 6H, CH$_3$), 6.7–7.5 (m, 13H, Ph; 1H, CH). MS m/z (I$_{rel}$, %): 328 [M]$^{+•}$ (<0.1), 175 (100), 158 (87), 144 (12), 131 (15), 116 (29), 102 (21), 91 (12),77 (10), 63 (12), 51 (4).

(1-Phenylprop-1-yn-3,3-diyl)bis(hydroxyl-4-chlorobenzene), Xc. Rf 0.24.

R_f 0.24. ^{1}H NMR (CDCl$_3$; δ, ppm, J/Hz): 6.90–7.48 (m., 13H, Ph, 1H, CH). MS m/z (I$_{rel}$, %): 368/370/372 [M]$^{+•}$ (<0.1), 175 (76), 158 (100), 144 (17), 131 (18), 116 (28), 102 (28), 91 (18), 77 (15), 63 (16) 51 (17).

Synthesis of compounds XIa and XIb. Sodium metal (0.012 mol) was gradually added to 0.1 mol of alcohol **VIa, VIb**. 2-Bromo-2-phenyl-1,1-dichlorocyclopropane **IX** (0.005 mol) was added to the prepared alcoholate and heated to 100 °C for 8–12 h. The reaction mixture was diluted with water, extracted with ether, and washed with water, and the solvent was evaporated. TLC analysis was performed on Silufol plates (Merk) using hexane–ethyl acetate, 9: 1, as an eluent. Preparative separation was accomplished by column chromatography on silica gel (hexane with ethyl acetate gradient from 5 to 100% as an eluent).

(3,3-Dibutoxyprop-1-yn-1-yl)benzene, XIa. Rf 0.27. ^{1}H NMR (CDCl3, δ, ppm, J, Hz): 0.95–1.01 (m, 6H, C4'H3, C4"H3), 1.38–1.49 (m, 4H, C3'Ha, C3'Hb, C3"Ha, C3"Hb), 1.72–1.81 (m, 4H, C2'Ha, C2'Hb, C2"Ha, C2"Hb), 3.43–3.50 (m, 2H, C1'Ha, C1'Hb), 3.66–3.72 C1"Ha, C1"Hb), 5.29 (c, 1H, C3H), 7.25–7.48 (m, 5H, Ph). MS m/z (Irel, %): 260 [M]+• (<0.1), 203 [M – C4H8]+• (1), 187 (17), 159 (3), 147 (6), 131 (100), 102 (36), 91 (6), 77 (22), 57 (36).

(3,3-Diallyloxyprop-1-yn-1-yl)benzene], XIb. R_f 0.30. ^{1}H NMR (CDCl$_3$; δ, ppm, J/Hz): 4.13–4.32 (m., 4H, C$^{1'}$H$_a$, C$^{1'}$H$_b$, C$^{1"}$H$_a$, C$^{1"}$H$_b$), 5.21–5.39 (m., 4H, C$^{3'}$H$_a$, C$^{3'}$H$_b$, C$^{3"}$H$_a$, C$^{3"}$H$_b$), 5.56 (s, 1H, C^{3}H), 5.91–6.04 (m., 2H, C$^{2'}$H, C$^{2"}$H), 7.27–7.50 (m., 5H, Ph). ^{13}C NMR (CDCl$_3$; δ, ppm): 66.29 (C$^{1'}$H$_2$, C$^{1"}$H$_2$), 83.97 (C^2), 85.78 (C^1), 91.02 (C^{3}H), 117.50 (C$^{3'}$H$_2$, C$^{3"}$H$_2$), 121.82, 128.29, 128,89, 131.96 (Ph), 134.05 (C$^{2'}$H, C$^{2"}$H). MS m/z (I$_{rel}$, %): 228 [M]$^{+•}$ (<0.1), 171 [M-$^•$OC$_3$H$_5$]$^+$ (18),142 (78), 128 (100), 115 (40), 102 (58), 91 (22), 77 (27), 63 (13), 51 (19).

25.4   CONCLUSION

In summary, the substitution of endocyclic chlorine atoms occurs in the reaction *gem*-dichlorocyclopropanes with alcohols and phenols. There is destroying of the carbocyclic fragment in brom-*gem*-dichlorocyclopropanes in these conditions.

KEYWORDS

- alcohols
- brom-*gem*-dichlorocyclopropanes
- diastereoisomers
- *gem*-dichlorocyclopropanes
- ketals
- phenols

REFERENCES

1. Yukel'son, I. I., Technology of Basic Organic Synthesis, Moscow: "Khimiya" ("Chemistry", in Rus.) Publishing House, 1968, p. 712.
2. Zefirov, N. S., Kazimirchik, I. V., Lukin, K. L., Cycloaddition of Dichlorocarbene to Olefins, Moscow: "Nauka" ("Science", in Rus.) Publishing House, 1985.
3. Günther, H., NMR Spectroscopy: An Introduction, New York: Wiley, 1980.
4. Bogomazova, A. A., Zlotskii, S. S., Kunakova, R. V., Proceedings of Higher Education (in Rus.), Ser. Chemistry and Chemical Technology, 2011, 54(4), p. 10.
5. Newman, M. S., Dhawan, B., Hashem, M. M., et al., Org. Chem., 1976, 41(24), P. 3925.
6. Fedorynski, M., Chem. Rev., 2003, 103(4), p. 1099.

INDEX